***Praise for* The**

"Whenever the Toxic Avenger and I need information about DVDs, or how to exfoliate those hard-to-reach areas, we turn to *The Digital Bits*. It is superlative and very entertaining, much better than watching a Troma movie!"

—*Lloyd Kaufman*
President of Troma Entertainment and creator of the Toxic Avenger

"My education in the ways of all things DVD began with *The Digital Bits*. *The Bits* was at the forefront of covering the dawn of the DVD industry. It remains an essential voice in reporting current industry developments and reviewing DVDs with the unabashed spirit of true film buffs."

—*Charles de Lauzirika*
DVD producer

"*The Digital Bits* has proven itself in its many enduring years of reporting on this industry. It has become an easy point of reference which in turn has made many of our lives in the DVD world vastly enriched."

—*Buena Vista Home Entertainment*

"*The Bits* is the best site on the Web for accurate, up-to-the-minute info on all the newest DVD releases. These guys are movie fans and it really shows. An A+ site for essential DVD information."

—*Donald May, Jr.*
President, Synapse Films, Inc.

"*The Digital Bits* has been an invaluable resource for me over the years. It's one of the best and most reliable DVD sites on the net. Every DVD collector should have it in their Favorites list."

—*Peter Staddon*
Senior Vice President, Marketing & DVD Development
Twentieth Century Fox Home Entertainment

"Without losing the flavor and 'aw shucks, we're cool' charm of their religiously updated web site, Bill and Todd have expanded their work into a book that gives DVD fans just the right amount of insight into how DVDs work, how Hollywood movies make it from film to disc, and how to get the most out of these shiny round things."

—*Jim Taylor*
Chief of DVD Technology and General Manager
Advanced Technology Group Sonic Solutions
Author of DVD Demystified *and*
Everything You Ever Wanted to Know About DVD

The Digital Bits:

Insider's Guide

to DVD

The Digital Bits:
Insider's Guide
to DVD

Bill Hunt and Todd Doogan

McGraw-Hill
New York   Chicago   San Francisco   Lisbon
London   Madrid   Mexico City   Milan   New Delhi
San Juan   Seoul   Singapore   Sydney   Toronto

The McGraw·Hill Companies

Cataloging-in-Publication Data is on file with the Library of Congress

1 2 3 4 5 6 7 8 9 0   DOC/DOC   0 9 8 7 6 5 4 3

ISBN 0-07-141852-0

The sponsoring editor for this book was Stephen S. Chapman and the production supervisor was Pamela A. Pelton. It was set in Helvetica by MacAllister Publishing Services, LLC. The art director for the cover was Anthony Landi.

Printed and bound by RR Donnelley.

This book is printed on acid-free paper.

McGraw-Hill books are available at special quantity discounts to use as premiums and sales promotions, or for use in corporate training programs. For more information, please write to the Director of Special Sales, McGraw-Hill Professional, Two Penn Plaza, New York, NY 10121-2298. Or contact your local bookstore.

DEDICATION

This book is dedicated to the loyal readers of *The Digital Bits*, who have stuck with us through thick and thin, and to film and DVD fans everywhere. Keep spinning those discs!

Contents

Acknowledgments

We could easily fill another book equal in size to this one if we were to properly thank absolutely everyone who helped us along the way. But there are definitely a few people we should acknowledge here.

For permission to use images from their films and DVDs in this book, and for allowing us to document the *Alien Quadrilogy* production, we'd like to thank the good folks at Twentieth Century Fox Home Entertainment, including Steve Feldstein, Peter Staddon, and Sven Davison. Special thanks also to DVD producer Charles de Lauzirika, who's been a good friend, and director Ridley Scott.

For their help with this book, our thanks to Rodrigo Brandao at Kino International, Michael Felsher at Anchor Bay, Danielle Garnier at Manga, Don May, Jr. at Synapse Films, R. O'Donnell at Criterion, Susan de Christofaro, Lloyd Kaufman, Rick Rhoades, Amy Friend, Beth Brown and her staff at MacAllister Publishing, Jim Taylor, and the folks at McGraw-Hill, including Anthony Landi and our editor, Steve Chapman.

We'd be remiss if we didn't acknowledge everyone who's contributed their writing and other efforts to *The Digital Bits* over the past six years. This includes Adam Jahnke, Matt Rowe, Robert A. Harris, Barrie Maxwell, Brad Pilcher, Sarah Hunt, Bob Banka, Greg Suarez, Rob Hale, Dan Kelly, Brian Ford Sullivan, Erin Lindsey, Jeff Kleist, Florian Kummert, Graham Greenlee, Dallas Ragan, Frank Ortiz, Chris Maynard, Andy Patrizio, Josh Lehman, Donald V. Day, and Drew Feinberg. Thanks also to Duane Leyva, D. W. Dunphy, John P. Dunphy, Brett Rudolph, Grey Cavitt, John Nelson, Dan Wolfson, Randy Stanley, Marco Passarelli, Joe Poropatich, and Robert Olsen of our *MusicTAP.net* affiliate.

Our sincere thanks to everyone at DVD Planet.com for their support in recent years, including Paul Ramaker, Holly Bell, and Vuong Hoang. Others in the industry to whom we owe thanks include David Prior, J. M. Kenny, Mark Atkinson, Michael Pellerin, Van Ling, Laurent Bouzereau, Alita Holly, Jeffrey Schwarz, Jeff Kurtti, Steve Gustafson, Mark Rance, Ken Crane, Pamela Fiorentino, Dave and Linda Lukas, Paul Prischman, Ron Epstein, Parker Clack and everyone at the *Home Theater Forum*, Devin Hamilton, Josh Dare, David Del Grosso, Rick Dean, Jeanne Cole, Lynne Hale, Jim Ward, Amy Jo Donner, Warren Lieberfarb, and all our peers in the online DVD community.

From the studios, special thanks to Garrett Lee, Spencer Savage and Marty Greenwald from Image Entertainment, Michael Mulvihill, Amy Gorton and Matt Lasorsa from New Line, Ronnee Sass, Janet Keller and Paul Hemstreet at Warner Bros., Urban Vision's Rhona Medina, Martin Blythe and Liz Haggar at Paramount, Ian Hendrie and the folks at Fantoma Films, Michael Stradford, Kavita Smith and Fritz Friedman at Columbia TriStar, Eric Maehara, Chris Bess and Bob Chapek from Buena Vista, Steve Biro at Unearthed Films, Laura Young at HBO, Cheryl Glenn and Missy Davy from DreamWorks, Matt Kiernan at First Run Features, Peter Becker from Criterion, Steve Wegner and Stacey Studebaker from MGM, and Craig Radow and Ken Graffeo at Universal.

Finally, on a personal note, Todd Doogan would like to thank every last person who ever read the site; Mom and Dad for a childhood filled with love, support, movies, and art; Corey for the cheap beer; Kari for love and support; Hiram Abif; Samit Choudhuri; the filmmakers who make DVDs worth owning; Bill for the voice of reason (three hours behind); Marvie for her warm belly; friends like Dallas and Brad for being there; and Erin for being the ying to his yang and filling up all the empty parts of his life with laughter, craziness, and love.

Bill would like to thank his parents Dan and Joan; his brother Jason (who rocks); Dave and Dale who are brothers too; his grandparents Louis and Vivian; all his friends and family, including Michael and Paul, George and Miranda, Lucile, Jim and Mary, Matt and Debbie, and Frank and Jill for all their love and support; Winfield for setting him straight; Kirby, Lucy, Winnie, Gracie, and Hilton for snuggles, scratches, and saving his sanity daily. Todd for his friendship and creative inspiration; and especially his wife, Sarah, for being absolutely everything that matters in this life.

Introduction

This is a book about DVD. *Oh sure*, you're probably thinking. *I figured out that much from the cover. Tell me something I don't know, Einstein.* Okay . . . how about this:

This book will teach you *everything* you need to know about DVD.

That's a bold statement, sure. But it's true. We've written *The Digital Bits: Insider's Guide to DVD* to make you a better fan—and a wiser consumer—of movies on DVD. In this staggering work of genius, we'll explain everything you need to know to get the most enjoyment from all your shiny little movie discs. Whether you have seven, seven hundred or even seven thousand in your collection (that means *you*, Roger Ebert!), we'll break it all down into plain, easy to understand language. Heck, you could give this book to your Grandma and she'd get it. And you know what? We're going to make it all fun. That's our promise to you . . . the reader.

After all, you've just shelled out your hard-earned cash for our spiffy trade paperback. Then again, maybe you're just standing there in the bookstore, minding your own business and browsing through all the books on film, and our snazzy cover caught your eye. Now you're paging through this book, trying to decide whether or not to take a chance on yet another video guide.

You're probably wondering, what are some of the things you need to know to be a better DVD fan? What wisdom can you expect to spill from these hallowed pages like honey? You probably already have a DVD player at home and a nice collection of discs. If you're even *half* as addicted as we are, you've probably got a *lot* of discs. So what else can we tell you that you don't already know?

Well, those are fair questions, and we're glad you asked 'em. Here's the deal . . .

First, we're going to explain how DVD works. We're going to tell you what those black bars on your TV are all about, and why some DVDs fill your screen and others don't. We'll tell you about a little something called anamorphic widescreen, and why it's very important. We'll explain the difference between Dolby Digital, DTS and THX (your first lesson: no, THX *isn't* a sound format). We'll tell you why, sometimes, you see a funny little pause when you're watching the film (don't worry it's *not* a problem with the disc). We'll teach you how to read the back of a DVD package, and what all those strange technical terms mean. We'll explain some of the different kinds of extras you can find on good DVDs, and tell you which are worth your time and which are just lame.

Oh, but we're not done yet—not even close. We're gonna teach you how to build a good home theater for a price that won't keep your kid from going to college or sink your retirement fund. We'll teach you what features you need to look for in a DVD player to make sure you aren't wasting your money on something that will be obsolete the day after you take it out of the box (and just because we're pals, we'll do that for TVs and audio equipment too). We'll give you suggestions on how to set all that gear up so you get the most from it. We'll tell you what the deal is with all these new fangled audio formats (like DVD-Audio and SACD) and we'll explain how they fit into your life (or not). We'll talk about the future of DVD. Where is the technology going? What's the deal with recordable DVD? Maybe you're thinking about buying one of those new high-definition Digital TVs. We'll tell you what it's all about, what you can watch right now, and when you'll be able to buy high-definition movies on disc. In short, we'll tell you what's coming next and how soon you have to start thinking (or worrying) about it.

Plus, we'll take a closer look at some of The Greatest DVDs Ever Made—more than a hundred in all. We'll tell you about DVD special editions that are so good, everyone should have a copy in their collection. We'll take a look at great movies of all kinds on disc: classics, action movies, comedies, scary movies, Sci-Fi, Westerns . . . you name it. We'll tell you about awful movies that have been turned into great DVDs. We'll point out wonderful music titles and documentaries on disc. We'll even run down a list of great DVDs that you might have missed in your trips to the video store.

And we'll give you a fascinating glimpse behind-the-scenes on the making of one of the biggest DVD releases of 2003: Twentieth Century Fox's nine-disc *Alien Quadrilogy* box set. The set includes brand-new special editions of all four films in the series (Ridley Scott's *Alien*, James Cameron's *Aliens*, David Fincher's *Alien³* and Jean-Pierre Jeunet's *Alien: Resurrection*), along with newly recorded audio commentary tracks, new interviews with the cast and crew, deleted scenes, outtakes, and tons of other bonus material, much of which has never been seen before by *anyone* but the filmmakers themselves. We think you'll be surprised at just how much work is involved in putting together some of your favorite special editions on DVD.

So how is it that we're able to tell you all of this? Who are we, these mysterious "insiders" who can make sense of all the complexities of DVD for you? And who or what is *The Digital Bits* anyway?

We, Bill Hunt and Todd Doogan, have been familiar fixtures in the DVD world since the format first debuted in the U.S. in March of 1997. Back then, most people really had no idea what to expect of DVD. Even most Hollywood studio execs believed that DVD was just another niche format that would amount to little and quickly disappear. But we knew better. We knew that DVD was the coolest thing to happen to movies since the projector and the VCR. And it was just a matter of time before DVD changed *everything*.

Bill, at the time a professional video producer, started *The Digital Bits* website with the idea of creating a place for people to go to learn more about DVD—one of the very first such places anywhere, online or in print. Todd, then a writer for TNT's *Rough Cut*, joined up a few months later. And as the site quickly grew, we discovered that not only were consumers from around the world reading what we had to say about DVD, many professionals within the Hollywood community were reading too. Almost inevitably, we fell into league with a rogues' gallery of film directors, DVD producers, and studio executives, all of whom shared our passion for the format and wanted to see it reach its full potential. We earned their trust. They shared their secrets with us. And we suddenly found ourselves in the strange and wonderful position of educating (and, even more importantly, *influencing*) the very people who were responsible for DVD.

So we did what anyone in that enviable position would do: we climbed up on our soapboxes. We talked up DVD to anyone who would listen. We spoke at conferences, we wrote magazine articles, we gave interviews to reporters, we served as judges for industry awards. We actively lobbied all of the studios to support the format, and to make important DVD quality features (like anamorphic widescreen) standard. We gave the world its first look at the dreaded Divx "pay-per-view" variation of DVD, sponsored by Circuit City, and then aggressively worked to make sure it never came to be—to make sure you didn't get robbed as consumers. (Imagine having to pay $3.25 every time you watched a DVD that you already owned. Ugh! It almost happened, folks.) We helped to lead a campaign that convinced George Lucas to begin releasing his *Star Wars* films on DVD. We've even helped to encourage the release of some of our other favorite films on disc as well. And, through it all, we've spoken our minds on *The Digital Bits*, every day, for more than six years now. It's been an amazing journey so far.

As for DVD . . . well, just as we expected, it's changed everything.

The bottom line is that we're in a unique position to be able to show you, firsthand, what DVD is all about. We've been in the trenches since Day One, fighting to help make this format as cool as it could be. But if you learn nothing else about us, know this: we're just like you. Sure, we have better agents and a killer web hosting deal, but we're still just a couple of guys who love movies and love watching them on DVD. *And we've got your back.*

So there it is. We've laid all our cards out on the table, and explained exactly what *The Digital Bits: Insider's Guide to DVD* is all about. And there you are, book in hand, about to come along with us as we take you deeper into the world of DVD. Seriously, just go bring this thing to the register and buy it won't you? We hope you do, 'cause we're gonna have a lot of fun.

Ready? Then hang on tight, folks! Here we go . . .

Bill Hunt and Todd Doogan
Editors of *The Digital Bits*
www.thedigitalbits.com

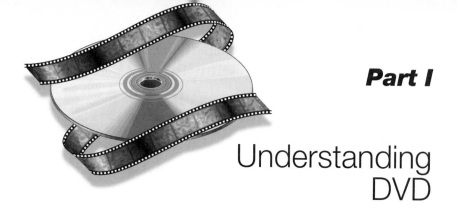

Part I

Understanding DVD

What Is DVD?

That's a pretty straightforward question, so we'll give you a straightforward answer. DVD is a 5-inch optical disc format that, when held in your hand, looks almost exactly like the music compact discs (or CDs, for those of us who like to abbreviate everything) that have been familiar to all of us since the mid-1980s. But the similarity ends there. A DVD disc can store up to 25 times as much data as a regular CD, and can be read some 9 times faster. In case those numbers make little sense to you, let's just say that DVD has a *lot* of versatility, and all that extra storage space allows the format to be used in a number of different ways.

There are three basic kinds of DVD discs. Let's take a look at what they're all about, shall we?

DVD-ROM discs, which are designed to be read by computers, can contain software programs like videogames and utilities. Writable versions of these discs (like DVD-R, which can be used for recording once, and DVD-RW, which can be recorded and erased many times) are a great storage medium for important digital information, such as text documents, spreadsheets, photos, and even video and sound files. Just think of them as CD-ROMs, with much more storage capacity.

DVD-Audio discs, designed to be read exclusively by DVD-Audio-compatible players, contain high-resolution music that rivals the sound quality of even the best standard CDs available today. Finally, music by your favorite artists can be heard in the same level of quality as the original master recordings—sometimes even better with the latest restoration techniques. And live music at last *truly* sounds live. DVD-Audio discs may also include video-based material as well, such as artist interviews, concert footage, song lyrics, and other bonus material.

Of course, we've saved the best for last: the DVD-Video disc. These are designed to be read by *all* DVD-compatible players and contain new and classic feature films, TV shows, documentaries, concerts, and other video-based material — all in picture and sound quality that's better than anything you've ever seen in your homes before. In fact, the video resolution of DVD-Video discs (while not quite as good as true high-definition video) is greater than VHS, laserdisc, and even current broadcast television. The format can also include the kind of multi-channel surround sound you hear in movie theaters. And DVD-Video allows for a wide variety of interactive supplemental features, such as filmmaker audio commentaries, alternate language tracks, subtitles, deleted scenes, multi-angle footage, and much, much more.

The bottom line is this: DVD is the best thing to happen to home entertainment since the introduction of the TV and the VCR. No kidding — it really is *that* cool.

DVD-Video is, by far, the most common type of DVD available today. So that's what we'll primarily be focusing on in this book. From now on, when we use the term DVD, we'll be talking about DVD-Video discs unless otherwise noted. But don't fret if you're a music fan, because we'll talk in more detail about DVD-Audio (as well as SACD, which is another new high-resolution audio disc format) later on. We've got lots of ground to cover.

But first, let's give you some history . . .

A Brief History of DVD

The idea for DVD was to create a simple, higher-capacity replacement for CD discs. The format originated with electronics manufacturers in 1993 and became a collaborative effort with Hollywood in 1994. It would take another two years for the various industry players to work out the specific format details and to agree on a copy-protection standard. By 1996, version 1.0 of the DVD spec was announced (covering DVD-Video and DVD-ROM), and at the end of that year, the very first players and movie titles were introduced in Japan.

DVD first appeared in the United States in March of 1997 in seven "test" market cities. At the time, only a handful of player models were available (priced at roughly $750 to $1,000) from RCA, Toshiba, Pioneer, Panasonic (Matsushita), and a couple of other hardware makers. And there were only about 15 or 20 movie titles available in those first couple of weeks (priced at about $30 each), primarily from Warner, New Line, Columbia TriStar, and MGM. As you'd expect of a new format trying to appeal to so-called "early adopters" — enthusiasts who pride themselves on being among the first to try new entertainment technologies — the first titles were primarily action and science fiction films like *The Fugitive*, *Blade Runner*, and *Eraser*, although classics like *Dr. Strangelove* and *Rocky* soon followed.

For those of us who were there at the beginning, it was a heady and exciting time. Sure, the prices were a bit on the high side and there wasn't a lot of variety yet in terms of movie discs. Early adopters, though, were quite pleased with what they were seeing. The introduction of DVD meant that, soon, *everyone* would have the chance to enjoy the kind of high-quality movie experience at home that had previously been reserved only for those select few technophiles with outrageously expensive home theaters and lots of money to burn. The promise of DVD was undeniable to us.

Strangely, though, not everyone was convinced. In fact, some of the most ardent naysayers were the very people in charge of DVD at the Hollywood studios. More than one conversation we had with studio executives at the time went something like this:

Q: "So when are you going to start releasing a wider variety of movies on DVD?"

A: "Well . . . it all depends. We're sort of testing the waters right now. DVD is cool, but we really don't see it becoming anything more than a niche product like laserdisc is."

Q: "You're kidding, right? Seriously, you don't really believe that, do you?"

A: "Our market research says people just aren't all that interested. Most people have never even heard of DVD. We don't think your average consumers are going to care."

Despite such negative attitudes, "early adopters" quickly flocked to the DVD format. Initial sales of movies and players in the test markets were encouraging and soon exceeded all expectations. More than 30,000 players were shipped in the first month. By the end of 1997, there were some 300,000 DVD players in the U.S. market, and the format had been launched nationwide. Another million players were added in 1998. Four million shipped in 1999. Eight million more followed in 2000 . . . 12 million in 2001.

Software sales, too, were surprising. Some 50,000 movie discs had been sold by the end of the first month of availability. DVD enthusiasts, quickly becoming hooked on the format, were voracious for *anything* new to watch on their players. It wasn't uncommon for people to line up at their favorite video retailers on Tuesday mornings (the day new movies and CDs are traditionally released to the public) to purchase every single DVD that came out that day. New titles were added each month, including *Batman*, *Twister*, *Air Force One*, and many more. In October of 1998, New Line's *Lost in Space* sold 200,000 copies in its first week, despite the fact that it was . . . well, a pretty miserable film. A year later, Warner's *The Matrix*—a much better offering—became the first title to sell over a million copies on DVD. Then, in 2001, DreamWorks' *Shrek* sold a whopping 5.5 million copies on DVD in its first week. After just 2 months, the film had sold 7.9 million copies on disc in the U.S., and another 2.1 million internationally.

By then, of course, one thing had become abundantly clear: The DVD format was succeeding beyond anything the industry had expected. In fact, DVD has become the most successful consumer electronics format *ever* introduced, finding its way into the living rooms of consumers far faster than either the VCR or the compact disc before it.

But don't let all this talk about success fool you; there were certainly roadblocks to that success. In 1998, Circuit City introduced the Divx pay-per-view variation of DVD. The Divx-enhanced player connected to a phone line, and users had to register an account with the company. You paid only $4.50 for each movie disc at the store, which gave you a 48-hour viewing period (starting from the time you first put the disc in your player). But if you wanted to watch the disc again after the 48 hours expired, you had to pay another $3.25 (via credit card) for another viewing period. In other words, your Divx movie collection became a de facto studio rental store, right in your very own living room.

The fact was Divx was doomed almost from the start. It wasn't so much that it was a bad idea. The technology itself was interesting and surprisingly clever. The problem was, it was very confusing to new consumers. Which should they buy—a regular DVD player or Divx-enhanced player? The ultimate downfall of Divx, however, lay in the fact that several studios that (at the time) had yet to announce "standard" DVD support, decided instead to aggressively release Divx titles. This included Buena Vista (Disney), DreamWorks, Paramount, Universal, and Twentieth Century Fox. In fact, Divx paid these studios millions of dollars in incentives to get certain titles as Divx exclusives, a fact that eager Divx executives were quick to tout to the press. The reaction from enthusiasts of standard DVD was as predictable and swift as it was furious. Uniting via the Internet, aggressive anti-Divx campaigns were mounted. Boycotts of Circuit City stores were organized, along with letter-writing campaigns targeted at the Divx-supporting studios. Consequently, only a year after its debut, the resulting terrible press and lackluster sales forced Circuit City to put Divx out of its misery, to the loss of more than $300 million in operating costs. Not long after this, the last of the remaining holdout studios finally jumped on the DVD bandwagon, and they were accepted with open arms.

Today, DVD is well on its way to supplanting VHS as the dominant home video format on the planet. As of the summer of 2003, the DVD Entertainment Group and the Consumer Electronics Association (or CEA —both industry trade groups that monitor such things) estimated that in just 6 years, some 60.9 million DVD players had been sold in the U.S., with some 48 million homes in the States having at *least* 1 DVD player. Those numbers don't even begin to include the many computers, laptops, and video game systems (like Microsoft's Xbox and Sony's PlayStation 2) equipped with DVD-ROM drives. Now consider this fact: The CEA estimates there are just over 95 million homes in the U.S. with VCRs—a market that took some 25 years to develop and has virtually stopped growing. It's not hard at all to foresee the day, in probably less than three years, when DVD becomes king.

But it's not just DVD hardware that's selling like crazy these days, and that's what has Hollywood so excited. More than 1.6 billion DVD discs have shipped to retailers since the format's launch in 1997. In 2002, consumers in the U.S. spent more than $11.6 billion buying and renting DVDs. Sales alone totaled more than $8.7 billion in 2002. Think about that. That's obviously a huge number, so let's put it into perspective for you. What else could you buy for $8.7 billion? Well . . . do you fancy a trip into orbit? Then how about 21.7 missions of NASA's space shuttle at roughly $400 million a piece? How about your own country? More than 90 countries around the world had a Gross Domestic Product (GDP) of less than $8.7 billion in 2002. Not that any of them are for sale, but you get the idea. $8.7 billion is a *lot* of money.

The fact is, there are major Hollywood studios whose balance books wouldn't be in the black without the profits they receive from DVD. And there are movies that make more money in their release on DVD than they do theatrically. The bottom line is simple: DVD is a huge success. It's big business and it's here to stay.

So how did DVD become so successful so quickly? To us, the answer is simple. DVD is a no-brainer. First there's the comfort factor. As we said before, DVD looks just like the CD discs that people have grown familiar and comfortable with for years. Heck, you can even play your CDs in your DVD player! Then there's the ease of use. The discs are a lot smaller and easier to handle than VHS tapes, and they're significantly more durable, as anyone who's ever had a videotape get stuck in their VCR can attest. You don't have to rewind them. You can skip to your favorite part of a film instantly. And finally, there's the quality. Let's say you have a 27-inch TV . . . immediately, you'll get better looking video with DVD than you ever saw with VHS. And, as you'll learn in this book, the quality you enjoy from the very same DVD will keep getting better and better as you upgrade to newer video and audio equipment.

All of that said, we haven't even mentioned one of the best parts of DVD: the extras you get. Some discs have many hours of bonus material to enjoy if you want to. All of these things mean *value*, folks. And all of this shiny goodness comes to you for an average price of about $25 a disc. The real question we ought to be asking is: How could DVD *not* succeed with all that going for it?

But the benefits of DVD aren't just to viewers at home and the studios' bottom lines. The success of DVD is encouraging Hollywood to revisit classic films that haven't sold well on VHS in years. Entire studio back catalogs are being re-released on DVD and, more importantly, many are being *restored* in the process. Some of the profits from DVD sales actually help to finance the ongoing and vitally important effort to preserve decaying film libraries . . . thus protecting the treasures of cinema history for future generations to enjoy.

DVD is, quite literally, the disc that saved Hollywood. But how does it work?

How DVD Works (in a Nutshell)

DVD stands for Digital Video Disc, although many people erroneously believe it means Digital Versatile disc, given that DVDs can be used to store and deliver more than just video content. Both types of discs are basically made up of layers of clear polycarbonate plastic. During the manufacturing process, the base layer is "stamped" with a long spiral pattern of microscopic "pits," that starts on the inside edge of the disc and winds around and around to the outside edge. This stamped plastic layer is coated with a thin layer of reflective aluminum, and then bonded (with an adhesive) to another layer of clear polycarbonate plastic to seal and protect the reflective surface. You can basically think of a DVD or CD disc as an aluminum sandwich with plastic bread – a crude but effective analogy.

Basic DVD Disc Composition

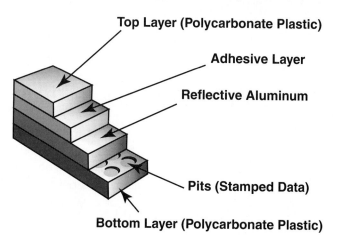

Top Layer (Polycarbonate Plastic)

Adhesive Layer

Reflective Aluminum

Pits (Stamped Data)

Bottom Layer (Polycarbonate Plastic)

DVD Disc – Laser Read Pattern

Digital information—which is basically nothing more than long patterns of *ones* and *zeros*—is stored on the disc in the particular pattern in which the tiny "pits" are arranged onto the disc's surface. When you put the disc in your player and press play, it starts spinning. A laser inside the player then reads that long spiral pattern of pits, starting on the inside edge of the disc and winding around and around to the outside edge, by focusing on the reflective aluminum layer. Whenever the laser passes over a pit, a *one* is recorded. If there's no pit, a *zero* is recorded. As the disc continues to spin, the laser reads on and the player's processor gradually builds up a digital picture of the data stored on the disc. Then it converts that data into either video or audio signals (or both) and sends that out to your TV and stereo equipment. This explanation is a simplification of the process, of course, but you get the basic idea.

As we said before, the difference with DVD is that it can store up to 25 times as much data as a regular CD. One way the format accomplishes this is to make the size of the physical pits on the disc that much smaller. By making them smaller, you can etch more of them onto the same surface area of the disc. DVD also employs a different kind of laser (one that uses shorter-wavelength red light) than current CD players do (which use longer-wavelength ruby lasers). The shorter wavelength of the red laser's light means that it can focus on a tinier area of the disc's surface and thus read the smaller pits.

But that's not the only method the DVD format uses to store more data than CDs. Other options available include recording data on *both* sides of the disc's surface (rather than just one side), or to physically sandwich more layers of recorded pits onto the same disc. With the latter option, the thickness of the disc is the same, but there are now *two* reflective

Pit Size Comparison

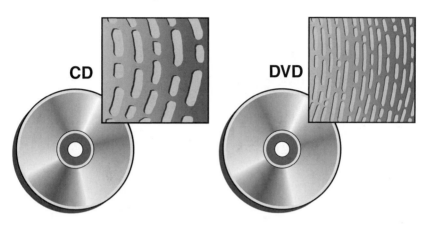

data layers rather than just one. When the laser gets to the end of the first layer on the disc, it simply refocuses onto the second layer and continues on reading the data as a continuous stream. This is called a *layer switch*.

If you watch a lot of DVD movies, you've probably seen a layer switch happen without even realizing it. Have you ever noticed how occasionally, when you're watching a movie on DVD, there will be a tiny little pause in the playback? It usually happens when the soundtrack is quiet, and when there's very little movement happening onscreen. Well, that's a layer switch. The pause results from the time it takes for the laser in your DVD player to shift focus from one layer of the disc to the other. See? Your disc *isn't* defective . . . it's working exactly like it should. This kind of switch usually happens with longer movies on DVD, or in situations where the DVD producers want to spread the data over two layers so that they have more room to work with. Depending on the disc and the model of player you have, this pause may be so brief as to be virtually invisible.

On the next page, you'll find a look at four different side and layer configurations of DVD discs. There are a couple of others as well, but these are by far the most common. The little triangle represents the direction from which the player's laser reads the data.

A DVD-5 disc has just one layer of data on one side of the disc (we say it's *Single Sided/Single Layered* or SS/SL). A DVD-9 disc has two layers of data on one side of the disc (*Single Sided/Dual Layered* or SS/DL). DVD-10 discs have two layers of data, recorded one on each side of the disc (*Dual Sided/Single Layered* or DS/SL). And DVD-18 discs have four

Common Types of DVD-Video Disc

DVD-5 (Single-sided, single-layered)

DVD-9 (Single-sided, dual-layered)

DVD-10 (Dual-sided, single-layered)

DVD-18 (Dual-sided, dual-layered)

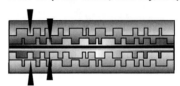

layers of data, recorded two per side (*Dual Sided/Dual Layered* or DS/DL). When a DVD disc has data recorded on both sides, you generally have to take the disc out of your player and flip it over to access the other side. In the early days of the format, there were discs where you had to actually flip the disc in the middle of the movie—half of it was recorded on each side (the discs were known as *flippers* for this very reason). These days, however, such discs are rare. Most DVD producers use the layer switch method to make flipping unnecessary. And some discs that are flippers have the movie on one side, and extras on another. Or, they might have the widescreen version of the movie on one side, and a full frame version on the other. There are also "hybrid" discs with DVD information recorded on one side, and CD information recorded on the other, but these are relatively rare.

So now you know, in a nutshell, how the DVD format works. That wasn't so bad, was it? We've covered the worst of the gear-head stuff already, and now we can get to the real fun of DVD . . . the video, the audio, and all those nifty extras.

Let's start with the picture first. So how is it that Hollywood can cram a whole movie onto a DVD in such high quality? Just how *does* the video part of DVD work? And what's the difference between widescreen and full frame anyway?

Telecine and Compression

Once the film is completed, a typical DVD production starts with something called *telecine*. This is the process of capturing the filmed image and converting it into a digital format. During telecine, the projected image of the film is recorded by a digital video camera, usually at high-definition resolution (or even better in some cases). Sometimes the filmed image is actually scanned, a frame at a time, by computer. In other cases, as with computer-generated films like Disney and Pixar's *Toy Story*, the film is generated by computer in the first place, so it's already available in digital format. The same also applies to films that are shot on digital video rather than actual film, like *Star Wars: Episode II* — something that's becoming more common these days.

In any case, once the film is available in a digital format, the video signal must again be converted into the TV format it will eventually be shown in — either NTSC, PAL, or SECAM, depending on the country in which you plan to release the final DVD. It might surprise you to learn that different countries around the world don't use the same TV system. In North America and Japan, TVs all use the NTSC format, which features 525 scan lines (lines of vertical resolution). In most of Western Europe and the rest of the world, PAL is the commonly used TV format, with 576 scan lines. And in France, parts of Eastern Europe, Russia and select other countries, SECAM is the standard, also featuring 576 scan lines. (Just by way of comparison, future High Definition TV, or HDTV, can have up to 1,080 scan lines.) It's worth noting here that if you have an NTSC DVD player, as do most of us in North America, you can't play DVDs with PAL or SECAM video on them – just NTSC.

No matter what TV format the video is converted into, the original digital files are still *way* too big to be included on a single DVD disc. A DVD can store up to 17 gigabytes of data depending on how many sides and layers are used. But a typical feature film, converted to a high-resolution digital video file, is many times larger than that — as much as 150 gigabytes or more. You'd need more than a dozen DVD discs to store the original digital file for just one 2-hour film. So how do the studios squeeze a whole film onto just one disc? The ingenious answer to this problem is *compression*.

DVD uses a compression format called MPEG-2 (which was established by the Moving Pictures Expert Group). The way compression works, on a basic level, is to analyze the original data and then remove any information that is considered to be redundant or unnecessary. The MPEG-2 encoder closely examines each frame of video in the source file

MPEG-2 Compression

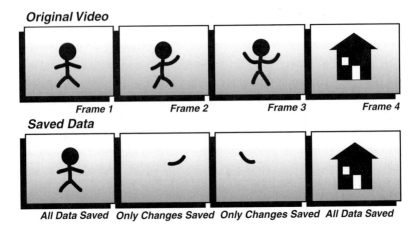

Original Video

Frame 1   Frame 2   Frame 3   Frame 4

Saved Data

All Data Saved   Only Changes Saved   Only Changes Saved   All Data Saved

(note that NTSC video runs at a rate of 30 frames per second). For the first frame analyzed, all of the picture data contained in the frame is fully saved. But if you've ever looked at a piece of actual movie film, you know that several of the subsequent frames are likely to be nearly identical to the first (film and video are similar in this respect). So, rather than saving identical data for each of the following frames, the encoder saves only the differences – the data that changes from frame to frame. When the video image changes significantly enough, another complete frame is stored as a reference. In this way, much of the data stored in the original file is removed (at a rate of about 30:1, or 30 units of data lost for every 1 stored). When the process is complete, the resulting MPEG-2 compressed digital video file is small enough to be stored on a single disc.

When it comes time for your DVD player to read the video file on the disc, it reads the first completely stored frame normally. Then, when it encounters a frame with only the changed data stored, the player looks back to the previous fully stored frame for the missing data, and then reassembles the complete image. This happens again and again, 30 times a second for the entire length of the film. If the encoding and decoding process is done properly, the eye of the casual viewer will never even detect that it's happening.

There's one other important thing we should mention here. The quality of the video encoding and decoding also depends on something called the *data rate*. Imagine that you're a fireman, and you're trying to put out a brushfire with a water hose. The more water the hose can transfer to the fire at one time, the more effective you'll be in dousing the fire, right? In the same way, the more video data your DVD player can read at a time, the more effective it is in reassembling the original video image. For water through a hose, this would be measured in gallons per second. For digital video data, this is measured in *megabits per second* (Mbps). We call this the data rate (it's also sometimes referred to as the *bit rate*). A DVD player can read video at a maximum data rate of 9.8 Mbps. But this is variable — the more complex the motion in the image, the higher the data rate needs to be to decode it properly. Generally, the video data rate will average around 4-5 Mbps.

Many DVD players allow you to monitor the video data rate on your TV screen while the film is playing. Consult your manual to find out if your DVD player allows this. If you *are* able to monitor the video data rate, play a scene from one of your favorite action films — one with lots of complex motion, such as explosions — and watch how the data rate goes up and down during the scene.

So that's how Hollywood is able to shoehorn an entire movie onto a 5-inch DVD disc. Pretty clever, huh?

All About Aspect Ratios

Perhaps the one thing we get asked most often at *The Digital Bits* is this: Why, on so many DVDs, are those damned black bars at the top and bottom of my TV?

The answer to that seemingly perplexing question has everything to do with something called *aspect ratios*. Simply put, a film's aspect ratio is a description of the shape of the image it presents — the relationship between the width of the image and its height.

The next time you go to a movie theater, take note of the shape of the screen before the show begins. Then, when you get home, look closely at your TV. You'll immediately notice that the shape of a theater screen is invariably much wider. The problem with watching movies at home stems from this very difference — the vast majority of films made today are shot in widescreen. In other words, they aren't the same shape as your TV. The funny thing is, though, that wasn't always the case.

Way back at the beginning of motion picture history, movies all looked roughly the same shape when projected in a theater. Starting in the late-1890s (and thru about the early-1950s or so), almost all films had a standard aspect ratio of roughly 1.33:1. In other words, the film image was

1.33:1

Academy Standard
NTSC Television (4x3)

1.33 times as wide as it was tall (another common way to denote this is 4x3, meaning 4 units of width for every 3 of height).

This ratio eventually became known as Academy Standard (when it was recognized formally by the Academy of Motion Picture Arts and Sciences in the 1930s. Almost every classic film you can think of from that period of time appeared in this ratio. The examples you see on the next page are actual screen shots (taken from DVD) of two films in their original Academy Standard aspect ratio.

When transferring films shot in the Academy Standard aspect ratio to video, there's no problem at all. Why? Well, you may have noticed that Academy Standard is shaped an awful lot like your current TV set. That's because when the time came for the television industry to decide what shape TVs would be (in the early 1950s), the National Television Standards Committee selected Academy Standard as the official aspect ratio for TV broadcasting here in the United States (as you may have guessed, the current TV format here in the U.S. got its name from this organization — NTSC). You'll remember that we mentioned 4x3 a few minutes ago — that's how many people in the industry refer to current TVs. This shape is also generally known as *full frame*.

But as the 1950s rolled on, and TV began capturing the imagination of American consumers, the Hollywood film industry was faced with a problem: So many people were buying TVs, and staying home to watch them, that theater attendance began to decline dramatically. In an effort to reverse this trend, Hollywood began making some changes to the look of their movies, releasing films in 3-D format and in *widescreen* aspect ratios.

Some of you may remember 3-D films, which required that you wear a pair of rather silly-looking cardboard glasses (c'mon, you know the ones . . . one of the plastic filters the glasses used for "lenses" was blue and

Gort and Klaatu (Michael Rennie) deliver Earth an ultimatum in *The Day the Earth Stood Still* (aspect ratio 1.33:1).

Lorelei Lee (Marilyn Monroe) sings *Diamonds Are a Girl's Best Friend* in *Gentlemen Prefer Blondes* (aspect ratio 1.33:1).

the other one was red). Experiments with both 3-D and widescreen in films had actually been going on since the early 1920s, but it was in the 50s that they really took off. Ultimately, 3-D was little more than a passing fad, but widescreen persisted. In 1953, Twentieth Century Fox introduced the world to the widescreen process known as CinemaScope, which was used by many studios between 1953 and 1967 (it eventually gave way to Panavision, which is the most commonly used widescreen process today).

In 1953, five films were released in a widescreen aspect ratio. By the following year, there were nearly 40. And by 1955, the number had exploded to more than 100. Today, widescreen dominates American filmmaking in a variety of aspect ratios. But there are two "standardized" ratios that are by far the most common: Academy Flat (1.85:1) and Anamorphic Scope (2.35:1).

1.85:1
Academy Flat

2.35:1
Anamorphic Scope
(aka Panavision/Cinemascope)

In the case of Academy Flat, at 1.85:1, the dimension of the image has 1.85 units of width for every 1 unit of height. In other words, the film is 1.85 times as wide as it is tall (this aspect ratio is often referred to today as simply "Flat"). Anamorphic Scope is even wider, at 2.35 times as wide as it is tall (it's usually called "Scope"). Some familiar films shot in the Flat aspect ratio include *The Right Stuff*, *All the President's Men*, and *Shrek*. Scope titles include *Star Wars*, *The Matrix*, and Peter Jackson's *Lord of the Rings* films. You can see more examples of Flat and Scope films on the next page.

There can be no doubt that widescreen films convey much more dynamic imagery, with the wider aspect ratio serving to enhance the dramatic impact of the film. But when it comes time to transfer such films to home video, there's a problem with those wider aspect ratios – they're too wide to fill your TV screen vertically if you're seeing the whole image horizontally.

Prior to DVD, there were two primary ways to deal with this problem on home video: *pan and scan* and *letterbox* transfers. The pan and scan process has the video camera, during the telecine process, literally scanning back and forth during the transfer to keep the most important action centered on your TV screen. On DVD packaging, this is rarely referred to as "pan and scan." More often you simply see the words "full frame" or "full screen."

There's a popular misconception these days that the words "full frame" on a DVD case mean that you're seeing the full (or entire) film image.

Detective Jimmy Doyle (Gene Hackman) in *The French Connection* (aspect ratio 1.85:1).

Twentieth Century Fox

Mary and Ted (Cameron Diaz and Ben Stiller) reunite in *There's Something About Mary* (aspect ratio 1.85:1).

Twentieth Century Fox

Twentieth Century Fox

Sergeant Keck (Woody Harrelson) and his soldiers come under fire in *The Thin Red Line* (aspect ratio 2.35:1).

Twentieth Century Fox

Chris and Rheya (George Clooney and Natascha McElhone) grapple with reality in Steven Soderbergh's re-imagined *Solaris* (aspect ratio 2.35:1).

Those who believe this also usually assume that the black bars you see on a widescreen DVD are actually *hiding* part of the film image. Nothing could be further from the truth.

The fact is, if you're watching a full frame version of a film that was originally shown in theaters in widescreen, you could be *missing* as much as 50 percent of the original film image. As you'll see in the next section of this book, not only are you *not* seeing the whole film, the beauty of the artistic composition of objects and movement within the frame (which was carefully crafted by the film's director and cinematographer) is lost completely.

For this reason, most serious film enthusiasts prefer the letterbox format, in which the *entire* film image is presented, and black bars fill the unused screen area at the top and bottom of the frame.

While some vertical picture resolution is sacrificed with letterbox presentation, the director's original widescreen composition is preserved — you're seeing the *whole* film, as you were meant to. Why would you want to see the film in any other way?

Why Widescreen Really Is Better

Let's take a look at some comparisons between full frame and widescreen presentations of films on DVD. As you'll soon see, being able to see the whole widescreen image can make a *huge* difference. There's just no comparison.

Widescreen vs. Full Frame (2.35:1 Ratio Films)

Since this 2.35:1 aspect ratio is the wider of the two common ratios in use today, it only stands to reason that you'll be missing out on the most picture area when watching a full frame version of a Scope film. All of the examples shown in this section are "frame grabs" of actual DVD video, taken from discs that include both full frame and widescreen versions of the film.

The following are screen shots taken from the film *Independence Day*. This is from a sequence early in the film, when the alien invaders' massive spaceships are arriving over major cities around the world. The whole point of this shot is to illustrate the vast scale of the ship and how it dom-

Twentieth Century Fox

ID4 (aspect ratio 2.35:1) presened in widescreen (above) and full frame (right).

Twentieth Century Fox

inates the expansive New York City skyline. Unfortunately, much of that sense of scale is lost in the full frame version.

On the next page, you'll find a scene from a less epic Scope film, *Life or Something Like It*, in which Ed Burns and Angelina Jolie play TV news staffers. They're having an argument here, and the dramatic tension of the scene depends on seeing them react to one another. But in the full frame version, you can only see one of them at any given time. You can't see the other person's reactions at all. As a result, the scene is much less effective in full frame.

Now let's try an action film, *The Transporter*. As you can see in the images on the next page, actor Jason Statham is involved in an intense battle with two thugs . . . except you can barely see the bad guys in the full frame version! Once again, the scene is much less effective and intense in full frame than it is in widescreen.

Twentieth Century Fox

Life or Something Like It (aspect ratio 2.35:1) presented in widescreen (above) and full frame (right).

Twentieth Century Fox

Twentieth Century Fox

The Transporter (aspect ratio 2.35:1) presented in widescreen (above) and full frame (right).

Twentieth Century Fox

Widescreen vs. Full Frame (1.85:1 Ratio Films)

Scope (2.35:1) films aren't the only ones to suffer from full frame presentation. Here's an example (below) of a film in Academy Flat (1.85:1) aspect ratio, *Ice Age*, in both widescreen and full frame versions. While the problem isn't as severe here as it can be with wider aspect ratios, the problem is still apparent. Once again, these images are actual DVD snapshots.

In this particular case, while the characters are all visible in the full frame version, the image is much tighter and more crowded. The sense of scale intended in the widescreen version is compromised.

Keep in mind that all of the examples we've shown you are just single shots from films with many hundreds or even thousands of such shots. If you think the difference is obvious, imagine the same effect multiplied, shot for shot, over the entire length of the film. The loss of dramatic impact widescreen films suffer when seen in full frame is tremendous.

Ice Age (aspect ratio 1.85:1) presented in widescreen (above) and full frame (right).

Other Options

Now that you've seen what a difference there is between widescreen and full frame, we should note that there are a few additional techniques that can be used to get around the problem of bringing widescreen films to home video.

Below, you'll find a look at the widescreen and full frame versions of a scene from *Cast Away* starring Tom Hanks. The film was presented in the 1.85:1 aspect ratio in theaters, exactly as the director intended. Using a special "spherical" lens during the filming process, however, director Robert Zemeckis was able to capture additional vertical image area in the frame. For standard full frame TV presentation, we're simply allowed to see more of that additional area. We've added the white box outline on the full frame image to show you exactly what portion of the picture was seen theatrically in widescreen.

Cast Away (aspect ratio 2.35:1) presented in widescreen (above) and full frame (right).

This can be an effective technique, but not every director chooses to film their movies in this way, for a variety of artistic and technical reasons. For example, when shooting with this process, it's very easy for unwanted objects like boom microphones, light stands, and other production equipment to stray into the frame. In addition, the dynamic composition of objects and motion in the frame isn't as dramatic in the full frame image. That's why directors who use this technique generally still prefer the widescreen version. This process is usually referred to as *35mm spherical* or *open matte*. A variation of it, using a special film format, is called *Super 35*. Director James Cameron has long used Super 35 to achieve this same widescreen and full frame compatibility for films like *Titanic* and *Terminator 2: Judgment Day*.

There's also another very new process that can be used to create more effective full frame presentations for home video. It only applies to computer-animated films, but we'll mention it here because, as more such films are released, the process is becoming more and more common. It involves recomposing and re-rendering the original widescreen image for the full frame format. All of the characters, props, and scenery visible in the widescreen image are still used, but they're simply repositioned to make better use of the full frame aspect ratio. To use Disney and Pixar's *Toy Story* as an example, in a scene involving a conversation between Buzz Lightyear and Woody, the characters might be moved a little closer together, so they're more in the center of the screen. It's still the same scene, with the same dialogue and dramatic effect — it's just been made more effective for full frame viewing.

The Anamorphic Advantage

Whatever the filming process or aspect ratio, we hope by now that you understand, from an artistic standpoint at least, why widescreen is almost always the better choice for watching films at home on DVD. To us at *The Digital Bits*, the full frame (and particularly pan and scan) presentation of widescreen films is as bad as colorizing a black and white film — it amounts to artistic butchery. Still, there are plenty of people who aren't convinced by the artistic argument for widescreen. The more pragmatic among you might be thinking: *So what? I've still only got a regular, 27-inch TV, and I want my whole screen filled when I'm watching a movie.*

Fair enough. We appreciate your opinion . . . but we're still not done. In fact, we've saved our *best* argument for choosing widescreen on DVD just for folks like you. It's a practical, financial argument, and it's very hard to ignore.

Most of you have probably heard of *Digital TV* (DTV) by now. If you've been to your local Best Buy or other electronics store in the last few years, DTVs are pretty hard to miss. The *Federal Communications Commission* (FCC) has mandated a full conversion of American television broadcasting to Digital TV by the year 2006 (although the realities of the marketplace will probably mean that the actual conversion will take somewhat longer). A similar process is underway elsewhere around the world as well. This conversion to digital broadcasting means that, in the next decade or so, the vast majority of you will have to buy new Digital TVs.

Here's the important thing you need to know about Digital TVs for the purposes of our argument: The vast majority of them are *widescreen*. That's because the chosen aspect ratio for Digital TV broadcasting is 1.78:1. In other words, DTVs are usually 1.78 times as wide as they are tall (this is also referred to commonly as 16x9).

However it's phrased, the bottom line is that the future of TV is widescreen. You can probably already imagine how much easier that will make it to watch widescreen movies at home. No longer will your TV's aspect ratio require the Hollywood studios to butcher the presentation of their widescreen films with pan and scan transfers.

Okay, you're now probably thinking, so *I'll start buying widescreen DVDs some day after I finally upgrade to a new widescreen Digital TV. What difference does it make whether I buy widescreen DVDs right now?*

It makes a *big* difference, folks. In the same way that the improved sound quality you currently enjoy on DVD will get even better when you

1.78:1
U.S. Digital Television (16x9)

eventually upgrade your audio equipment to multi-channel Dolby Digital and DTS surround sound, so to will the picture quality of your current DVDs improve with a new Digital TV. But here's the catch: You'll *only* enjoy that improved picture quality with *widescreen* DVDs. If you've been buying all your discs in full frame, you're out of luck.

Let's use the example of a 1.85:1 aspect ratio film (see the images on the next page).

If you have a full frame version of the film on DVD, it might look fine on your current TV, but it won't fill your new widescreen DTV. Think you hate the black bars now? Just wait until you have to watch all your DVDs with gray bars (generated by the TV) on the sides of your screen because the image on the disc doesn't fill the widescreen frame! On the other hand, if you're wisely buying widescreen DVDs now, you'll have to deal with black bars on a current TV, but the image will fill your new DTV screen completely. Cool, huh?

We should note here that 2.35:1 films will still have slight black bars on the top and bottom of a DTV (because 2.35:1 is a wider aspect ratio than 1.78:1), but they'll be *much* less annoying, as you'll see in a moment.

The bottom line is that the vast majority of widescreen DVDs will look better than ever on DTVs. On the other hand, those of you with full frame DVD collections will have to replace all your discs. How's *that* for a good argument for widescreen on DVD?

The reason that most widescreen DVDs will look better on DTVs has to do with a special feature of the DVD format known as *anamorphic enhancement*. Simply put, the anamorphic enhancement of a film on DVD means that the video on the disc is of the highest resolution possible (short of true high-definition), while maintaining the film's original widescreen aspect ratio. Naturally, the best way to view this anamorphic widescreen video on the disc is on a widescreen display, like a new Digital TV.

So how does anamorphic enhancement work? It all starts in the telecine process. When transferring the film to a digital video master for home video, the film is recorded in anamorphic format. The resulting video image appears to be "squished" horizontally. In an anamorphic image, for example, people look unnaturally tall and thin. This transfer process often involves nothing more than digitally recording the exact image on the film print—most widescreen film prints feature an image that, if you looked at a single frame, appears squished horizontally. What happens in theaters

Full frame version of *Ice Age* (aspect ratio 1.85:1) on a standard 4x3 TV (right) and a future 16x9 DTV (below).

Twentieth Century Fox

Twentieth Century Fox

Widescreen version of *Ice Age* (aspect ratio 1.85:1) on a standard 4x3 TV (right) and a future 16x9 DTV (below).

Twentieth Century Fox

Twentieth Century Fox

is that when the film is projected, a special lens is used that "unsquishes" the image you see on screen so that it looks normal.

Once the transfer is done, and the video is being prepared for DVD, the authoring technician simply inserts a code into the disc's software instructions indicating that the video is in anamorphic mode. Your DVD player reads this code when playing the disc and recognizes that the video is anamorphic. Your widescreen TV then "unsquishes" the image so that you can view it normally, much like the lens on a projector in the theater. The only difference is that, on your widescreen TV, the process is electronic.

To help you understand how anamorphic enhancement works, and what a difference it makes, let's take a closer look at how your DVD player and TV handle anamorphic and non-anamorphic video. We'll use examples of both 1.85:1 and 2.35:1 films, and show you how they look on both standard 4x3 and widescreen 16x9 TVs.

Anamorphic vs. Non-anamorphic (1.85:1 Film)

For this demonstration, we've chosen to use an infamous scene from the film *There's Something About Mary* (aspect ratio 1.85:1). Here's the video recorded on a non-anamorphic DVD. Notice the black bars at the top and bottom of the frame. These are actually present in the signal.

Twentieth Century Fox

Now here's the video recorded on an anamorphic DVD (at the top of the next page). Notice that the image appears "squished" horizontally, while retaining its full vertical resolution, and that there are virtually no black bars visible in the signal. Normally, you would never see the video in this state. The only time you would see this "squished" picture is if you were watching the disc on an improperly set-up DVD player, using a standard 4x3 TV - the player thinks you have a widescreen 16x9 TV. A quick adjustment in the player's set-up menu would correct this problem.

Here's non-anamorphic video as it appears on a standard 4x3 TV. This is the familiar letterboxed image you're used to (below).

Now here's anamorphic video as it appears on a standard 4x3 TV. The DVD player performs a mathematical down-conversion on the video signal, in effect combining every 4 lines of vertical resolution into 3 until the correct aspect ratio is achieved. The black bars at the top and bottom of the image are generated electronically, completing the image. Visually, it's nearly indistinguishable from a non-anamorphic (letter-boxed) DVD image.

Here's where it gets interesting. This next image is non-anamorphic video as it appears on a widescreen 16x9 TV. The gray bars are generated by the TV to fill in the unused portions of the screen. Using the TV's "zoom" mode, you can magnify the image to fill the screen electronically, but at the cost of degrading the image quality significantly.

Twentieth Century Fox

Now here's the payoff (below). This is anamorphic video as it appears on a widescreen 16x9 TV. The "squished" image recorded on the disc (seen at top) is sent directly to the TV, which stretches the video signal horizontally until the correct aspect ratio is achieved. As you can see, the image fills the frame, while retaining its full vertical resolution. The picture quality is stunning.

Twentieth Century Fox

Anamorphic vs. Non-anamorphic (2.35:1 Film)

For our second demonstration, we've chosen a scene from the film *X-Men* (aspect ratio 2.35:1). Here's the video recorded on a non-anamorphic DVD. Notice that the black bars at the top and bottom of the frame are somewhat thicker than in a 1.85:1 presentation. Since the 2.35:1

aspect ratio is wider than the TV's 1.78:1 (16x9) ratio, the thicker bars are necessary to maintain the proper composition. These are actually present in the signal.

Now here's the video recorded on an anamorphic DVD (below). Notice that the image appears "squished" horizontally, while retaining nearly its full vertical resolution. In addition, black bars are now visible at the top and bottom of the frame. Since the 2.35:1 aspect ratio is wider, the bars are necessary to maintain the proper composition. These are actually present in the signal. Normally, you would never see the video in this state. The only time you would see this "squished" picture is if you were watching the disc on an improperly set-up DVD player, using a standard 4x3 TV - the player thinks you have a widescreen 16x9 TV. A quick adjustment in the player's menu would correct this problem.

Here's non-anamorphic video as it appears on a standard 4x3 TV (at the top of the next page). This is the familiar letterboxed image you're used to.

Twentieth Century Fox

Now here's anamorphic video as it appears on a standard 4x3 TV (below). The DVD player performs a mathematical down-conversion on the video signal, in effect combining every 4 lines of vertical resolution into 3 until the correct aspect ratio is achieved. Electronically generated black bars are added to the existing ones (to fill in the remaining screen area), completing the image. Visually, it's nearly indistinguishable from a non-anamorphic (letterboxed) DVD image.

Twentieth Century Fox

Once again, here's where it gets interesting. This next image is non-anamorphic video as it appears on a widescreen 16x9 TV. The gray bars are generated by the TV to fill in the remaining screen area. Using the TV's "zoom" mode, you can magnify the image to fill the screen electronically, but at the cost of degrading the image quality significantly.

Twentieth Century Fox

And again, here's the payoff. This final image is anamorphic video as it appears on a widescreen 16x9 TV. The "squished" image recorded on the disc is sent directly to the TV, which stretches the video signal horizontally until the correct aspect ratio is achieved. As you can see, the image fills the frame, while retaining nearly its full vertical resolution. Since the 2.35:1 aspect ratio is wider than the TV's 1.78:1 (16x9) ratio, thin black bars are still necessary to maintain the proper composition. Nevertheless, the picture quality is stunning.

Twentieth Century Fox

As you can see from these demonstrations, anamorphic enhancement will make a huge difference when you get that new widescreen TV. To knowledgeable DVD fans, seeing the terms "16x9" or "anamorphic" on a disc's packaging is reassuring. It tells them that, in today's blistering consumer electronics marketplace, where change and obsolescence can happen all too quickly, the DVDs they're spending their hard-earned cash on are effectively future proof. And that's important.

How Do I Know a DVD Is Anamorphic?

The first thing you should know is that only widescreen DVDs can have anamorphic enhancement. So if you have a full frame disc (1.33:1 aspect ratio), you know right away that it's not anamorphic.

If your DVD is widescreen, chances are good that it is anamorphic, particularly if it's a new disc from one of the major Hollywood studios. Virtually all of the Hollywood studios take advantage of DVD's anamorphic feature whenever they can these days, but smaller, independent studios occasionally do not use this feature. The reason is that releasing a DVD in anamorphic requires a new anamorphic film-to-video transfer. Smaller companies might not have access to the original film elements to do this, or they might not be able to afford a new transfer. In addition, some of the very earliest DVD releases from the major studios aren't anamorphic either. It took a year or so for all of the studios to realize the benefits of releasing widescreen discs with anamorphic enhancement.

Anamorphic enhancement is labeled in a variety of different ways on the back of a DVD's packaging. Look for key words or phrases like "16x9," "anamorphic widescreen," "enhanced for widescreen TVs," or "enhanced for 16x9 displays." They're all referring to the same thing. You might also find symbols like those shown below on the back of a DVD package.

But what if you've got a widescreen DVD and you can't find any markings about anamorphic on the packaging? It does happen. Many of Columbia TriStar's widescreen DVDs are anamorphic but are not labeled as such. If you only have a 4x3 TV, how do you tell? Simply go into your DVD player's setup menu and tell it that you have a widescreen TV (it may be labeled simply "16x9"). On your standard TV, if a disc is anamorphic, the image displayed by your player will look squished. If it looks normal, on the other hand, the disc is non-anamorphic. Don't forget to switch your DVD player back to standard "4x3" TV mode when you're done!

So now you should know everything you need to about how the video part of DVD works. You should understand aspect ratios and anamorphic enhancement, and why widescreen really *is* the better choice on DVD.

With that in mind, it's time to look more closely at the audio side of the DVD format.

Common
anamorphic
packaging symbols

What Is Surround Sound?

If you've spent any time in a movie theater, then it's a good bet that you're already familiar with the concept of surround sound. But what exactly is it? How does it work?

Simply put, surround sound is the use of multiple channels of audio, delivered to multiple speakers positioned around the listening space, in order to create the illusion of a three-dimensional sound environment. The desired effect is to immerse the listener in the fictional world of the movie, with ambient and directional sound cues that draw you into the images unfolding onscreen. In other words, surround sound is one of the ways that filmmakers get you to forget your own reality when you're watching the one they've created to entertain you.

One of the common myths of home video is that the sound you heard in the movie theater is the same sound you're hearing in your home. Very often, they're actually completely different sound mixes, with different design philosophies in mind. When mastering movie surround sound for DVD, the same basic processes used for theater surround sound are applied, except that special consideration must be given to the fact that home theaters are fundamentally different listening environments than theaters. Home theaters aren't likely to have a dozen, massive, professional-quality loudspeakers, specifically designed to fill stadium-sized rooms with sound. Whereas movie theaters are often designed with special acoustic treatments and materials used to enhance the sound quality, home theaters are often built in living rooms and other typical home spaces, filled with furniture, windows, bookshelves, doorways, and other surfaces that can reflect sound in unique and undesirable ways.

Sound engineers typically work from the film's original digital audio stems when mixing surround sound for DVD. In some cases, particularly for new movies, the same engineers who worked on the film are involved in the new mix for DVD. Taking care to remain faithful to the original theatrical audio experience, they create new configurations of surround sound specifically designed for home theater use. This is typically done via a variety of common surround sound formats, including Dolby Surround, Dolby Digital, and DTS. We'll talk more about those in a minute, but first we need to teach you how to read and understand typical surround sound notations—the kind you'll often find on the back of a DVD package.

At some point, if you've been watching DVDs for any length of time, you've probably heard the numbers "5.1" used to describe DVD surround sound. These numbers are nothing more than a numerical way to describe the optimal speaker configuration that the audio track is designed to be played back on. The "5" refers to the number of speakers in the system. These typically correspond to standard left and right speakers (usually the largest in the system, positioned to the left and right of your TV), a center channel (usually placed above your TV), and two surround sound speakers (located to the side or slightly behind the listening position, one on the left and one on the right). The ".1" number indicates that the track also includes a low-frequency effects channel (sometimes referred to as an LFE) that is usually delivered to a powered subwoofer — a particular kind of speaker that is specially designed to play the lowest portions of the audio track. People often think of this as the "bass" sound. The sub-woofer can generally be positioned almost anywhere in the sound environment because the longer wavelength of low-frequency sound means that it's difficult to determine the direction it's coming from.

While 5.1 is the most common configuration of DVD sound, it's by no means the only one. Older movies, made before the advent of stereo recording, often feature mono audio, noted as 1.0 (meaning that the sound is delivered to the center speaker only) or 2.0 mono (meaning that the same sound signal is sent to both the left and right speakers instead of the center). Stereo audio is usually noted as just 2.0 (once again indicating that the sound is delivered to the left and right speakers, but each speaker gets a different portion of the soundtrack — the left half and the right half). Surround sound configurations are even more complex. In the figure on the next page, you'll find some examples of typical DVD sound configuration icons that you may find on the back of a disc's packaging.

Now that you understand how to read these sound icons, and you know what the numbers mean, it's time we got back to those sound formats we mentioned earlier.

Dolby Surround, Dolby Digital, DTS, and PCM

In the same way that a film's video must be encoded for DVD using MPEG-2 compression, the digital audio data must also be prepared for DVD. In most cases, it too must be compressed to fit in the space available on the disc. There are three different digital audio compression formats that are commonly found on DVDs today, plus another common format that's uncompressed. Let's take a closer look at them.

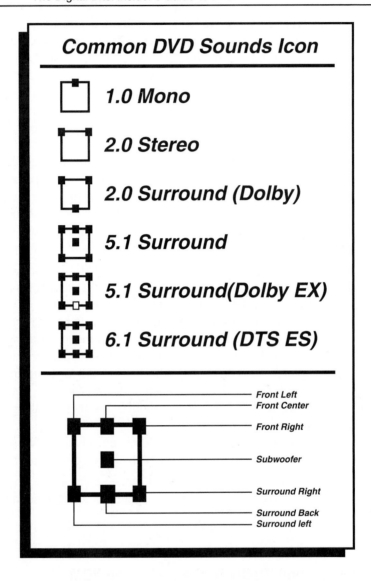

Dolby Surround

Dolby Surround (also sometimes known as Dolby ProLogic) is one of the oldest audio encoding formats found on home video. Dolby Surround is an analog audio format and has been used since the days of VHS (it was also common on laserdisc). It's a proprietary format, created by Dolby Labs, which takes a normal stereo signal (containing just two channels of audio—a left channel and a right channel) and extrapolates center and surround channels from it to give you a virtual surround sound effect.

In order to take advantage of Dolby Surround, you need a Dolby Surround-compatible receiver, along with five speakers (front left, center, and right speakers, along with left and right surround speakers). Most home theater receivers today feature Dolby Surround decoding, which is one of the reasons Dolby Surround audio tracks are commonly included on DVDs.

Dolby Digital

Dolby Digital (also sometimes referred to as AC3) is a more advanced proprietary format, also created by Dolby Labs, which features 6 channels of discrete digital audio (noted as 5.1, or front left, center, and right channels, along with left and right surround channels, and a channel of low-frequency sound). By encoding the surround audio in six discrete channels, greater clarity is achieved, meaning that the surround effect is often much more dynamic and natural sounding.

In order to take advantage of Dolby Digital, you'll need a DVD player and an audio receiver with on-board Dolby Digital decoding capability . . . or a DVD player with on-board Dolby Digital decoding capability and a receiver that is "5.1 ready" (meaning that the 6 channels of audio are run in separate cables from your DVD player into your receiver, which then delivers them to your speakers). Of course, you'll also need at least five speakers and a subwoofer.

A new variation of this format, called Dolby Digital EX, uses a process similar to Dolby Surround to create a virtual surround back (or center back) channel as well, creating an even more dynamic effect. Note that Dolby Digital EX playback requires that your receiver be specifically Dolby Digital EX compatible. You must also have an additional speaker (which is placed directly behind the listener) to deliver the center back audio.

Dolby Digital is the standard audio format for DVD (meaning that all DVD players *must* be compatible with it), and it's also part of the audio standard for future HDTV broadcasting.

DTS

DTS is another proprietary audio format, created by Digital Theater Systems, which also features 6 channels of discrete digital audio (also 5.1, once again including front left, center, and right channels, left and right surround channels, and a low-frequency channel). You can think of DTS as something of a competing format for Dolby Digital. It's not as common as Dolby Digital on DVD-Video discs, but it uses a higher audio data rate than Dolby Digital does. Remember our explanation of video data rate? Audio data rate works in much the same way, just for digital audio information. Because DTS uses a higher audio data rate than Dolby Digital, it's generally less compressed. Many home theater enthusiasts therefore believe it sounds better.

Our experience at *The Digital Bits* is that DTS *can* often sound "smoother" and more natural than Dolby Digital; although this is not always the case (the quality of the audio signal depends on many additional factors). That said, it's important to note that the difference between the two formats is going to be very hard for most people to appreciate, particularly if you're not a serious sound enthusiast. Think of it this way: The difference between Dolby Digital and DTS is much like the difference between a Porsche and a Lamborghini in the world of cars. If you've been driving a Ford all your life, both high-end cars are going to blow you away. Likewise, if you're used to hearing stereo or mono sound from your TV's speakers when watching movies, or even Dolby Surround for that matter, either of these new digital sound formats is going to represent a *huge* improvement. Any way you slice it, Digital and DTS are both terrific surround sound formats.

We should mention that a new variation of DTS, known as DTS-ES, is actually a 6.1 format. It features the same six channels as regular DTS, along with a seventh channel containing discrete surround back (or center back) audio information. The addition of the center back channel (the speaker for which is placed directly behind the listener), helps to smooth out the surround field even more, creating an even more natural listening environment. After all, you're effectively immersed in the center of 360 degrees of active movie audio — a very cool experience.

DTS playback requires that you have a DVD player equipped with a special DTS-Output, plus a receiver that features on-board DTS decoding capability. As with Dolby Digital, you will also need at least five speakers and a subwoofer. DTS-ES will require at least one additional speaker (for the surround back channel) and a receiver with specific DTS-ES compatibility.

DTS is an *optional* part of the audio standard for DVD (in other words, DVD players are not required to feature DTS compatibility, although most do these days).

PCM

One other digital audio format is occasionally available on DVDs in Region 1. It's called PCM (which stands for *Pulse Code Modulation*). This is the same basic audio format used on current music CDs, and it's also part of the audio standard for DVD (meaning that all DVD players must be compatible with PCM). PCM is generally a stereo-only format, meaning that it doesn't encode surround sound information. In addition, it's different from Dolby Surround, Dolby Digital, and DTS, in that it's an uncompressed audio format.

DVD Special Features

Obviously, the most important aspect of any good DVD is the movie itself. After all, if the movie isn't any good, and if it isn't presented in the best video and audio quality possible, then what's the point of watching the disc in the first place?

But now that we've looked at the video and audio aspects of DVD, it's important to let you know that there's often much more on a DVD than just the film. That's where special features come in.

Once you're finished watching the movie, you might want to explore further and learn more about it. You may have questions, such as: How was the film made? How do the filmmakers feel about their work? What was it like on the set? One of the big advantages of DVD is the format's capability to include many different kinds of interactive bonus materials on the disc, along with the movie itself, to add value to the viewing experience.

Not all DVDs have extra features, but most have at least a few offerings, like behind-the-scenes footage and trailers, and many have quite a bit more on top of that. Good DVD special editions will have *lots* of interesting bonus material — as much as many hours worth in some cases.

Special features on DVD can make a great movie an even more incredible experience. If the movie is all you want, then these extras are just the icing on an already delicious cake. On the other hand, if you *are* looking for more, it's probably right here, wrapped up inside a nice shiny package. Special features are a big reason why so many people love DVD — it's all about the possibilities.

Let's run down some of the most common kinds of DVD special features and explain what they're all about.

Audio Commentaries

These are alternate audio tracks that you can choose to listen to during the film, instead of the actual film audio. Imagine sitting in a room with the director, the lead actor, or a film historian, watching the movie as he or she explains various aspects of the production, and tells you interesting behind-the-scenes stories and anecdotes. That's the basic idea of an audio commentary track. Sometimes there's just one person on the track. Other times, there will be several people talking in an ongoing conversation. A few DVDs even have multiple audio commentary tracks — one with the director, one with the actors, one with the writers, and so on.

Mostly, what you'll find are interesting bits of trivia and other information you didn't know. They can be thoughtful . . . even revealing. And they might even make you view the film in a different light. Listening to director Francis Ford Coppola's track on the *Godfather* DVDs, for example, is an absolutely amazing experience. On Warner's *Citizen Kane* DVD, film critic Roger Ebert discusses the various themes of the story, the unique ways in which director Orson Welles composed the onscreen images, and even the film's history. For the *Fight Club: Special Edition* DVD, director David Fincher and star Brad Pitt crack wise about their experiences together on the set. And that's just scratching the surface.

To be fair, some audio commentaries aren't exactly riveting. Actually, the bad ones can be downright dull. What you often get, on a bad commentary, is a rambling play-by-play: "Oh, yeah . . . here I am walking down this dark alley. The bad guy's gonna jump out in a second. There he is . . . cool. You know, I remember this day of shooting. I think I had a tuna sandwich for lunch that afternoon."

That said, each commentary is a new and different experience because of the unique personalities and perspectives of the people involved. Great audio commentaries are one of the reasons that DVD special editions are often referred to as "film schools on disc." A good audio commentary track will not only entertain you; it'll teach you a few things too. If you love film, an audio commentary is like a personal tour into the minds of the filmmakers themselves. And that's a pretty cool thing.

Deleted Scenes

Sometimes, during the making of a film, whole scenes and sequences are created that are never seen by theater audiences. The scenes might not be as effective as they were thought to be originally, they might be confusing, they might give key plot points away too soon, or they might slow down the pace of the film. Some are cut for no other reason except that the film is running longer than desired by the studio.

Scenes like these are ultimately edited out of the final cut of the film. As the expression goes, they're "lost on the cutting room floor." Nonetheless, they're often fascinating to see. After all, a lot of effort was involved in creating them in the first place. They may represent whole unseen subplots of the story. They can even contain the entire performance of a supporting actor, who wasn't otherwise seen in the film at all.

Cutting these scenes is often a very difficult decision for a director to make. And so, very often, directors will save these trimmed scenes and include them on the DVD versions of their films for fans to enjoy. The DVD version of *Star Wars: Episode I* is a good example of a disc that features a number of such deleted scenes. In this particular case, all of the scenes were completely finished with special effects and sound just for the DVD release.

When you're watching these scenes, you can often easily understand why these were deleted. But every now and then, you'll see a deleted scene that's so good, you'll be surprised it wasn't in the film. Dream-Works' *Gladiator* DVD has a few like this.

Occasionally, a DVD will give you the opportunity to view these scenes in the context of the film, using something called *branching*. Basically, the film will play until the point where the scene was cut. Then the DVD player will be instructed to jump to that particular deleted scene on the disc and play it. When it's done, the player will resume playing the rest of the film normally. In some cases, this is done in such a way that you don't even realize the branching has occurred. We call this *seamless branching*. Artisan's *Terminator 2: Ultimate Edition* allows you to watch no less than three different versions of the film using this technique.

Every now and then, a director will even decide to restore some of these scenes back into the film for the DVD release. New Line's 4-disc "extended" edition of *The Lord of the Rings: The Fellowship of the Ring* features more than 30 minutes of new scenes restored back into the film by director Peter Jackson, complete with new music and special effects. The resulting film would have been far too long for theater showings, but on DVD, fans can delight in new moments of action and character development that significantly heighten the drama.

No matter how they're presented, deleted scenes are a fascinating look at the way films are crafted and shaped in the editing room. And they're a treat for fans.

Documentaries

Most of you are probably already quite familiar with documentaries, particularly if you watch a lot of public television or cable TV. Documentaries are basically nonfiction films created to look at a particular subject in great detail. In the world of filmmaking, these are often devoted to the making of a particular film, some aspect of the history of film, or the work of an individual actor or director.

On DVD, you access and play documentaries just like you do the film itself. On elaborate, multidisc sets, documentaries are often the most important component featured on the extras discs. Sometimes a documentary will even get a disc all to itself. A perfect example of this is Warner's *Stanley Kubrick Collection*, which not only features DVDs of eight of the director's best films, it also includes a separate, bonus disc with the marvelous documentary, *Stanley Kubrick: A Life in Pictures*.

Documentaries are as important to good DVD special editions as audio commentaries and deleted scenes. They can literally take you behind the scenes and allow you to glimpse aspects of the filmmaking process that you otherwise might never get to see. Some documentaries are as long (or longer) than the films they're about. And the best documentaries contain a wealth of information that provides valuable context and can deepen your appreciation of the subject matter.

Featurettes

Featurettes are similar to documentaries, except that they're much shorter — usually 30 minutes or less. Whereas a documentary generally covers a wide range of topics relating to the making of a film, a featurette might focus on just one particular aspect of the production in greater detail. On the other hand, some featurettes are made by the studio in advance of a film's release to help promote it and get people into the theaters. These are usually fluffy and lacking in substance. They're often referred to as *EPK* featurettes, because they were created for the film's Electronic Press Kit.

In many cases, a DVD might not have a single documentary, but rather a whole series of featurettes that together add up to a documentary. You can often choose to view them one at a time, or all at once via a "play all" option. This practice is becoming more and more common on discs these days, for reasons that might surprise you.

As the DVD format has become bigger and bigger business, and studio earnings continue to increase, actors and other talent involved in the creation of films have become more aggressive in their efforts to be paid their fair share of the DVD profits. Currently, there are union rules in Hollywood that say actors must be paid for documentary participation . . . but participation in a film's *promotion* is included in their original contracts. That means that actors usually don't get paid for appearances in featurettes. What's the difference between a documentary and a featurette? Well, remember how we said that featurettes are usually 30 minutes or less? Anything over 30 minutes is considered a documentary. So the next time you're watching a disc with no documentary but lots of featurettes, now you'll know why.

Alternate Audio Tracks

Like an audio commentary track, alternate audio tracks are those that you can choose to listen to instead of the film's original audio. In fact, commentaries *are* alternate audio tracks. But there are many other types as well.

Imagine listening to your favorite film, with all of the dialogue performed in a foreign language and "dubbed" to roughly match the actors' lip movements onscreen. Or imagine listening to a music-only track that allows you to appreciate just the film's score, without any dialogue or sound effects. On the other hand, how about listening to an effects-only track that illustrates the process of creating all the interesting little sounds you hear in a film—all the little bumps and bangs and crashes that make up a film's sound effects.

Alternate audio tracks let you experience films in a variety of new and interesting ways.

Image Galleries

An image gallery can take many forms on a DVD. Basically, they're composed of a number of still images that you can browse with your remote. You might have a gallery of photographs taken on the set by the cast and crew. You might find images of production sketches and other artwork that was made to help visualize the costumes and sets and characters in pre-production. Or you may find storyboards—little drawings, much like the kind you see in comic books, that are done to help lay out the various camera angles and shots to be used during filming. MGM's *Hannibal: Special Edition* includes a gallery of unused designs for the film's poster artwork—artwork that was rejected for various reasons and normally never gets seen by the public. Some DVDs have literally hundreds of images for you to browse through. There are many possibilities when it comes to image galleries.

Subtitles, Captions, and Trivia Tracks

Most DVDs offer a variety of text tracks in addition to their audio and video options. Text tracks are exactly what they sound like. If you select one, lines of text will appear on your TV screen (usually generated electronically by your DVD player) while you're watching the film.

Subtitle tracks are almost always included for foreign films, originally produced in another language. Film purists often prefer to listen to the original language audio, while viewing subtitle text that allows them to read and understand the dialogue.

In much the same way, captions are often included on a DVD so that viewers who are hearing-impaired can read and understand the dialogue as well as other audio elements of the film. For example, when a phone rings on the film's soundtrack, text will appear onscreen describing this: *(phone ringing).* Many, but unfortunately not all, DVDs also include official *closed captioning* as mandated by U.S. government law (specifically, the Television Decoder Circuitry Act of 1990). These tracks are usually created for TV broadcast in the United States, and are actually part of the video signal. In this case, your TV decodes the captions from the video signal and electronically displays them at the bottom of your screen.

Another text option occasionally found on DVDs today is trivia tracks. These work just like subtitles, except that, rather than containing the film's dialogue, they feature interesting facts and other behind-the-scenes information related to what you're seeing onscreen. Paramount's special editions of the *Star Trek* feature films on DVD all feature text trivia options.

Trailers and TV Spots

You've all seen these at one time or another. You simply can't go to a film in theaters these days without having to sit through a few (and sometimes many) trailers for other, upcoming films. Trailers have one purpose . . . to get butts in theater seats. They're designed to sell you on the promise of a good time at the movies. In the same way, shorter versions of the trailers you see in theaters are often assembled to be shown on TV during commercial breaks in your favorite TV shows.

Because of their promotional nature, they can sometimes be quite fascinating. But they're also a double-edged sword. Many times, studios are so desperate to get you into the theaters that they'll edit the trailers to make the film look a lot better than it really is. Sadly, trailers also often give away the best parts of the film—you'll feel like you've already seen the movie. In any case, fans often enjoy these sneak previews, and so they like to have them included on the DVDs of their favorite films.

Multi-Angle Video

Haven't you ever wanted to be able to switch to a different camera angle during, say, sporting events? It's also fun to be able to switch to different angles during live concerts, to see the different musicians performing. Many DVDs of sports and concerts feature just this ability. You simply press the "angle" button on your remote at any time during the program and you can switch to a different view of the action.

Some feature film DVDs use this option in their special features to let you watch, for example, an important stunt sequence from the film from different camera angles. That's easy to do, because expensive stunts are often filmed with more than one camera to be sure that the action is captured adequately on film. Other uses for the multi-angle feature might be to let you view a special effects scene in various stages of completion — the raw footage, the rough animation, and the final completed scene from the film. Or, you might be allowed to see the film's storyboard drawings compared to the final scene.

It's worth noting that, usually, you don't find multi-angle video during the main program of a feature film DVD. The reason for this is that the film's director and editor deliberately select the camera angle you're seeing onscreen to achieve the dramatic effect they wish to convey. Allowing you to choose a different angle would work against their efforts. For this reason, multi-angle video is a less common DVD special feature, but it can be quite interesting when it is available.

Easter Eggs

Sometimes fun little bonus items will be deliberately hidden on a DVD for you to discover as you navigate through the disc's various menu pages. They're not listed on the back of the disc's packaging — they're meant to be surprises. They might include outtake footage, gag reels, additional deleted scenes, trailers . . . you never know what you might find. Some are easy to find, while others are quite complicated to access, involving hidden codes that must be entered with your remote. Not all DVDs have Easter Eggs, but many of them do and they're generally quite popular with fans.

Interactive Games

Some DVDs, particularly those made for a younger audience, feature different kinds of interactive games that you can play with your player's remote control. These can include trivia questions, hide-and-seek games . . . the list of possibilities is almost endless. Usually, they're relatively easy to play and are themed to the various characters and situations seen in the film. Many Disney DVDs include such games, some of them quite elaborate, featuring new animations and lots of fun surprises as rewards for playing.

It should probably be noted that most adults, as well as more serious film fans, will probably have little interest in interactive games, but parents will certainly appreciate the entertainment value they offer to children.

So that's a look at some of the most common kinds of special features that you can find on DVDs today. In many ways, the kind of extras a DVD includes is only limited by the imagination of the DVD producer and the amount of space that's available on the disc. Of course, depending on the disc, there are other types of features that you might be able to access in addition to those we've listed here. But this gives you a pretty good idea of what extra features are all about.

It's also important to note that what we've talked about here covers only the features you can access via a regular DVD player — the kind you have in your living room or home theater. Some discs also contain extras that can only be accessed by playing the disc in your computer's DVD-ROM drive.

The ROM Side of DVD

These days, most PC and Mac-based computers equipped with DVD-ROM drives allow you to view movies on disc. In fact, if you've done any traveling by airplane lately, you've probably noticed that a lot of folks on the go like to watch DVDs on their laptops to pass the time. But there are other kinds of DVD features that can *only* be experienced with a DVD-ROM drive. Many discs these days, particularly those released by the major Hollywood studios, contain a variety of interactive DVD-ROM features, in addition to those you view with a regular DVD player in your living room — features specifically designed to be run on computers.

In our experience, some of these ROM extras are genuinely interesting and are well worth your time. The best can include things like interactive script readers, which allow you to view the original screenplay text while watching the film, storyboard art, and more. Occasionally, a studio will include the film's promotional web site on the disc. If your computer has Internet access, you can also sometimes connect to a special, online web site that features additional supplemental content, such as deleted scenes and other materials that are not actually on the disc itself. This allows the studio the opportunity to continually offer new bonus content to its customers. Another interesting ROM feature for Internet users is the capability to participate in special live events. MGM, for example, once hosted an online DVD-ROM viewing of *Ronin* with the film's director, who provided live, running audio commentary and answered select questions from participants.

One of the coolest DVD-ROM features we've seen at *The Digital Bits* can be found on DreamWorks' *Shrek* disc. It's called the ReVoice Studio, and it allows you to dub your own voice into various scenes from the film, replacing the original character dialogue (note that you need a PC equipped with a microphone, sound card, and speakers to use this feature). You'd be surprised how much fun it is, and the process is very similar to professional *automatic dialogue replacement* (ADR), which allows actors to improve upon their lines in post-production, even after they've been recorded on the set.

Unfortunately, though, most DVD-ROM material tends to be of little real value to movie fans. You'll often find a ton of promotional material for other titles available on DVD from the studio, or links to online stores where you can spend money on things like T-shirts and action figures. There may also be "trivial" items such as screensavers, wallpapers, or interactive games aimed at younger viewers. Since these are often software programs that need to be installed on your computer to be used, there's no guarantee that they'll work (and they often don't). You might find these things interesting for a while, but they generally have little replay value, and they typically contribute nothing to your understanding of the film.

There are other problems, too. When studios include genuinely valuable material on an online web site, accessible via DVD-ROM, the material often doesn't stay on the site indefinitely, so if you want to access it a year later, you're out of luck. What's worse, for a variety of reasons, most of these DVD-ROM extras are generally only available to PC-based computers, so you Mac users are out of luck when it comes to this sort of material.

All in all, there's no denying the potential of DVD-ROM extras. But our experience is that Hollywood has a very long way to go before they'll even come close to exploiting this potential in a way that will hold real interest to movie fans. Bottom line . . . let's just say that if you don't happen to have a PC computer equipped with a DVD-ROM drive, you're usually not missing all that much.

Menus and Authoring: Making It All Work

But DVDs aren't just made up of video, audio, and supplemental material. They are, in a very real way, computer software. Each disc contains software code—digital instructions that tell your DVD player how to read and access the information stored on the disc.

When you put a tape in your VCR and press play, it just starts playing. There's no special instructions the VCR needs—it just begins mechanically reading the electronic information on the tape. But in order to watch a DVD movie, your DVD player needs to know where to start. It needs to know what format you want to watch the movie in, what language you wish to hear, and so on. In short, it needs instructions.

So, encoded onto the DVD is a set of those instructions in a software format. You put the disc in your player and these instructions tell your player to, for example, go to the main menu page of the disc. From there, you can select Dolby Digital 2.0 stereo or 5.1 surround sound. You can choose to view the film in widescreen. You can choose, say, French subtitles. And you do all this by navigating through the DVD's onscreen menus. Each time you highlight and select an option with your remote, software code on the disc instructs the player to enable this option during playback. You are literally programming your player to do what you want it to.

The menu pages themselves are often quite lively, with animation, full-motion video, and even music and sound playing in the background as you make your choices. In many cases, these menus are *themed* to the film on the disc. For example, the menus use images or poster art from the film in the background. Sometimes the menus are even made to look like three-dimensional environments from the "world" of the film itself.

Twentieth Century Fox

The main menu screen from Fox's *Die Hard: Five Star 1.5* DVD special edition.

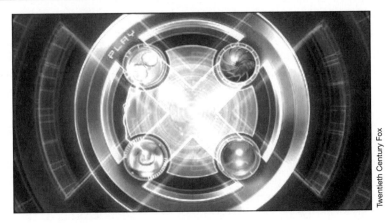

Twentieth Century Fox

The main menu screen from Fox's *X-Men 1.5* DVD special edition.

The basic idea of such onscreen menus is to allow you to make your playback choices and to access bonus material. But a secondary benefit is that they can actually help get you in the mood to watch the film. And the very best menu pages are intuitive and easy to navigate quickly. After all, who wants to watch a behind-the-scenes featurette on the disc if it takes 10 minutes to find the thing?

An interesting and important point to make here is that because DVDs involve an element of software, they must be programmed just like regular software is. This process is called *authoring*. DVD producers and technicians, using special computer programs, carefully assemble all of the necessary instructions to make the disc work. (Pressing this button plays the movie, pressing that button takes you to the special features page, and so on.) But there's more. Not only must DVDs be programmed, they also have to be *tested* just like computer software is. Before DVDs are released to the public, they undergo a rigorous testing process to make sure they work on as many existing models of DVD player as possible. This QC (or *Quality Control*) period can last weeks or even months in some cases. And if a defect is found that causes a disc to malfunction on a certain kind of player, it gets rejected and the authoring technicians must go back and correct the error. Sometimes it isn't the disc that's faulty, but rather a particular model of DVD player, in which case the manufacturer must "upgrade" the player in order for it to work properly. In the very early days of DVD, such player upgrades happened a lot, although they're much less common now. Over the years, manufacturers have gotten better at anticipating such problems and eliminating them before the players appear in stores.

In any case, DVD authoring and QC work is a complicated trial-and-error process. So, as you see, there's a *lot* more to a DVD than meets the eye.

Region Codes and Copy Protection

One of the biggest concerns that the Hollywood studios have with regard to releasing their films on DVD, or in any format for that matter, is protecting the content. The *Motion Picture Association of America* (MPAA) estimates that the U.S. filmmaking industry loses some $3 billion annually from piracy—the illegal copying and distribution of their films. So during the development of DVD, a number of copy protection features were deliberately built into the format.

The first of these is something called *region codes*. Basically, the entire planet is divided up into geographic regions (seven in all). The U.S., for example, is in Region 1, along with Canada and Mexico. All DVD players and DVD-ROM drives are coded for a specific region. Likewise, most movie discs contain software instructions that specify a specific region. In order to play a movie coded for Region 1, you have to have a DVD player coded for Region 1. If you try to play a DVD coded for a different region (say, a movie sold in Japan, which is in Region 2), you'll get a message saying that you can't view the disc. On the next page, you'll find a map of the different regions created for the DVD format.

The main reason for region codes is that many films are released on DVD in the U.S. before they even appear in theaters in other countries. The fear is that, without region codes, people in other countries could simply import DVDs of the latest films from the U.S., which would hurt profits from the film's international theatrical release. Given that these international consumers would also already own a copy of the film on DVD, it would potentially hurt the home video release in that country as well.

Many DVDs indicate their region code on the back of the packaging, using specialized icons. These usually include some sort of graphic representation of a globe, along with a number that designates the specific region the disc is coded for (generally 1–7, or "0" or "All" for discs that are region free). There may also be a notation of the television format the disc is compatible with (NTSC, PAL, or SECAM).

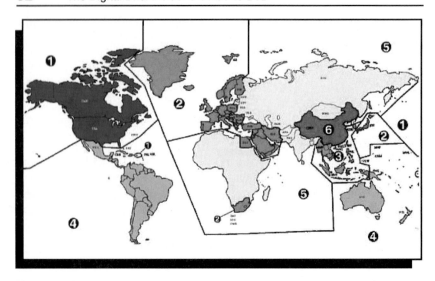

Region code map

Sample region code icons

Another feature built into the DVD format is an electronic copy protection scheme called *CSS* (which stands for *Content Scrambling System*). CSS is designed to thwart digital piracy – of great concern given the high quality of DVD movies and the potential to make perfect digital copies. Simply explained, it works by encrypting various portions of the data recorded on the disc. The data can only be properly decrypted by a DVD player or DVD-ROM drive, which then outputs the video and audio signals to your video and audio equipment. At no point is an unencrypted digital video signal ever allowed to leave the DVD player or ROM drive. New (and future) DVD recorders are designed to recognize a DVD's CSS encryption and disallow copying of the disc.

Finally, on many Hollywood studio DVDs, the video signal itself is encoded with *MacroVision* signals. These work to help discourage analog copies of movies, on VHS tape, by interfering with the VCR's recording process. When you view a movie on tape recorded from a DVD encoded with MacroVision, you'll generally see the brightness level of the video change randomly from bright to dark. The idea is that it's annoying enough that it will discourage pirates from making (and selling) lots of illegal copies.

All of these copy-protection schemes aside, we believe that the best way to prevent piracy has absolutely nothing to do with technology, but rather price and availability. After all, why would anyone want to buy a pirated copy of a film on DVD, in terrible quality and with no extras, when you can get a high-quality, authorized copy with lots of extras for a low price? The answer to that question, as we say at *The Bits*, is a no-brainer.

DVD Packaging and Labeling

DVDs come in many different kinds of packaging types, made of different materials in all manner of unique shapes and sizes. But there are four basic types that are the most common: the Snapper, the keep case, the Super Jewel Box, and the Digipak.

The Keep Case

The keep case has become the de facto standard for DVD-Video packaging and is by far the most common type of case that you'll find today. Most titles released by Columbia TriStar, Twentieth Century Fox, DreamWorks, Universal, and Disney come shipped in various versions of the keep case. It's also the case we tend to prefer, for both its simplicity and durability.

Keep cases are manufactured by several companies, including Amaray and Alpha, come in a variety of colors, and can be configured to hold a single disc or several depending on the title (the more discs contained, the thicker the case). The single-disc version consists of an all-plastic shell that snaps open like a book to reveal space for an insert on the left, and a snap-on holder for a disc on the right. The outside of the case is always wrapped with a clear plastic sleeve that holds a paper sheet featuring the DVD cover artwork.

Sarah Hunt

Examples of Digipak (left), keep case (center top), Snapper (right), and Super Jewel Box (center bottom) DVD packaging.

The Snapper

The Snapper case is also quite common, as nearly all titles released by Warner Bros. use this style of packaging. They're also found on many older New Line and Image titles. The Snapper is our least favorite DVD packaging type, as its cardboard surface is easily dented, creased and otherwise damaged.

Snappers are generally available only in single-disc configurations. Multi-disc sets using this packaging type require a cardboard slipcase to contain multiple Snappers, making them somewhat bulky. The Snapper basically consists of a plastic disc holder wrapped in a cardboard shell that features the DVD cover artwork. It's held closed on the right side of the package with a plastic snap tab, thus the name.

The Super Jewel Box

The Super Jewel Box, and variations of it, should be fairly familiar to most of you already. That's because it's basically just an oversized version of the same clear plastic jewel case that's been used to hold music CDs for decades. The dimensions have simply been altered to make them similar to the keep case and the Snapper.

The Super Jewel Box is generally made entirely of clear, hard plastic. When opened, the left side is basically a lid, which holds a booklet or insert containing the cover art (this can be seen through the case itself). The right side holds the disc and also contains the back and side artwork insert. The Super Jewel Box is a bit prone to scratching and cracking, but it is generally very durable and secure.

The Digipak

The Digipak comes in many different shapes and configurations. It's similar to the Snapper in that it's mostly made of cardboard. The Digipak is commonly used for multidisc special editions by Twentieth Century Fox, New Line, and more recently by Warner. In this application, the package generally consists of an outer cardboard slipcase, featuring the cover artwork, which in turn encloses a gatefold style cardboard inner package. This folds open to reveal multiple panels of artwork. Some of these have clear plastic disc holders affixed to them, while others have sleeves for booklets and other inserts.

We're a bit torn about the Digipak at *The Bits*. On one hand, they generally make very attractive packages. But, like Snappers, they're not as durable as other packaging options and can easily be damaged.

Custom Formats

In addition to the types of packaging we've discussed, DVDs also come in a whole variety of custom cases. These are generally more rare but can include plastic boxes, embossed metal tins, and even wood boxes (see Figure 1-61). DVDs have been released in book-like packages and, in at least one instance, foam latex cases. Nonstandard packaging types have the advantage of standing out visually on store shelves. Unfortunately, some are also very difficult to store on your DVD library shelf with other discs.

DVD Features Labeling

When it comes to listing a disc's special features on DVD packaging, there is unfortunately very little in the way of standardized labeling. Over the years, this has resulted in a lot of confused customers. Fortunately, a few studios have perfected a system of labeling that works quite well. Universal pioneered the "features grid" concept for its DVD releases, and

Sarah Hunt

Figure 1-61   Examples of nonstandard DVD packaging, including large box sets, clear plastic cases, metal collector's tins, and even a foam latex package (far right).

the idea was quickly adopted by DreamWorks as well. Recently, Twentieth Century Fox has also begun to use a version of it.

The grid is quite handy, in that you can learn pretty much everything you need to know about the disc at a glance, from its video aspect ratio to its audio surround sound format. There is information about subtitles, captions, languages, and more. If only all studios made their packaging this easy to read! You can see an example of a features grid on the next page.

You've probably noticed that the features grid says little about the supplemental material included on the disc — the documentaries, the audio commentaries and the like. That's because studios will often list these extras as marketing bullet points on the back of the DVD case. Some discs have few, if any, extra features, while others have so many that they couldn't possibly all be listed on the package.

The variety of different features' labeling formats can be confusing, but if the labeling is at least accurate, you should get a pretty good idea of

LANGUAGES	English	Español	Français		Dolby Digital and DTS soundtracks contain up to 5.1 channels of discrete audio. Dolby Surround soundtracks contain up to 4 channels of encoded audio. Playback from 2-channel DVD outputs is compatible with stereo and Dolby Pro Logic reproduction. DTS Digital Surround play requires the use of a DTS decoder.
dts	5.1 SURROUND			**2.35:1**	
DD DOLBY DIGITAL	5.1 SURROUND		5.1 SURROUND	ANAMORPHIC WIDESCREEN	
DD DOLBY SURROUND	2.0 DOLBY SURROUND				
CAPTIONS SUBTITLES	Captioned & Subtitled	Subtitulos	Sous-Titre	COLOR	Dual Layer · 1 Hr. 32 Mins. · 90626

The features grid from the back of DreamWorks' *Old School* DVD.

what's in store for you on the DVD. And the more practice you have in reading DVD packaging, the more sense it will start to make.

What Is THX?

This is another question we get a lot at *The Digital Bits*. You've probably seen one of the many different THX trailers in your local movie theater — they generally run just before the movie begins. If you've been buying or renting movies on disc for any length of time, you've probably seen them on your favorite DVDs too. In fact, DVD special editions will often display the THX logo right on the front cover, usually accompanied by the words: Digitally Mastered for Superior Sound and Picture Quality. But what the heck *is* THX anyway? And what does it stand for?

The first thing you should know is that THX is *not* a sound format. It's not like Dolby Digital or DTS. THX is nothing more than a quality assurance certification. It's kind of like a "Good Housekeeping" Seal of Approval for movie viewing.

THX was first developed by engineers working for filmmaker George Lucas in the 1980s. Lucas wanted a way to ensure that the movie theaters he was showing his films in were providing the best possible presentation quality for audiences. He wanted to know that the picture and sound experience of his films would be the same each time, no matter where a film was shown. THX was actually named after Lucas' first film, *THX-1138*, but it also unofficially stands for Tomlinson Holman eXperiment. Tomlinson Holman is the name of the Lucasfilm technical director who led the team of engineers that first established the THX guidelines for theaters.

Basically, these guidelines set rules about things like how much noise you should hear from the lobby and adjacent auditoriums, how projection and sound equipment should properly perform, how bright the screen should be, where the speakers should be placed, and so on. THX

engineers would work with theater owners to correct problems and create the best theater environment possible. In 1983, *Return of the Jedi* became the first film to be shown in a THX-certified theater. Today, there are more than 3,000 THX-certified theaters around the world.

The THX concept was so successful that the same basic idea was later applied to film production equipment, home theater and stereo equipment, and even DVDs. For home theater equipment, a THX certification means that the device has been designed to perform to specific standards, allowing you to enjoy the best possible video and audio performance at home. When it appears on a DVD, THX means that the picture and sound included on the disc have been mastered according to specific production guidelines. Once again, this allows you to enjoy the best possible video and audio experience.

Of course, the quality you enjoy will only be as good as your specific equipment allows. And even good equipment can produce lousy quality if it isn't set correctly. For this reason, many THX-certified DVDs include a feature called an *Optimizer*, which is basically a set of text instructions and test patterns. The Optimizer will guide you, step by step, through the process of setting up your TV and sound system correctly, in order to deliver the best quality (we'll talk more about this in the next section of this book).

The next time you're watching a movie with your friends and the THX logo appears, drop that little Tomlinson Holman eXperiment factoid on them. They'll either be impressed . . . or they'll decide that you've got *way* too much time on your hands.

Conclusion

So there you have it. We've now recounted some of the DVD format's history, explained the technical ins and outs of how it works, and described to you some of the many treats the format has in store for you as movie fans. You should all have a pretty solid understanding of what DVD is all about at this point.

But, in order to take full advantage of all the shiny plastic goodness DVDs can offer, you need something to play them on. In other words, you need to know how to put together a decent home theater. Even more importantly, you need to know how to assemble a decent home theater for a price that won't break your bank accounts, thereby robbing your kids of their college education and plunging your golden years into abject poverty.

Worry not, folks. You *don't* have to be Steven Spielberg to watch movies at home in style. It is absolutely possible to build a home theater you can be proud of on a budget. Don't believe us? Well . . . *The Digital Bits* Home Theater School is now in session. If you'll just continue on to the next section of this book, we'll show you how it's done.

Part II

Building a Home Theater Made Simple

So You Want to Build a Home Theater . . .

What do people mean when uttering the words "home theater"? Are they talking about their living room, with a rear-projection TV and kid's toys strewn about the floor in front of a great big cuddling couch? Or do they mean that old room in the back of the house, with a TV screen, five speakers, and a DVD player? Maybe they're speaking of a massive room with elaborate lighting sconces on the wall, a massive projection screen that rolls down from the ceiling, and seating for a dozen people?

All of these are correct.

Quite simply, a home theater is a space in your house in which you attempt to recreate the experience you get when watching movies in a full-fledged theater.

It doesn't matter whether you have theater-style seats or a frumpy old couch to sit on. It doesn't matter if there are curtains that open on cue to reveal your screen, or if you have to pull curtains shut over your windows to block out the light. What matters is how well you're able to experience the film. Are you drawn into a big, dazzling movie image? Does multi-channel surround sound swirl around you, putting you right smack in the middle of the action? When you watch a DVD in your home theater, do you get that same sense of being transported that you do in a real movie theater? That's what matters.

You should all know right now that it is absolutely possible to build a good home theater on a budget.

Thanks in large part to the success of DVD, home theater equipment has never been cheaper — quality and features that used to be reserved only for those with the deepest wallets are now available at surprisingly affordable prices. And new features are being added all the time.

If you've decided that you want to build your very own home theater . . . the first thing we want to say to you is, "Congratulations!" We're here to tell you that you're in for a world of thrilling entertainment. There's nothing like curling up with your family and enjoying a great movie in your own home theater. There's no traffic to deal with, there's no idiot talking on his cell phone, and the popcorn doesn't cost eight bucks a bowl. A home theater is like your own little slice of cinema paradise.

But before you start buying all that spiffy equipment, you first need to ask yourself a few questions. For example, what features do you want in your home theater? Do you want the very best performance in terms of video and audio quality? Do you want your system to be upgradeable and expandable? Do you want your home theater to double as a good music listening environment? These are just a few of the things you need to consider. The way to answer all these questions (and many others) is simple: Do your research.

Research is the most important thing you can do when you're planning your home theater.

This one step can save you not only headaches down the line, but also huge amounts of money when you're buying equipment. Thankfully, when it comes to research, there are more resources available to you today than ever before. Let's take a look at a few of them.

Consumer Guides

Consumer product review magazines can offer valuable advice, particularly about specific models of equipment. Be aware, however, that with technology products (such as DVD and computers), they're often a bit behind the curve with regard to advice about what's considered state of the art. As a result, referring *only* to consumer magazines means that you're not always going to be getting the whole picture. If you want to make a good purchase decision — one that you'll be happy with for years to come — you need to consult other sources as well.

Home Theater Magazines

Another good place to turn is home theater-specific magazines. These will be packed with information on a wide variety of models and equipment types, and they do an excellent job of looking ahead at new developments in technology. Unfortunately, they can often be very difficult to read, brimming with terms you've never heard of and performance test numbers that seem almost meaningless to the uninitiated. Even technology-savvy consumers will have trouble understanding it all. They also often focus on very high-end equipment — equipment that's either too expensive for most consumers, or simply overkill in terms of features offered. So again, if you want to make good purchase decisions, you should consult additional sources than just home theater magazines.

In-Store Visits

Of course, there's always a simple visit to your local electronics store. Nothing beats being able to see the particular DVD player or TV you want firsthand. Salespeople are usually there to answer your questions, and they'll probably even hand you the remote and let you take the device for a test drive. Even this avenue of research has drawbacks, however. Salespeople aren't always the experts they claim to be on the technology. More than once, we've gone into a store to look at gadgets and been told *completely* inaccurate information by well-meaning salespeople. Additionally, many salespeople are paid on commission, so you have to keep that in mind when considering their advice.

"Okay," you're probably thinking, " . . . so product review magazines, home theater magazines, and salespeople are all problematic. How the heck are we supposed to do our research?" That's another good, old-fashioned, straightforward question. So we're going to give you another easy answer: *The very best thing you can do is to seek advice from other consumers just like yourself.*

There are *lots* of people out there who have successfully built great home theaters — people who were once beginners just like you may be, and who are willing to offer you the benefit of their experience. Even longtime home theater enthusiasts find it valuable to consult with their peers from time to time. Let's say you're thinking about a particular model of DVD player . . . somewhere out there, there are people who have already purchased it and who can answer every question you may have. Do they like it? Have they had any problems? How does it work with the rest of their system? What's the performance like? This same idea applies to virtually every topic or aspect of home theater. Looking for the best way to position your surround sound speakers? Someone out there has already figured it out. Looking for the best price on a receiver? Someone out there has *already* found a good deal on exactly the model you want.

The Internet

Thanks to the Internet, seeking the advice of fellow home theater enthusiasts has never been easier. There are a number of great online discussion groups and forums you can visit where you'll find literally thousands of people chatting about home theater. These range from the complete beginner to some of the most knowledgeable technologists there are. Even manufacturer reps and other industry experts check in to offer advice.

A simple search of Yahoo or Google using the keywords "home theater discussion" will reveal a long list of links to good home theater discussion forums. We can immediately recommend the following:

The Home Theater Forum (www.hometheaterforum.com)

AV Science Forum (www.avsforum.com)

Home Theater Spot (www.hometheaterspot.com)

There are many others as well. Registration with these forums is usually free and the discussions are well moderated. There are often numerous discussion areas for different topics (like movie software, surround sound equipment, high-definition television, and so on) — all you have to do is find the topic you're interested in and then read the existing threads. If the particular question you have isn't already answered, just post it on the board. Soon, you'll likely have a number of replies from people willing to help you.

Not only is the Internet a good place to do your research, there are a number of reputable online retailers that specialize in home theater equipment. While some of you may not be comfortable buying high-ticket items like DVD players online, you can often get a significantly better price than you would at your local electronics store. Sometimes you don't even have to pay sales tax, which can save you a lot of money, even with the cost of shipping factored in. If you're interested in the possibility of buying equipment online, we recommend you visit the forums mentioned above — many have specific areas where members can rate their experiences with online retailers, and offer their recommendations about those they trust.

Another good way that you can sometimes save money is to buy models of equipment that are soon to be discontinued. Manufacturers introduce new models of DVD players, receivers, and TVs every year. When they do, retailers typically lower the discontinued models in price — sometimes dramatically — to make room for the new inventory. There's generally nothing wrong with these models — they just might not have the very latest, state-of-the-art features. But if you do your research, and find that the features these discontinued models do offer are well matched to your needs, there's a bargain to be had.

Our Goal

What we're going to do in the pages ahead is to provide you with an understanding of the components that make up a good home theater system, and give you practical guidance and advice that will help you make sound purchase decisions. We'll give you an idea of the different features you might want to consider, and those that might not be worth your money. We'll also give you some *general* tips on how to set up and calibrate your system so that you'll enjoy the best video and audio performance your equipment allows.

What we can't do is tell you how to set up and calibrate your system in *specific* detail. There are just too many possibilities involved — too many models of equipment, possible configurations, and the like. As a general rule, we suggest you refer to the instruction manuals that will come with each component of your system for specific instructions on how to make connections and so on. Still, there's a lot we *can* tell you that will make things easier as you move into the world of home theater.

Before we get started, however, there's one other very important rule you should keep in mind when buying home theater equipment . . . or *any* technology product for that matter. It's the rule of diminishing returns. What it basically means is this:

You're generally going to get what you're willing to pay for . . . up to a point. But once you reach that point, the more money you spend, the less value you get in return.

Building a good home theater is all about finding the right compromise between the features you want and the price you're willing to pay. That may sound complicated, but if you break it down, it's not so difficult. Let's begin by identifying the general price range of the home theater you wish to build.

How Much Can You Spend?

For the purposes of this book, we're going to identify four very general price ranges that apply when building a home theater. These aren't meant to be specific but are simply designed to give you an idea of the different levels of home theater you can go for, based on how much money you'll probably end up spending.

The "I Don't Want to Spend More Than $2,000 Bucks" Home Theater

If all you're interested in is seeing a good movie on DVD in your home, this is the simplest solution. It's certainly not the best option, but who's to fault you for wanting to watch DVDs on a regular old 27-inch TV?

That's right, for less than $2,000 you can probably put a good standard TV (a tube set), or possibly even a lesser quality digital-ready TV, together with what's known as a "home theater in a box." A typical "home theater in a box" system (so named because all the components are *literally* packaged in the same box) will include a combination DVD player and surround sound receiver, along with a set of five small bookshelf speakers and a subwoofer. Keep in mind, however, that while this *sounds* like a cool idea, the receiver, speakers, and subwoofer are often quite low powered and are therefore unsuitable for use in any room larger than, say, a typical dorm room or a very small office. We can tell you from experience that they're just not powerful enough to fill a medium to large room with surround sound, certainly not at any reasonable volume. In addition, the DVD player is generally lacking anything more than basic features.

Our advice to those of you who *can* be swayed from this option is this: Rather than trying to buy *all* your equipment for this price, we recommend that you simply buy one or two components of higher quality. Then, in a few months, when you've got a little more money available to spend, buy a couple more of the components you need. That way, over time, you can assemble a genuinely good or even great system, all while still managing your budget. Just because your current home theater budget is less than $2,000 doesn't mean you have to settle for a mediocre system.

The "What Am I Going to Spend My Tax Refund Check On?" Home Theater ($2,000–5,000)

For those of you who are assembling your first home theater, or for the more casual home theater enthusiasts, this option is probably the best way to go. For under $5,000, if you're a savvy enough shopper and you put in your research, you should easily be able to buy a fully featured DVD player, a good solid surround sound receiver, a decent set of matched speakers with a subwoofer, and a fair-sized, widescreen Digital TV (tube or rear projection). Properly set up and calibrated, that's all you need to enjoy a good theater experience in your home. Just add popcorn.

The "I've Got the Money to Spend So Why Not?" Home Theater ($5,000–15,000)

If you're a little more serious about your home theater, then you will probably want top-of-the-line video and audio performance from your system. For about double the cost of a good system, you can assemble a truly great home theater—one that includes a fully featured DVD player (capable of playing the latest high-resolution audio formats in addition to regular DVDs and CDs), an equally well equipped surround sound receiver (capable of rendering up to 7.1 channels of movie audio), an excellent set of matched speakers with a powerful subwoofer, and a huge, state-of-the-art, widescreen, rear-projection Digital TV. If you're a *very* savvy shopper, you may even be able to upgrade to a true video projection system with a very large, separately mounted screen. Properly set up and calibrated, a system in this price range will deliver tremendous quality—the kind that will have all your friends just begging to come over and watch movies at your house. And that's half the fun right there, isn't it?

The "I've Just Got Way Too Much Money to Burn" Home Theater ($15,000 and up)

This is just way too much home theater for most folks to appreciate. Remember that rule we told you about, where you get what you're willing to pay for up to a point, but after that point the extra money buys you less in return? Well, if you're spending more than $15,000 on a home theater, you've pretty much crossed that point.

Don't get us wrong—the quality you're going to get for this kind of payola will be truly amazing. But it's more than only the most trained aficionados will really be able to appreciate. You see . . . home theater is one of those hobbies where you can spend as much money as you want to. The sky is really the limit. Want a $5,000 DVD player with all the internal video and audio circuits made of gold? They're out there. Want a fully automated touch screen built into your coffee table that will let you dim the lights, lower the projection screen and start the movie all with the touch of a single button? No problem, but it'll cost you. You can buy rows of actual, ultra-plush theater seats, complete with cup holders. They're very cool – definitely not cheap, but cool. Hell, there are even ultra high-end speaker cables that will run you $1,000 a foot. No kidding.

Unless your last name is Spielberg, Lucas, Cruise, or Ebert . . . and/or you've got money coming out your ears . . . spending this kind of money on your home theater is overkill. On the other hand, if your name *is* Spielberg, Lucas, Cruise, or Ebert . . . *and/or* you've got money coming out your ears . . . let us simply say, "Thanks for sending one of your people out to buy our book."

For the rest of you, however, let's see what we can do about setting you up with a good ($2,000 to $5,000) or great ($5,000 to $15,000) home theater system, shall we?

Picking the Right Room for a Home Theater

One of the keys to success in building a high-quality home theater system is to pick the right room, and then to prepare the room properly. Unless you live in a large house, and have a spare room to devote to a home theater, most of you will be using your living or family rooms to watch movies. That's not a problem, but there are a few things to consider.

A home theater room should allow you to precisely place your surround sound speakers. It doesn't matter if they're on stands or they're wall-mounted. There are general rules to how you should place them to get the best sound, and your room needs to be able to accommodate this.

A home theater room should have carpeted floors. If the room you've chosen has hardwood, tile, or laminate floors, you should be prepared to use a large area rug in your home theater. The reason is that hard flooring surfaces tend to reflect sound in unusual ways, ruining the quality of the surround sound effect you're trying to achieve.

A home theater room should also allow you to precisely control the direct and ambient lighting in the room. Are there windows in the room? You'll need to be able to block their light with curtains or other coverings. Are there lamps in the room, or light coming from other rooms? You'll need to be able to turn them off, or close doors to keep light from getting in. This is very important. In the same way that real movie theaters are darkened when the film is playing, you should also be able to darken your home theater room. Unwanted light sources can cause reflections or otherwise compete with the brightness of the video display, preventing you from seeing the movie in full detail.

Finally, keep in mind that in order to enjoy the most comfortable viewing in your home theater, your seating — your couch or favorite comfy chair — should be positioned a certain distance away from your video display. As a general rule, for a standard 4x3 TV, take the diagonal measurement of your video display's screen and multiply it by three. The result, in inches, should be the distance between your display and your seating. If you're planning to use a widescreen 16x9 TV, multiply the vertical measurement of the screen by four to get the correct distance.

Once you've addressed these concerns, you need to measure your room's dimensions. This will be important, particularly when you go to buy speaker cables. But more on that in a little bit.

Now let's take a look at the various components of a home theater.

The Video Display

The video display is the heart of your home theater. The entire system, from speakers to seating, is going to be arranged so as to allow you the best view of the video display. The choice of display device you make will make or break your system more than any other single component.

There are three basic types of video display we'll consider here: tube, rear projection, and front projection.

Of these types, *front projection* is generally the most expensive. This involves a video projector, mounted in the rear of your home theater environment (often on the ceiling), projecting the video image onto an actual, separate screen, very similar to the arrangement you find in a real movie theater. In addition to being the most expensive option, front projection is also the most demanding technically. It not only requires that the projector and screen be precisely aligned so as to render the correct image, it also requires that you have the greatest degree of control over ambient light. The projected image in this case is generally not going to be as bright as a tube or rear-projection display, so any unwanted light in the room will tend to ruin the effect. While front projection is arguably the ultimate way to display movies at home, for the purposes of this book, we'll consider it to be overkill.

The other two types of video display are *tube* and *rear projection*. Each has their own strengths and weaknesses in terms of their use for home theater purposes. Tubes tend to be brighter than rear projection sets and offer a somewhat wider viewing angle, but they're also much heavier. In addition, there are fewer tubes in widescreen format, and only the very biggest tubes will give you good home theater performance. Remember, the goal is to recreate the movie experience in your home . . . and that means a large screen is critical.

Rear-projection sets, on the other hand, offer some unique advantages over tubes. While good rear-projection sets are a bit more expensive than tubes, they're lighter, are often available in widescreen aspect ratios, and come in sizes upwards of 65 inches diagonal. While not as bright as tubes, in a properly darkened home theater environment, they're capable of rendering images of stunning quality. New rear-projection sets also have much wider viewing angles than early models did, so if that's your objection to rear projection, you'd do well to take another look.

In addition to the choice of tube or rear projection, there are two formats of video display to consider: *standard* and *digital*. Standard TVs are the same basic analog TV sets you've known for years. They're also going to become obsolete before the end of the decade. As we've explained earlier in this book, the FCC has mandated the conversion of American television broadcasting to Digital TV by the year 2006. The promise of DTV is higher quality video and audio, and even true high-definition broadcasting. But more important than just the potential quality improvement, this conversion means that, in the next few years, the vast majority of you will have to buy new Digital TVs. So if you're currently considering the purchase of a new video display, the only reasonable choice is a Digital TV. At the very least, make sure that you get a digital-ready TV, which can be made fully compatible with digital broadcasting by adding a separate DTV receiver/decoder at a later time (these currently sell for less than $200).

In terms of digital program availability, there's a lot for you to watch right now — all you need to do is connect your Digital TV's decoder to a small broadcast antenna. Virtually every television market in the country now offers at least some free, over-the-air content in digital format. All of the major broadcast networks (ABC, CBS, NBC, and FOX) offer programming in digital format, including sporting events and many of your favorite primetime shows. This varies in resolution from 480 progressive all the way up to true high-definition in 1080 interlaced format. PBS stations have been particularly aggressive in broadcasting high-definition content.

In addition, digital satellite services (such as DirecTV) now offer special high-definition channels to their customers, such as HD Net, Discovery HD, and ESPN HD. If you're a subscriber to HBO and Showtime, both offer special channels of high-definition feature film programming. Even some cable companies are now starting to jump on the bandwagon with at least a few channels in high-definition, although they've been somewhat slower to make their TV signals compliant with digital (since the FCC mandate did not specifically apply to cable operators).

In addition to allowing you to watch great quality broadcast, satellite, and cable signals, Digital TVs also offer an additional advantage for home theater use . . . their compatibility with progressive scan. As we've mentioned before, video runs at a rate of 30 frames per second. Most current TV sets scan each frame of that video on the display screen in two passes, called fields. The first field contains all the even-numbered lines of resolution in the image, while the second contains the odd-numbered lines. Together, they form the complete image (this is known as *interlaced* video). The fields are scanned so fast that your eyes, and brain, are fooled into thinking that the complete image is always present. But a slight flickering can sometimes be seen. Digital TVs equipped with progressive scan, on the other hand, render each frame of video on the display screen in a single pass. This is the same process that happens on a computer screen. The result is a smoother image. When used in home theater for watching DVDs, progressive scan renders a much more film-like image. If you ever have the chance to compare DVDs projected side by side, both with and without progressive scanning, you'll be stunned at the difference progressive can make.

For a good system, we recommend selecting a small to medium rear-projection Digital TV as the centerpiece of your home theater. For a great system, look for a larger rear-projection Digital TV of higher quality. Widescreen is definitely the preferred aspect ratio, so you can take advantage of not only widescreen movies on DVD, but current and future HDTV broadcasting in widescreen as well.

When selecting the particular model that's right for you, be sure to choose one that has a wide variety of audio and video inputs, so it will accommodate not only your DVD player, but also VCRs and other equipment you might have, both now and in the future.

Before we move on, we want to mention one other important note. Many people these days are becoming more interested in thin-screen plasma TVs. Plasma TVs are certainly attractive, in that you can literally hang them on your wall. They're also becoming more affordable. However, there are two important things to consider. First of all, plasma TVs are not good at rendering the darkest areas of a video image with great detail. The blackest areas of an image tend to look "crushed" or muddy. In addition, current plasma TVs are notoriously fragile. The screens are easily damaged . . . and the screens are the single most expensive component of the display—literally more than 80 percent of the cost of the TV. Damage the screen on a plasma TV, and you'd do just as well to chuck the TV and get a new one as have it repaired. For both of these reasons, plasma TVs are not suitable for use in home theater. Someday thin-screen technology will probably be the way to go. But that day is still a ways off.

The DVD Player

If the video display is the heart of your home theater, a DVD player is its soul. And the great thing about DVD players is that you can get a terrific DVD player, loaded with the features that will give you the best home theater performance, for well under $500. In fact, we've seen truly excellent players priced at about $250. It's true that you can even buy DVD players for less than $100, but these tend to be off brands, with questionable reliability records. For our money, we recommend a good midrange model from a major manufacturer.

In terms of features, DVD players have a lot to offer. To start with, nearly all DVD players these days allow you to play music CD discs as well. Some (but not all) will also play "burned" CD-R discs full of MP3 music files.

Most players these days offer compatibility with both Dolby Digital and DTS audio formats. Some (but not all) have built-in Dolby Digital decoding. This can be useful if your receiver is 5.1 ready, but doesn't have Dolby Digital decoding built in. Our recommendation, however, is to let your receiver do the surround sound decoding (it's much more likely to do a better job of it).

Some (but not all) DVD players will output progressive scan video through their component video output jacks. Progressive scan is a *must* in any good DVD player, even if you don't currently have a Digital TV that can take advantage of it. Someday you will have one . . . and then you'll be very glad that your DVD player is progressive scan compatible. Trust us on this.

DVD players generally come in two varieties: single-disc players and multi-disc changers. There are many people who like the convenience of changers, which allow you to load several different discs at the same time — you can switch from one to another at the touch of a button. For home theater purposes, however, we tend to recommend the single-disc variety. Because of the need for a rotating disc carousel, changers have more mechanical parts than single-disc players do . . . and more mechanical parts mean more things that can wear out over time. The multi-disc capability is also a little more expensive, so unless you really appreciate the feature, we recommend that you spend the money on other features that are more directly useful for home theater.

Some new DVD players offer the ability to record your favorite programs on blank, recordable DVD discs in addition to just playing your favorite movies. Recording is a very nice feature to have if you really want it, but again, you'll typically have to pay a premium for it (they currently cost well over $500). And there are other issues to consider. First, there are a variety of DVD recording formats available, and only a few of these will work on the majority of existing DVD players. Also, none of these recordable players are compatible with high-definition video, which is going to be more common in the next few years. Eventually, there will be high-definition disc formats that will allow you to record in true high-definition resolution. Until then, however, we recommend that most of you pass on DVD recording technology unless you really want or need it. You'll find more on recordable DVD later in this book.

Some new DVD players also feature compatibility with one or both of the new high-resolution audio formats: DVD-Audio and *Super Audio CD* (SACD). Each format offers your favorite music in quality that's far superior to current CDs. Since both formats seem to be receiving the support of many different record companies, and with most titles available in only one format or the other, we recommend that you get a combo player that's compatible with *both* SACDs and DVD-Audio discs. Again, you're going to pay a premium for this feature, but not nearly as much as with recordable DVD (you can get a few such models of combo players for as little as $250). You'll find more about high-resolution music later in this book as well.

There are other features offered by DVD players, of course, but these are by far the most important. When you're looking for a DVD player, whether you're building a good or a great home theater system, our advice is to find one that's compatible with both Dolby Digital and DTS audio, as well as progressive scan video. Beyond that, it's all gravy.

The Receiver

In many ways, the surround sound receiver is both the nerve center and the muscle of your home theater system. *Every* device in the system connects to it in some way. The receiver generally does most or all of the surround sound decoding in your home theater, and it routes that decoded audio information to all your speakers. The receiver's built-in amplifier also provides the power that runs those speakers, with the sole exception of the powered subwoofer. In addition, a receiver also allows you to tune in and listen to over-the-air broadcast radio signals.

Because of all this, it's important to get a high-quality receiver. Virtually all receivers these days feature built-in Dolby Surround decoding. Whether you're building a good or great level system, you *definitely* need on-board Dolby Digital and DTS decoding, and you'd do well to see that the receiver is also compatible with the new 5.1 EX and 6.1 ES variations. We also recommend that you check to make sure that there are plenty of inputs and outputs to accommodate all your current and future equipment needs. For example, if you're interested in high-resolution audio, you'll need your receiver to have dedicated 5.1 inputs (in both SACD and DVD-Audio, the player does the decoding and sends the decoded signals to your receiver, which then routes them out to your speakers). If you buy a combo DVD player that plays both SACD and DVD-Audio, you'll just need one set of 5.1 inputs, but if you buy separate SACD and DVD-Audio compatible players, you'll need separate 5.1 inputs for each. Some high-end receivers even have 7.1 inputs to allow you to eventually upgrade to future surround sound formats.

It's also very important to get a receiver that is rated with more than enough power to run all your speakers. Keep in mind that the *amount* of power delivered to your speakers determines, in large part, the *quality* of the sound your speakers reproduce. So be sure to check that your receiver's power rating is high enough to meet the power demands of your speakers, and then some. It's definitely better to have *more* power than not enough.

We should also note that in addition to simply decoding audio signals and powering your speakers, the receiver can also allow for the easy control of all your equipment. By routing all your video and audio signals through your receiver, before sending them to your TV and speakers, you can often use a single, universal remote (the receiver's) to control everything. There is one drawback of doing this, however. More on that in a minute.

The Speakers

Whether you're listening to stereo music or 5.1 surround sound on DVD, it stands to reason that you need to have a speaker to reproduce each of the channels that make up the audio signal. In home theater surround sound, different types of speakers are used depending on which channel they're intended to reproduce.

The main speakers (the front left and right) are generally large, floor-standing speakers (sometimes called "towers") that are designed to reproduce a wide portion of the audio spectrum. This is important, because the main speakers are typically used to reproduce both movie audio and music. You can also get smaller, bookshelf-sized speakers if your budget can't afford larger ones, but at the cost of a loss of performance.

The center channel speaker is often similar in size to the front left and right speakers. The center channel speaker typically handles more than 50 percent of the sound in a film's soundtrack — including almost all of the dialogue. It's therefore important to make sure the center channel speaker is capable of reproducing the same full range of sound as the mains.

Surround speakers are generally smaller than the main or center channel speakers, but these also need to be full-range whenever possible. Surround speakers are typically placed on speaker stands or are mounted on the walls of your home theater.

Subwoofers are different sorts of speakers, in that they aren't designed to reproduce the full range of sound. Rather, they deliver sound only in the low end of the spectrum, generally lower than the rest of the speakers in your system are designed to handle. This gives your home theater the capability to reproduce rich, deep bass — the kind that's important when, say, there are explosions happening in the 5.1 movie soundtrack. But bass isn't all subwoofers are good for. By taking the low-frequency load off of your main speakers, those speakers are able to use their power to more efficiently reproduce the midrange portion of the soundtrack. Most subwoofers are powered (meaning you have to plug them into a separate power source), which allows your receiver to operate more efficiently as well.

For home theater purposes, it's very important to purchase a set of *voice-matched* (or timbre-matched) speakers. This means that all the speakers in your system (with the exception of your subwoofer) have the same tonal characteristics. Different speaker manufactures use different materials in the construction of their speakers. Different types of wood used for the cabinets, for example, have different sound properties that can color the tonal qualities of the sound you hear. The same is true of the cones and other parts of the speaker. In addition, manufacturers often have different lines of surround sound speakers tailored to fit different budgets — for example, an entry-level set, a midrange set, and more expensive, high-end sets. The basic difference between each of these lines is the type of materials used in the speaker construction.

The reason this is important is that the whole goal of home theater surround sound is to create a smooth, natural sound environment. If one of your speakers sounds different than the rest, that smooth quality will be disrupted. The speaker will call attention to itself, ruining the natural effect you're trying to create.

For this reason, whether you're building a good or a great system, you need to purchase your speakers as a matched set. That's not to say that you can't purchase the speakers at different *times* . . . just that you generally need to make sure you're buying them all from the same manufacturer and the same quality line. Once again, the subwoofer is an exception to this. Because it generally reproduces a portion of the spectrum the other speakers in your system do not, there's no need to voice-match.

It's also very important to make sure that the center channel speaker in your system and possibly the mains as well (depending on how close you plan to position them to your TV — within a foot of the display is the general rule of thumb) are video shielded. Speakers use magnetic drivers that, if left unshielded, can badly distort the video signal being displayed by your TV. Unshielded speakers can even permanently damage your TV over extended periods of time. You can generally assume that if a set of speakers is designed for home theater surround sound, the center channel speaker at least is video shielded. Still, it's worth double-checking to make sure, especially with cheaper or lesser-known brands.

Cables

Cables are an important and commonly overlooked part of your home theater system. They are often the largest hidden cost of your system. More than one newcomer to home theater has been surprised to discover how expensive they can be: "Fifty dollars for a set of component video cables? You're kidding!"

You should remember that if you plan to build a 5.1 or better surround sound system, you're going to need a lot of cables — not just for all the video and audio connections for your DVD player, TV, and receiver, but also for each of your speakers as well. That adds up fast. It's not uncommon to spend $500 or more on cables for your system.

The quality of the cables you buy is important . . . but this is one of those areas where the law of diminishing returns applies in force. In general, we recommend that you buy a *good* set of cables from a good manufacturer, so that you can be sure you're getting high-quality video and audio signals to and from all your devices. In particular, cables need to be shielded so they don't pick up unwanted interference. However, you don't need to buy the *best* cables. It's only really advisable to buy the best cables if you intend to use them with very high-end equipment. That said, the simple reality is, most of you will never be able to differentiate the slight quality differences that the best cables can deliver. Even experienced home theater enthusiasts have difficulty with this. For this reason, it's best not to go overboard when buying cables. You can spend silly amounts of money on them, but if your eyes and ears aren't refined enough to appreciate the differences, you're just throwing that money away.

Make sure the cables you're buying are *long* enough. This is where those room measurements you took come into play. You need to make sure that the cable lengths allow you to properly place your components and speakers, particularly the surround speakers, which are going to be the farthest away from the rest of your equipment. You also want to make sure to allow for some extra length in your cables so that you have room to move the speakers or components a foot or two if necessary, and so that the cables aren't pulled too tight. Remember too that if you intend to hide your speaker cables under the carpet or along the walls of your room (or even *inside* the walls), you need to allow enough extra length for that as well.

In addition, be sure to check your equipment to see *what type of cables* you need, and *what type of connectors* those cables should have. Typical video connections come in component, S-video, and composite types (note that component and composite connection cables both usually feature RCA connectors). Audio connections also come in a variety of types, including Toslink (optical digital), coaxial, and line-level (coaxial and line-level connection cables also usually feature RCA connectors). And speaker cables feature several different types of connectors, including pin, banana, spade, and bare wire. Make sure to buy the right cables, with connectors that match your equipment's needs.

We generally don't recommend brands, but it's fairly easy to do so with cables, because there are two brands that really stand out as good choices for their quality and reliability: Monster Cables and Better Cables. Monster Cables are by far the most common — you can find them virtually anywhere you can buy video and audio equipment, and they come in a wide variety of price ranges to fit most budgets. Better Cables also come in a variety of price ranges, and tend to be a little cheaper, but they're generally not available in stores (you can find them online at www.better-cables.com).

Positions and Connections

The positioning of all the equipment in your system is critical to getting the best performance. Thankfully, there are some general rules to follow.

As we mentioned earlier, your primary seating position should be directly in front of the video display, at a distance of roughly three times the diagonal measurement of the screen (for standard 4x3 TVs) or four times the vertical measurement of the screen (for widescreen 16x9 TVs).

All your components (your DVD player, your surround sound receiver, your VCR, and so on) should be grouped together, ideally in an equipment rack that matches the style of the video display, and placed a short distance away from the display. Keeping them together allows you to use shorter cables to connect one device to another (shorter cables are generally less expensive). Similarly, keeping your components relatively close to your video display allows your video cables to be shorter.

Your left and right main speakers should always be positioned directly to the left and right of your video display, at equal distances from the centerline of the video display. In a surround sound system, the center channel speaker should always be positioned on the centerline of the video display, either above or below it. Many people choose to place it directly on top of the display. Your left and right surround speakers are typically placed to the left and right of the primary listening position, either mounted slightly above the position on the wall, or on stands at about ear level. Some people like to position these slightly behind the primary listening position as well. 6.1 systems add an additional speaker – the surround back – positioned directly behind the primary listening position. 7.1 systems split the center back channel into separate left and right speakers. Because low-frequency sound is omni-directional, you can usually place the subwoofer almost anywhere in the room. Many people choose to position it on one side of the home theater, often near a wall, which will tend to reinforce its low-frequency sound. Refer to the illustrations on the following pages for examples of typical speaker positioning.

In terms of connecting all your equipment together, on the video side we *always* recommend using the component video connections whenever possible. Component splits the video signal into its three parts (red, blue, and green), and sends each to the display device separately. This prevents them from interfering with one another, and allows for both the clearest image and the most vibrant color reproduction. In addition, if you have a DVD player and video display that are compatible with progressive scan, this feature is typically *only* available through the component con-

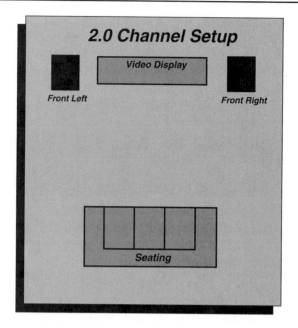

Typical speaker positions for 2.0 surround sound systems

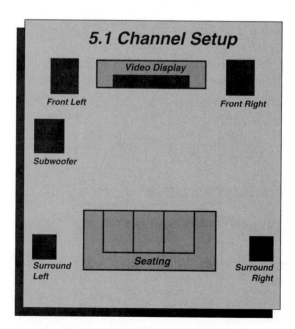

Typical speaker positions for 5.1 surround sound systems

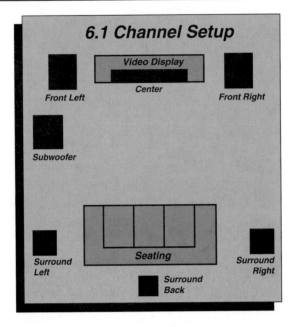

Typical speaker positions for 6.1 surround sound systems

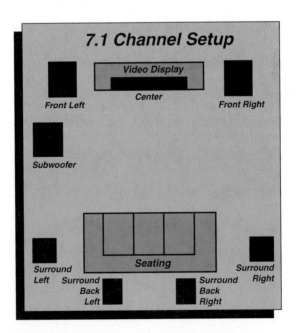

Typical speaker positions for 7.1 surround sound systems

nections. If you can't use component, the next best option is S-video. You should only use composite video connections as a last resort, because they deliver the poorest-quality video performance.

On the audio side of things, most of your connections between devices will be made using standard, RCA-ended cables, which separate the audio signal into its left and right channels. When connecting your DVD player to your receiver, it's best to use either the Toslink (optical digital) or coaxial digital connection. Both deliver excellent quality, although some believe that the fiber-optic data transmission of Toslink is more error free, and therefore lends itself to better quality. Keep in mind, however, that if your DVD player is compatible with either SACD or DVD-Audio (or both), you'll have to use the analog 5.1 connections between your player and receiver to take advantage of these formats—you can't deliver the high-resolution audio signals contained on SACD and DVD-Audio through the digital outputs. This is a deliberate measure that's intended to prevent pirates from making digital copies of the high-resolution music.

There's one other thing you should consider when making your home theater connections. As we mentioned earlier, by routing all your video and audio signals through your receiver, you can often control the entire system with the receiver's remote. You must *always* connect the audio signals through the receiver; however, it's sometimes preferable *not* to do this with the video signals. The reason for this is that, when you pass the video through the receiver, even the best receivers can sometimes introduce unwanted noise or other interference into the video signal. The very best video connection you can have is one that goes *directly* from your DVD player to your display device. So you need to consider what's more important to you: the very best video quality or the ease of not needing a pile of remotes to control your system. Also, keep in mind that when you pass the video signal through your receiver, you'll need two sets of video cables—one going from the DVD player to the receiver, and another leading from the receiver to the display. If you'd rather not spend your money on a second set of component cables, running the video straight to your TV is probably the better option.

Tuning Makes a Big Difference

Just setting up your equipment properly, of course, is no guarantee that you'll be getting the best performance from it. Both your surround sound system and your video display will need to be properly adjusted or calibrated in order to experience DVDs in the highest possible quality. None of these adjustments is terribly difficult, but you'll need a few inexpensive tools to help you out.

First of all, you'll need a device called a *sound-level meter*. This is an inexpensive item that you can purchase from your local RadioShack store for less than $35 (look for catalog number 33-2050). It allows you to monitor the level of sounds in your home. When you use this meter, in conjunction with special test tones played from the individual speakers in your surround sound system, you'll be able to balance the output levels from each speaker to match all the others. This is critical to your home theater's ability to reproduce movie surround sound accurately. If one speaker is louder than others, your ear will be drawn to it, ruining the natural effect of the sound field.

So where do those test tones come from? Many high-end receivers can produce them. Or, you can use any one of a handful of good DVDs designed specifically to help you calibrate your home theater into top performance. Among them are the *Avia Guide to Home Theater*, *Sound & Vision Home Theater Tune-Up*, *Video Essentials*, and *Digital Video Essentials*. These discs sell for between $20 and $50, and are available at good home theater stores. You can also buy them online from a variety of DVD retailers. Each of these discs provides not only the proper test tones you'll need to calibrate your surround sound system, but also simple, step-by-step instructions to talk you through the process.

Even more importantly, these discs will talk you through the process of calibrating your video display as well, again with the proper video test patterns and step-by-step instructions. This is particularly critical because most TVs ship from the manufacturers with their video settings terribly out of adjustment. For one thing, most have their *contrast* and *brightness* settings turned way too high, under the assumption that most people believe a brighter picture is a better picture. Nothing could be further from the truth. Remember, TVs and other video display devices aren't *meant* to compete with sunlight or other bright sources of light. But since most people use these displays in living rooms and other places with unshielded windows, shipping them properly calibrated would mean that people wouldn't see a bright picture out of the box, and so might think they were defective.

Have you ever read your TV's instruction manual and seen the warnings about the potential for images to be burned onto the screen? You often find this in reference to the use of videogame systems in particular. The fact is, if you simply turn the contrast and brightness settings of your display device way down, adjusting them to the *proper* settings, there's very little danger of burn-in, even with videogames. As is the norm in a movie theater, and as we've already advised you with your home theater, the key is to dim the light in the room somewhat so it isn't competing with your TV. Just turn off that lamp that's reflecting off the screen, and close your curtains over the windows. Not only will you enjoy much improved video quality and greatly reduce the risk of burn-in, you'll also significantly extend the life of your TV.

The contrast and brightness aren't the only settings that are generally wrong when you first bring your new TV home. Many display devices have what's called a "red-push," meaning that the *color* is often set too high, and the *tint* tends to be shifted too far toward the red end of the spectrum. What's more, the *sharpness* setting is almost always set way too high. In fact, the correct sharpness setting when viewing DVDs is to have it almost all the way off. Sharpness is designed to artificially enhance the edges of a video image by adding noise, particularly with analog video sources of lower quality, such as over-the-air broadcast programming or VHS videotapes. But good DVDs provide the clearest, cleanest video images your TV can produce. Why would you want to add noise into an already perfect picture?

For all of these reasons and more, it's absolutely vital that you calibrate your video display and surround sound system properly. We *definitely* recommend that you obtain both the sound meter and one of the test discs we've listed — tools that are absolutely *indispensable* for any good home theater enthusiast to whom quality is important. Once you've seen how good your new TV can look when properly calibrated, for example, you'll have a very hard time looking at an uncalibrated set without cringing.

If, for some reason, you have a hard time finding one of the test discs we've recommended at your local retailer, and you just can't wait until the copy you ordered online arrives in the mail, don't fret. We've got good news for you. Just go to your DVD library and look through the movie discs there. Find a recent title that features the THX stamp on the packaging — the *Star Wars* movies, for example, or almost any recent title from Twentieth Century Fox (and a few other studios as well). Then put the disc in your player. Chances are you'll find a feature called a THX Optimizer. It features a handful of the most important audio and video test signals, along with simple instructions on how to do the proper calibrations of your home theater equipment. The THX Optimizer isn't as good as a full-fledged test disc, of course, but it'll do the trick in a pinch.

See? We've got you covered, folks.

Conclusion

With any luck, you should now have a better idea of what's involved in putting together a home theater you can be proud of. By doing your research, balancing the features you want against the amount of money you wish to spend, and making smart buying decisions, there's absolutely no reason that you can't build yourself a good or even great system for enjoying DVD movies at home. There's really nothing like the sheer joy you get from watching your favorite movies in a home theater you planned and built yourself. All you need now are a stack of great DVDs!

But what goes into the making of a great DVD? The best special editions are literally loaded with extra features in addition to the film itself. Those features don't just fall out of a Hollywood studio's vaults—they have to be *created*. Someone's got to craft those features and pull them all together. As you're about to see, that involves a *lot* of work.

In the next section, we're going to give you a glimpse behind the scenes on one of the biggest DVD productions in recent years—Twentieth Century Fox's nine-disc *Alien Quadrilogy* box set. Then, later in this book, we'll give you our handpicked, ultimate list of the best discs the DVD format has to offer. We'll review more than a hundred discs, in a variety of genres, that we feel deserve both your attention and a place in your movie collection.

After all, DVD isn't just an excuse to play with the latest technology . . . it's about presenting movies in the best possible quality, and teaching you a little something along the way. So let's take a look at an example of the hard work that makes all that possible.

Part III

Inside the
Alien Quadrilogy

Introduction

Charles de Lauzirika understands the craft of DVD production as well as just about anyone in this business. Among the many titles he's produced are Columbia TriStar's *Black Hawk Down: Deluxe Edition*, Universal's *Legend: Ultimate Edition*, MGM's *Hannibal: Special Edition*, DreamWorks' *Gladiator: Signature Selection*, and Twentieth Century Fox's *Speed: Five Star Collection*. Lauzirika brings to each and every one of his projects the insight and vision of a true fan of film. He's also a DVD purist, who wants the special edition material he produces to serve the *film* above all else. This sort of dedication to the medium shows, because many of the discs he's produced push the capabilities of the format and rank as great examples of "film school in a box." This isn't just a coincidence, because it was film school that brought Lauzirika to DVD.

Looking at the list of titles he's produced, you've probably noticed that they tend to have an element in common — most of these films were directed by Ridley Scott. While still studying film at USC, Lauzirika found a position at Scott Free, the production company founded by legendary filmmaking brothers Tony and Ridley Scott. Though he started as an intern, he soon found himself working in development and eventually went on to direct a number of commercials and music videos.

In mid-1998, while reading *The Digital Bits*, Lauzirika learned that Twentieth Century Fox was planning a DVD box set of the *Alien* series. As the original *Alien* was one of Ridley Scott's best early films, Lauzirika brought the DVD plan to the director's attention. After a few phone calls and meetings with Fox, the DVD became a full-fledged special edition with the involvement of Scott Free. Ridley was just about to leave the country to shoot *Gladiator*, however, so he asked Lauzirika to supervise the project in his absence. Thus began Lauzirika's involvement in DVD, and a partnership between director and DVD producer that's lasted to this day.

Twentieth Century
Fox's original *Alien:
20th Anniversary
Edition*

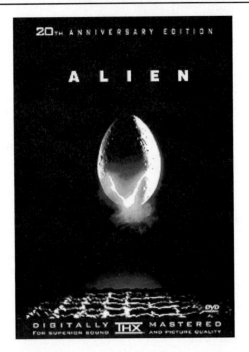

Twentieth Century
Fox's *The Alien
Legacy* box set

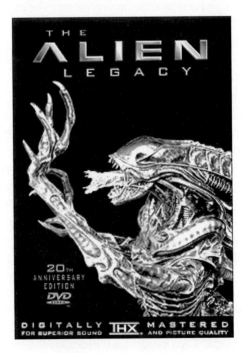

As most of you know, *Alien* is an important film not just to Twentieth Century Fox as a studio (the *Alien* series was their second blockbuster franchise, but the first they actually owned and controlled), but also to fans of the sci-fi and horror genres as a whole. *Alien* is a modern classic of filmmaking, and its technique and visual style have influenced virtually every genre film that has followed it. The *Alien* series also has a rich history, well chronicled in books, comic books, magazine articles, and previous special edition releases on laserdisc. The films remain so popular, in fact, that there's always talk in the air of another sequel—something fans of the franchise would no doubt love to see.

As the twenty-fifth anniversary of the original *Alien* approached, Fox was eager to revisit the series with a much more definitive special edition treatment on DVD than the previous release. Out of this desire, the *Alien Quadrilogy* concept was born. The *Alien Quadrilogy* would be a nine-disc box set of all the films, complete with all new, never-before-seen special features. And who better to create these features than Lauzirika, the man who supervised the original disc for Ridley Scott, and who has since become a talented and prolific DVD producer in his own right?

When we at *The Digital Bits* learned that *Alien Quadrilogy* was going into production, we knew it would be a great opportunity for this book to look more closely at the work that goes into creating a DVD special edition. Lauzirika and Twentieth Century Fox were kind enough to allow us along behind the scenes, giving us access to such things as commentary sessions and interview shoots. We observed the film restoration and editing process. We visited the Fox Archives with the producer and even spoke with director Ridley Scott. Trust us when we tell you that the process of creating a great DVD special edition is incredibly challenging.

What you're about to read is an in-depth interview with the man in charge of this massive project. In the pages that follow, Lauzirika talks about the art of DVD production and provides the kind of insights into the craft that only an experienced producer can. Not only will you learn specific details about the *Alien Quadrilogy*, you'll also get a fascinating glimpse at the tremendous effort involved in creating your favorite discs. And with that, off we go . . .

The Digital Bits (DB): Your first DVD experience was on the original *Alien*, but you weren't technically the producer of the disc were you?

Charles de Lauzirika (CL): No, my credit was project supervisor for Ridley Scott, which meant that—you know Ridley had left to go shoot *Gladiator*, so he put me in charge, on his end, of the disc. I was supervising a lot of stuff. It was the first disc I'd ever worked on and I didn't know a lot of the details. It was kind of a trial by fire. Primarily, I focused on the menus and the new transfer of the film, along with how the supplements would be laid out and navigated. I also worked with Fox on the packaging. I was involved in pretty much everything but the actual supplements.

DB: How did that experience lead you to other DVD projects?

CL: When I started working on the first *Alien* disc, I was very open to working with everyone and being very supportive. I said, "Anything you need from Ridley's office, I'll help you with. Use me." And everyone pretty much did use me, except for the guys doing the supplements. As a result, I formed great friendships with everyone else, and so that's how I ended up continuing to work for Fox on other projects. Then, at some point, the work began speaking for itself. More and more studios started coming to me, so now I'm producing DVDs for other filmmakers, in addition to Ridley and Tony Scott.

DB: So the *Alien: 20th Anniversary Edition* DVD gets finished and Fox releases their *Alien Legacy* box set. Obviously, a few years go by. We remember you telling us a few times since then that you were interested in going back and doing a special edition of *Alien³* — something more elaborate.

CL: For *Alien³*, yes. I had no serious urge to revisit *Alien*, although I'd spoken with Fox off and on about it. I remember, maybe about a year after the first release, we were speculating about the twenty-fifth anniversary of *Alien*, or the high-def DVD version of *Alien*. That was kind of a joke for a couple of years. Then it became less and less of a joke. I'd get a call, literally once every six months, saying, "Fox wants to revisit the original *Alien* as a new special edition. Do you think Ridley would be on board? Do you want to do anything for it?" I was busy with other projects at the time, and I thought the first disc was fine. So why go back to it?

But the seed had been planted in my mind after the first DVD release that the movie that really needed the most exploration was *Alien³*. I remember we were at Complete Sound working on the first *Alien* disc, and I remember mentioning, "You know, *Alien³* has all these great deleted scenes that no one's ever seen before." And the powers that be at Fox back in 1999, who aren't there anymore, basically said, "It's not your movie." So that was really the last time I mentioned it. But over the years, I kept thinking, *that's* the film that really needs exploration, not only because Fincher went on to become such a prominent director, but because it was a hellish production, you know — sort of on par with *The Abyss* and *Apocalypse Now*. It was a really troubled production.

DB: Tell us a little more about that production for some of our readers who might not be familiar with it. There were a number of different directors that worked on *Alien³*, weren't there?

CL: Sure. It actually starts with the number of different writers. Off the top of my head, I think there were seven or eight different writers, and a ton of different drafts done. It was always changing. It started with William Gibson's script, which was more militaristic like *Aliens*. That's probably what the fans wanted. Then David Twohy, Eric Red, Rex Pickett, John Fasano, and Vincent Ward—all these guys took a stab at it. Eventually, they started to hone in on this sort of monastic society, which finally got turned into prisoners with religious convictions. And it was a mess. It was a cobbled together train wreck of a story. But I found it fascinating.

The film was fascinating too, because there were three different directors on it. Renny Harlan worked on it for a year; then Vincent Ward took over. He took it almost all the way to production. They were actually building sets while he was still on it. Ward had this idea that the whole film would take place on a wooden planet, and he was talking about casting Gary Oldman in one of the parts. He was getting ready to pull the trigger on it. Finally, though, Fox and the producers basically said, "These are terrific ideas, but they're just way too out there for what we need for a franchise movie." So enter David Fincher . . .

DB: This was Fincher's first movie.

CL: It was his first film. I think he was 27 at the time. He was an amazingly talented commercial and music video director, but he stepped in and inherited this shipwreck. He had to do with it the best he could. And that was just the beginning. [laughs] It got worse.

DB: So your whole motivation was to come back for *Alien³*.

CL: Yeah. Fox would come to me and ask, "Would you like to do *Alien*?" and I'd come back with, "No, but I'd love to do *Alien³*." That went back and forth. Then on about the fifth go-round, they asked, "How would you like to do all four?" And I dropped the phone. It was definitely a case of be careful what you wish for. Honestly, I didn't have a burning desire to do *Aliens* or *Alien Resurrection*, because *Aliens* had been done pretty well in the past, and frankly I thought it was territory that Van Ling might want to revisit at some point.

DB: Because of his previous DVD and laserdisc work for James Cameron?

CL: Right. And frankly, *Resurrection* . . . I'm not a big fan of as a movie. So I really wanted to do *Alien³*, and I would be happy to go back and do the original *Alien* for Ridley. But in a moment of insanity, I agreed to do all four. I thought, "Well, if we can pull it off, it'll be a hell of an achievement." And it would be great to have a unified vision behind them all. So that's why I agreed to do *Alien Quadrilogy*, against my better judgment. [laughs] Even as I sit here today, I still wonder whether I should have said yes to this project.

DB: Once you've said yes, and you've recovered from the shock, do you go home and start putting together ideas? What's your next step?

CL: That was absolutely the first thing I did. I started working on a proposal for the kind of content I thought would be great to go on the discs. I was pretty well aware of what already existed. What I thought would be the challenge was to find things that hadn't been previously released. With *Alien³* and *Alien Resurrection*, that would be pretty easy, because they hadn't been fully explored. But with *Alien* and *Aliens*, it was significantly more difficult. Almost immediately, though, when I started going through the Fox Archives, and when I started calling people and checking out materials — even going through Ridley's private collection of *Alien* memorabilia — I started finding more and more material that no one had ever seen before. Frankly, I was shocked that this stuff hadn't been put together for the laserdisc or the previous DVD. So there was a kind of cache of treasures that I knew we could go into and show people.

With that wealth of original material, I never even *had* to look back at the previous laserdiscs or DVDs. I could start with a clean slate. So all of the extras on Discs One through Eight of the *Quadrilogy* are either all-new, never-before-seen material, or completely remastered archival material. All of the photos and artwork have been rescanned from either the original Fox transparencies, or from the original artwork itself, so they look better than ever. All of the deleted scenes have been remastered from the original negative. All of the research for the supplemental materials was done from scratch, and since we interviewed over 80 people for this set, we got all of our information straight from the horse's mouth. The only exception to that rule was that we found *Cinefex* magazine to be an invaluable resource. It really is the gold standard of visual effects journalism, and Don Shay was kind enough to help me out with advice and additional support materials.

DB: Was the plan for *Quadrilogy* originally to have two discs for each film, plus a ninth disc of bonus material?

CL: No, I think originally the plan was for an eight-disc set — two discs per film. But the problem came out of my anal retentive desire for symmetry in the way that supplements are presented. Maybe to my detri-

DVD producer Charles de Lauzirika searches through boxes of original production artwork from the *Alien* films in the Fox archives.

ment, I'm very anal about having a consistent style in terms of how things are presented navigationally. So in looking at the available material . . . you know, there are a few different documentaries on *Alien*, there are some bits on *Aliens* that exist, there are different featurettes and different levels of already produced materials for the other films, but they didn't line up in a way that I thought was symmetrical across all four supplement discs for each film. So I thought, rather than trying to shoehorn all of the old material onto the supplement discs, why not have a bonus disc that's kind of a catch-all repository for everything that had already been done? That would include *Alien Evolution*, previous laserdisc content—most everything that had come before. We can put all that onto the final disc for the hardcore fans to enjoy. That would give me the freedom to start from scratch in creating all new content for the supplement discs. So right from the start, we decided to interview everyone we possibly could, to dig up everything we could find, and basically to really put this together not only in a new way, but with all new content that people had never seen before.

A production design sketch of from *Alien Resurrection*.

Even with the extra breathing room on Disc Nine, however, there were still some sacrifices that had to be made. For instance, due to space limitations, Fox unfortunately chose to once again omit the *Alien Legacy* documentary. I told Fox that this was a bad idea, but given that we were physically out of space on Disc Nine, there's not much you can do. When faced with that kind of limitation, you have to focus on unique material that's not redundant with the rest of the set. The same material — and much, much more — is covered elsewhere and in far greater detail, so its omission is probably not the end of the world. But I was still hoping there would be some way to include *Legacy* for completists. Ultimately, it was Fox's decision, not mine.

DB: So you have your proposal, and you submit it to Fox. How much freedom do they give you? Do they say, "This is great; how much money do you need?"

CL: No. I wish. Generally, Fox is very supportive of putting together the biggest, best, most cutting-edge disc you can do. But when it comes to scheduling and budgets, well . . . [laughs]

DB: That kind of support from the studio is pretty rare these days, isn't it?

Quadrilogy Facts

More than 20 people worked on the creation of the new supplements for these DVDs, including the DVD producer, three coordinators, five editors, two or three fill-in editors, two or three people scanning photos, a handful of researchers, and eight or nine people to go through boxes of film and identify footage. This *doesn't* include the many technicians and studio personnel who worked on the menus, the packaging, disc authoring, the new special effects footage, and other necessary work.

All told, literally thousands of hours of production time were involved in the creation of the DVD supplements, spanning an entire year (from mid-summer 2002 to mid-summer 2003).

Approximately 30 new effects shots are being completed for the extended cuts of the films that are available on these DVDs. Some of these are digital composites of original optical elements, while others are being created from scratch using a variety of live action plates, original pre-visualizations, and other internal documentation.

Five separate companies were involved in finishing the new visual effects shots and editing them seamlessly back into the films to create the new special edition versions.

Four separate camera teams were hired to gather new interview footage in the Los Angeles area, and additional teams were hired to do the same in the U.K., France, Australia, India, Canada, the Czech Republic, and elsewhere around the world.

More than 80 people involved in the production of the *Alien* films were interviewed for this DVD production, at an average of about 45 minutes per interview. (The shortest interview—Alan Ladd, Jr.— clocked in at under 15 minutes, while the longest—Ridley Scott— was over 2 and a half hours long!)

New audio commentary tracks were created for these DVDs, compiled from newly recorded sessions with nearly 40 of the original filmmakers and actors.

The supplement disc for each film will feature more than three hours' worth of short documentaries, which can be viewed in a "play all" mode. These have been culled together from hundreds of hours of new and previously created footage, including interviews, on-set footage, and other production-related video. The vast majority of this material has never been released before.

(continued)

The final discs are expected to incorporate more than ten thousand still images, either in the menus, the documentaries, or in still gallery format. This includes production photos, conceptual art, storyboards, original script text, and more.

You'll need more than 40 hours to go through everything that will be included in the *Alien Quadrilogy* box set. That includes watching each of the four films three times (the theatrical version, the seamlessly branched alternate version, and the film with audio commentary), plus roughly three hours of material on each of the supplement discs and the bonus disc. We suggest that you book your vacation time now.

CL: Very rare. There are a couple studios that give me everything I need, but it always varies. It depends on the DVD regime at the studio, what mood they're in that week . . . what their latest marketing analysis tells them. When it comes to Fox, they give me a lot of creative freedom. The downside is, we had a very limited budget in terms of what we had to stick to and how far we could push the envelope. That was tough considering how much work that we really felt needed to be done to make this a worthy re-release. You know, this is a set of films that fans have already purchased on DVD, and now we're asking them to buy them again. We're doubling the content, sure, but it needs to be content that's worthy of their time and money.

So as a DVD producer, you can sit all day and dream up this amazing proposal for the ultimate disc, which is what I always do on any title I produce. I take a day and think up what I would do if there were no limitations — if there were no political, legal, or financial restrictions. But then reality sets in, and you have to start asking, "Okay, what can we *really* get away with?" That was very difficult on this set.

DB: Do the studios generally arrive at the budget for the project based on your proposal, or do they have a number in mind before that?

CL: A lot of times, they'll give me the number first and say, "Here's what we have to spend. What can you do?" And I come back with my best proposal based on that. In this case, there was a number that I felt was too low for the work that needed to be done. So I submitted my proposal, and a proposed budget to Fox.

DB: When was this?

CL: Probably August or September of 2002. Maybe a little earlier. My proposed budget was more than Fox wanted to spend, so we went back and forth a little bit and agreed on a number that wasn't unworkable in terms of how low it was, but wasn't very comfortable in terms of how high it was either. That was the challenge.

DB: When did you get the green light to start working?

CL: I would say it was late summer of last year. I basically knew I going to do it; it was really just a matter of the details of what and how it was going to be done. What was the budget? What was the timeframe? Those were the question marks. And I was willing to do anything just to make it for *Alien³*.

DB: Obviously, this is bigger than anything you've attempted before as a DVD producer—four films all at once from the ground up. How did you organize a project of that size?

CL: The first thing we did was to put together a list of the people we wanted to interview. Actually, the original list I put together wasn't as comprehensive as it turned out being. I started out with the usuals—Ridley Scott, James Cameron, David Fincher, Jean-Pierre Jeunet—we went right down the list. Then at some point during early production, AMC ran that *Alien Saga* documentary, in which they really interviewed a lot of people. I saw that and thought, "Okay, now we have to go even bigger with this." So we started trying to go beyond the usual suspects, trying to dig up people that you wouldn't normally hear from—the visual effects people, the sound design people. We even interviewed Veronica Cartwright's stand-in for *Alien*, because we heard she had some really great stories to tell. But it's like, who would think to talk with a person like that for a DVD? Our grand total is over 80 all-new interviews for this project.

DB: These are on-camera interviews?

CL: Yeah. And a lot of them are for commentaries too. So that was really the starting place—figuring out who were going to be the voices that would tell the stories behind the making of these films.

DB: How do you approach actors and filmmakers when it comes to doing on-camera interviews and audio commentary? What are you looking for as a DVD producer?

CL: There are two different approaches. For on-camera interviews, I just like to have a conversation with people. I don't like having a list of questions that I have to keep looking down at. I know all this material pretty well, so I feel like I can have an intelligent conversation about it. I like the eye contact with people. You tend to get better material when

Lauzirika and his crew prepare to shoot on-camera interviews with *Alien* cast members Harry Dean Stanton and Veronica Cartwright. These will be used as part of new documentaries for the *Quadrilogy* DVDs.

Harry Dean Stanton

Veronica Cartwright

you're having a direct one-on-one conversation. In terms of audio commentary, it's the opposite. I like to have as much paperwork around as I can, so that if the participant blanks on something, I can check my notes or we can look up something to refresh their memory.

You have to remember, though, that *Alien* was made 25 years ago. I was interviewing Michael Seymour, the production designer on *Alien*, and I was asking about the details of the sets—really film geek stuff. And he looked at me and said, "Do you remember what you were doing 25 years ago?" I couldn't tell you what I had for dinner 3 days ago, and here I am asking this guy what he was doing 25 years ago, and what the doorknob on the Nostromo looked like. I realized just how ridiculous this can be sometimes. When I was interviewing Tom Skerritt, he suggested that some of the questions were of the "get a life" variety, you know? And he's right, of course, but that's what fans want to hear.

DB: Do actors tend to be a little different, in terms of how much they remember about a film, compared to production people?

CL: No, you really can't divvy it up that way. You might have one actor who went on to do a million films after the one you're talking about, and so they don't remember anything. It was just a job. Then you might have another actor who remembers everything. I was stunned when we interviewed Veronica Cartwright, because she had an encyclopedic memory. She remembered every little detail, whereas Skerritt was very nice to talk with but didn't recall as much.

DB: Does it help to trigger their memories if you show them the film during a commentary, or if you bring a group of people into the commentary booth together?

CL: Sometimes. Unfortunately, though, if they haven't seen the movie in a long time, they often end up watching it. And that's when I really have to earn my money and pop in with questions. "Hey, do you want to talk about this person, or what do you think of that scene?" That's why, personally, I hate commentary sessions. Nine times out of 10, you're trying to draw blood from a stone. Occasionally, you get someone who just has so much to say that you can sit back and enjoy. In the case of Ridley, for example, he remembers a lot of stuff. He'll just keep talking. I have to pop in with questions occasionally, but that's fine, because he knows how to pick up the ball and run with it. But that's rare.

Cast members Harry Dean Stanton (left), Veronica Cartwright, and Tom Skerritt react to a memorable scene from *Alien* while recording audio commentary for the DVD.

DB: We would imagine commentary is a little different for newer films.

CL: People definitely have more to say, because it's fresh in their minds. But there's no historical perspective, so it's a double-edged sword. *Die Another Day* is a perfect example. I did the commentary with Pierce Brosnan and Rosamund Pike, and they remembered everything. But the film hadn't even been released in theaters when we recorded it. So there's no sense of what the film's place in the world is yet. What is there to talk about other than the production itself? What's interesting about this film? We don't know yet. It's very strange.

A Commentary Tale

March 3, 2003—1 PM. It's an unusually overcast afternoon in Southern California, and we've just arrived at the Santa Monica offices of P.O.P. Sound. After parking in a small, off-street lot, we enter the building to find *Alien Quadrilogy* DVD producer Charles de Lauzirika waiting in the lobby. With him is with Jon Mefford, his lead coordinator on the project. On the agenda today is a full afternoon of DVD work—the reason for our visit. This will include an audio commentary recording session with three members of the cast of the original *Alien*, as well as separate, on-camera interviews with each.

P.O.P. Sound, a division of Pacific Ocean Post (which is located just across the street), is a state-of-the-art, audio post-production facility. Its two levels contain numerous recording booths and mixing stages, in which the soundtracks of many different feature films and DVD releases have been tweaked to perfection. On this day, the *Quadrilogy* production crew has occupied two stages. A fully equipped suite upstairs, Studio G, will host the commentary work. Studio B, on the first floor, is being converted into a makeshift interview set by a two-man crew. A Sony DVCAM is already standing at the ready on its tripod, along with various lights, microphones, and other equipment.

While we wait for the actors to arrive, Lauzirika and Mefford take a few moments to confer and go over their notes for the day. There's a degree of nervous anticipation in the air, and for good reason. "You never know what's going to happen on days like this," Lauzirika says. "Commentaries and interviews are always a challenge. You just never know what you're going to get, and given the budget and the schedule, you've really only got one shot."

(continued)

Of course, this is only one session of many set during the year-long *Quadrilogy* production. Several have already been completed and many more will be scheduled before the end of the project. But today's session is an important one. First on the docket is an on-camera interview with Tom Skerritt, set for 2 PM, followed by another with Harry Dean Stanton at 2:30. Veronica Cartwright is then expected to arrive in time for the group to begin recording their audio commentary together at 3 PM. An hour or two later, Cartwright will round out the afternoon with her own on-camera interview. Naturally, the unexpected happens almost immediately. Harry Dean Stanton strolls in at 1:45 PM, fully 45 minutes early, and Skerritt is nowhere to be seen.

Stanton seems not so very different from Brett, the character he plays in the film. His clothes are blue collar, his manner is laid back, and he's rarely without a cigarette. With the clock ticking, the decision is quickly made to shoot Stanton's interview first. So he's led into Studio B, the door is closed and the videotape starts rolling. Lauzirika begins the interview, and it takes Stanton a little while to get going. A lot of years have passed, and he doesn't remember a great deal about his work on *Alien*. Stanton, it seems, is a man of few words, about 1 in every 10 of which is an expletive wrapped with a smile. As a result, the stories he does have to tell are both colorful and entertaining.

By 2:15, Stanton's footage is in the can, and we discover that Tom Skerritt has just appeared in the lobby. Skerritt, who played Dallas in the film, has flown in from out of state for the day. He's dressed in a smart-looking suit and he's all business. So after a few minutes of preparation, the tape is rolling once again. Like Stanton, Skerritt's recollection of specific details about his work on *Alien* is limited. But he knows his craft and he's able to tell a few interesting anecdotes. In all, the interview continues for about 20 minutes.

At this point, we're fast approaching 3 PM. Everyone heads into the lobby to regroup and to grab something to eat or drink. As Skerritt and Stanton catch up, you can tell that the *Quadrilogy* team is a little concerned about the commentary session to come. Given the interviews thus far, how much will the actors remember? Will they get enough good, usable material? These concerns evaporate, however, when Veronica Cartwright arrives moments later, full of smiles. Cartwright's a fireball and her enthusiasm is infectious. She's clearly thrilled to be involved in the project, and to see her fellow cast members again after so many years. Energized, the group moves upstairs into Studio G.

Studio G is typical of the kind of space used to record audio commentaries. On one end, there's a sound-isolated booth in which the actors are seated together in front of microphones. This is separated by a glass window from the control room, where all the mixing and recording equipment is located. A recording engineer and an assistant are already there, ready to go, as we enter. Elevated in the back of the room are a desk and couches for the production team to sit and supervise the process. As the recording session gets underway, *Alien* begins playing on video monitors mounted throughout both rooms. The idea is that actors can watch along and comment on what they're seeing.

Right from the start, it's obvious that the energy is different than in the earlier interviews. Cartwright, who played the character of Lambert in the film, picks up the ball and runs with it. It's not overstating things to say that she seems to remember everything as if it happened yesterday. She's got lots of interesting behind-the-scenes stories to tell, and Skerritt steps right in line with her. A few times during the session, Lauzirika has to pop in over the intercom to prompt the actors to talk about various subjects, but the commentary goes very well. Even Stanton has opened up more with Cartwright seated beside him. You can't exactly call him chatty and his comments are still laced with expletives, but he's absolutely hilarious and he's got us all laughing out in the control room. At one point, when a newly restored deleted scene appears on the screen (we're watching the new special edition cut of *Alien*), the actors realize they're seeing new material. "I didn't know they cut any of my scenes out," Stanton quips. "Bastards."

When it's all over, the group moves downstairs again, and Skerritt and Stanton say their goodbyes. We chat for a few minutes with Cartwright as her makeup is touched up for her on-camera interview, and then she's ready to start. She picks right up where she left off, with animated answers to all of the questions Lauzirika asks of her. She even recalls a funny moment during the filming of the Chestburster scene, when she was hit in the face with a spray of fake blood and fell over backwards during the shot. We've seen the footage—she disappears behind the table so that all you can see are her boots sticking up in the air. Moments later, she just gets up and keeps on

(continued)

acting. Cartwright is thrilled to learn that the outtake has been recently found, and that it will be included on the DVD (in fact, *every* take shot during the filming of the infamous scene will be included on the disc in a multi-angle scene deconstruction).

It's nearly 6 PM when Cartwright finally concludes her interview and, with yet another smile, makes her exit. And that, as they say, is a wrap. Both Lauzirika and Mefford agree, it's been a good afternoon's work. There's plenty of usable material for both the documentaries and the commentary track. Cartwright, in particular, has contributed some great stories. So it's no surprise then, that as we gather our things and get ready to leave, Lauzirika seems a little more relaxed. "See what I mean?" he says with a smile as we make our way toward the parking lot. "You never know what to expect."

DB: Aside from the unusually large number of new interviews and commentaries, what's the other major challenge you've had on *Alien Quadrilogy*?

CL: Well . . . you have to remember that simultaneous with all this, we had worked out the idea with Fox that not only were we going to do all new supplements, we were also going to restore the films, and do special edition versions of *Alien*, *Alien³*, and *Alien Resurrection*. We were going to fully restore all the deleted scenes. In the case of *Alien³*, which had such a sordid history in terms of how badly it had been butchered before its original release, we felt we absolutely had to go back and find the long lost cut of the film. So in addition to coordinating all the supplements, we also had to do restoration and finish visual effects for the movies.

DB: The idea is that the final DVDs for each film will give you a choice between watching either the original theatrical cut or a new special edition/alternate version, via seamless branching.

CL: Correct.

DB: Whose idea was it to go back and create longer versions for this release?

CL: Well, it was my idea to do *Alien³*. It was always part of the plan to include deleted scenes on the discs, and to give people the opportunity to see all of that. But what we quickly realized was that, with *Alien³* and *Alien: Resurrection*, in order to fully restore some key sequences, the visual effects had to be completed. There were a number of shots that had never originally been finished. This was much like the DVDs of *Star Wars: Episode I* and *II*, where Lucasfilm went back in and finished some of the special effects for the deleted scenes, because otherwise they would have been just actors standing in front of a green screen. We had the same problem here. So we spoke with Fox about the idea of completing the effects, and they were excited about it. I was actually shocked they were willing to spend the money!

DB: So once you knew you had to restore *Alien³* and *Resurrection*, it then made sense to do all four?

CL: Yes . . . well, three of them anyway. The great thing is we didn't have to do anything for *Aliens*, because that had been done previously. There was already a new special edition version, with all of the special effects completed, done for the previous DVD and laserdisc. So initially we really only had to focus on *Alien³* and *Alien Resurrection*, because *Alien* didn't originally need much. Eventually, towards the end, several shots in *Alien* were tweaked or enhanced more to Ridley's liking. Mostly just starfields and tiny little details most people wouldn't notice. I always cringe when faced with digital revisionism, but the changes made to *Alien* are very slight. Nothing even close to the controversial changes made to the *Star Wars* films or *E.T.* So that was all happening simultaneously while we were producing the supplements for the discs.

DB: Given the vast amount of work, how do you assemble a production team to get it all done?

CL: Well . . . the biggest expenditure in DVD supplement production is usually editorial. So it was a matter of figuring how many editors I needed. There are a couple I've been using pretty consistently over the years. One is David Crowther; the other is Will Hooke. David has been my "A" editor for a long time, but he was getting burned out on doing documentaries and behind-the-scenes material. So I thought, since he has a film background—he was assistant editor on *Titanic* and *True Lies*—maybe he should focus on the restoration of the films and the special effects. So David really dove into going through all the boxes of negative, looking through all the code books and lined scripts to try and figure out what footage was missing and what needed to be restored. So right from the start, one of my editors was lost to that for almost the whole length of the project.

DB: That's an archeological process, isn't it?

CL: Absolutely. But it's great, because as David's looking through all the boxes, he has an eye towards, "Oh, this outtake or this lost scene

Editor David Crowther at work on the new special edition cut of *Alien³*

might be cool for the supplements." So Dave focused on that. Will took on the task of editing the special features for *Alien*. Then I had to find three other editors for the other films, which I did through this company I used to have office space at called Sparkhill. They have in-house editors, and because I was bringing a pretty sizable project through, the editors came as part of that package. The problem is, Sparkhill had their own projects going on in addition to mine and those of another DVD producer, David Prior, who also had an office there. So it becomes a scheduling challenge. Thankfully, Jon Mefford, who is an in-house producer and coordinator at Sparkhill, became sort of the first coordinator involved on the *Alien Quadrilogy*. He tracked people down, set up interviews, made sure all the camera crews were going where they needed to be . . . stuff like that. Then, along the way, it became necessary to bring in a couple of additional coordinators to pick up the slack. Alex Close set up several of the interviews, and even conducted a couple of them, while Cory Watson tracked additional people down and then focused on organizing the huge image galleries.

DB: Since you're the DVD producer, we would imagine that logistics isn't what you want to be spending your time on.

Assistant editor Curtis Bisel assembles new documentary featurettes for *Aliens*. In case you hadn't already guessed, long hours spent in dark rooms in front of glowing video monitors are an occupational hazard when you work in the film industry.

CL: I really need to be focusing on the creative side. For example, once the interviews are scheduled, I usually conduct them if I can be there. Out of the 80 or so interviews for this project, I probably conducted about 85 percent of them myself. If I can't be there, I'll have one of my coordinators conduct them. If they're in another country I'm not in at the time, we'll have the remote crew conduct them. I'll write up the questions and e-mail them over.

DB: So you've got people looking through film, you've got interviews being done, and you've got footage being collected. Basically, it's a big gathering process.

CL: Yes, and it's really daunting when you think about it, all the material that has to be found and gone through and cataloged so we know, basically, what we have to work with. Frankly, I'm in awe of this whole project — that we've even gotten this far in the time we have. I think, back in the laserdisc days, this is a project that would have taken three or four years to do. We're going to do it in just under a year. It shows you how much the demand for DVD has accelerated the whole process.

DB: Do you have some kind of list that helps you know what you're looking for?

CL: For the most part, yes, but it's a curse and a blessing. As you continue on with the project, doing more interviews and digging through more boxes, new things always come to light. For example, late in the game, you'll learn that they shot something you didn't know about, because it wasn't documented during the production. There's just too much film to go through every single frame on all four movies. But someone you interview might say, "Hey, did you ever find that scene?" And we didn't know about it. So we'll look for it and find it, but it's frustrating. You'll wish you'd known about it earlier, so you could have asked the people you've already interviewed about it. It's really tough to ask people back once you've already interviewed them. I think there are only two instances where we've been able to do a second round of interviews with people — one with Jean-Pierre Jeunet and the other with Gale Anne Hurd. So it's a continuously evolving process. It's always fluid. That makes for an interesting project, but it's also scary as hell, because you're afraid you're going to miss something.

DB: It must be a challenge. You have an idea what you *want* to do, but until you've collected all the material available, you really have no idea what's *actually* going to be on the discs.

Aliens producer Gale Anne Hurd records new audio commentary for the *Quadrilogy* DVDs.

Hurd (center) and *Aliens* creature designer Stan Winston during a break in the commentary.

CL: You know, pretty much every studio now, when you start doing a special edition DVD — they want to see a really detailed proposal of every-thing you're going to create for them. In addition to the timeframe and the budget, they want to know what the running time of each piece is going to be. And you do it, just to go through the motions, but it's a joke. I like writing proposals for myself, because they help me focus on the struc-ture, but I never want them to be in a contract. When that happens, the studio comes back and says, "Okay, you have to deliver this and this and this now." That's ridiculous. What if you don't find that material, or the tal-ent involved doesn't want to participate . . . or what if something better comes along? Producing good supplemental materials is an organic process, and you *never* want to lock yourself into something from the start, because that's limiting yourself creatively. Even on this project, Fox will occasionally say, "We need a spec list. What are the titles of each piece? What are the running times?" They want that because they're try-ing to create the packaging and the menus, which is understandable. But I literally don't know yet. I won't know until I deliver it all, because until the very last moment, we're going to be tweaking and working to make it all the best we can.

DB: It also seems like, on a project of this size, there's just so *much* to deal with.

CL: That's another big challenge, because I need to go through every-thing multiple times at different stages to approve things and make tweaks and adjustments. And now I have to do it for *four* films — it's four times as much as you usually have to deal with. There's just not enough time in the day to get through it all, but I *have* to do it. I know I'll get done. It always gets done. But man . . . of all the projects I've worked on . . . if one project was ever going to break me, this was the one. It's just so huge.

DB: We know that one of the biggest issues you've struggled with here is the lack of participation by certain key people.

CL: Definitely the one issue with this project, other than legal problems, is the fact that some key players just didn't have the time or the interest in participating. And we're still talking to these people. We're still trying to get them on board, and I'll keep trying until the very last minute.

Obviously, the big person who's MIA is David Fincher. That pains me, because the whole reason I took on this project was to put his work in its best light, and to try to salvage as much as I could from the wreckage that the film ended up being. That said, I don't feel like it's for me to single-handedly rescue this film — not that I could, even if I wanted to. That's not what I'm going to do. But as a fan of Fincher and his work, I felt like I

Lauzirika examines outtake footage from *Alien³*.

wanted to really try to show what his original vision was for this film — to show people what he *wanted* to do, and to preserve that for all time. But, without Fincher involved, that's not necessarily going to happen. We're going to show you what he was working on and show you some of the alternate ideas he was working on. We're going to show you the footage he shot and later abandoned — you're gonna see all that stuff. But it's not going to have that extra level of authorship that it would've had if Fincher been a part of this project.

DB: If Fincher *had* been a part of this effort, you'd get more of a sense of perspective. It wouldn't be just, "Here's what could have been" or "Here's what was in progress." It would have been, "Here's what I wanted to do. Here's why this ended up the way it did."

CL: Exactly. You know, there are very few directors out there who do commentary better than David Fincher. As a fan, I would just *love* to have him do a commentary for this, as much as I know he would've hated to do it. This, of all of his films, is the one that most needs his voice in terms of what went wrong . . . and what went right, perhaps. I know the overwhelming majority of his thoughts will be negative, but that's *interesting*. It's a cautionary tale for young filmmakers out there. People, who want to follow in Fincher's footsteps, want to know *why* his first feature film went wrong. It would be fantastic to see what he had to say about that.

DB: For fans, the question of what went wrong is not only a part of the history of *Alien³*, it's history that's never really fully been told before.

CL: We've gotten people to talk about it in the new interviews we've done, but I'm not sure we've gotten one hundred percent honesty from everyone. Again, it was an incredibly difficult project. Most people either don't want to talk about it, or they want to forget about it, or they have forgotten about it, or they want to whitewash the whole thing. We've only had a couple of interviews that I would really consider brutally honest. But my final cut of the documentary, which *did* go into some interesting detail and was initially approved by Fox, eventually scared the hell out of some Fox executives and lawyers. So they went and made several cuts without my participation, most of which made absolutely no sense to anyone working on the disc. I've actually taken my name off of the documentary because of it. I've disowned it and it's truly a shame because the primary reason I signed on for this project was to create an in-depth documentary on *Alien³*. So for those people who are expecting this DVD to really be the tell-all, all the dirt you've always wanted to hear about *Alien³*, it's not going to be that. It's not going to be *Hearts of Darkness* for *Alien³*. But it *was* that . . . before, much like the film itself, studio politics ruined it.

DB: Tell us more about each of the other directors for this series — the degree of willingness they've had to come back and revisit their films for the new DVD release.

CL: Obviously, Ridley Scott has been very helpful and enthusiastic. In fact, he's become more enthusiastic as the project's progressed. At first, he was busy working on *Matchstick Men*, his new movie, plus he was prepping another film that he's trying to get off the ground, called *Tripoli*. So he didn't really have the time to focus on *Alien* for a long while. But now that his schedule is starting to free up a little bit, he's been very helpful and supportive. He really wants to get in there and be part of the project. We recorded all new audio commentary with him twice, once by himself and then again with Sigourney Weaver. Considering what an incredible schedule he's got, the fact that he's so supportive of what we're trying to do . . . well, it's why I love working with him. Ridley isn't just a great director; he's a great person. He's really supportive of the way we're trying to immortalize not just his work, but the efforts of *everyone* who worked on all four of these films. Ridley was a slam dunk.

A true *Quadrilogy* moment. *Alien* director Ridley Scott (left) confers with Lauzirika about the DVD production, just as *Aliens* star Michael Biehn happens by down the hall (he's the one looking at the camera).

CL: James Cameron . . . is also very busy. And he's busy doing things that don't seem to lend themselves to having him be available to participate in this DVD. I read an interview with him, related to *Solaris*, in which someone asked him what he's been up to since *Titanic*. And his reply was basically, when he's 70 years old, he can still make a movie. But he won't be able to go down to the bottom of the ocean. He's still in his prime right now, so now is the time for him to be doing all the adventurous stuff. And that's perfectly reasonable. It doesn't really help us on the DVD, but that's okay. He's done lots of interviews in the past and we can use those. We've been in contact with Lightstorm, his company, and they've been very helpful whenever they can. When I went to Lightstorm, I did a little bit of a pitch and said basically, "You know, the laserdisc has a lot of stuff on it, which is all fine. But it doesn't really have a human voice. I want to hear from everyone who worked on the film, not just read a bunch of notes." So I told Lightstorm that I wanted to go out and interview as many people as I could and they were very supportive of that.

Fortunately, at the eleventh hour — actually more like 11 hours, 59 minutes and 59 seconds — Cameron finally came through for us with a commentary track. We went down to Lightstorm the day before he was going to leave on another expedition to the *Titanic*, and he gave us a fantastic solo commentary. I don't know why he was hesitant to do one before *Solaris* or the *T2: Extreme Edition* DVDs, but I think he's got the bug now. It was a very informative and entertaining track.

DB: You've mentioned that Fincher ultimately declined to be a part of this project, but how did you go about trying to get him involved in the beginning?

CL: I wrote a very long letter to Fincher, explaining exactly what I wanted to do and why I wanted to do it. I was very passionate about it. I basically said, "I'll do anything, just please be a part of this." I've never actually spoken with him directly, but I was told by his people that my letter at least got him to consider it, and they said they'd entertain the idea. So that's when we really dug in and started looking for material. Eventually, we found this long lost, 150-minute cut of the film. So we sent it over to his office.

The thing you have to understand is, Fincher was circling around three different projects for his next film at the time. Of the four directors, he's probably the closest to actually going into production on something, so his schedule is tight. Plus, it was a very negative experience for him. This film was hell for him. So to come back and talk about it must be painful. If you look at his filmography on the *Panic Room* DVD, *Alien*³ isn't even on there, so he's obviously disowned the project. All these things kind of combined into a very polite reply from his office: "No, he won't participate, but good luck. You're free to do whatever you want; you just can't call it a director's cut."

I'd like to think my letter had some effect, but frankly, it may just be that he didn't care. But I wanted him to know that even if he didn't care, I *would* care. I would try to do the best I could – to put his work in the best light I could. Now, I don't know that we've done that. I think that what we *have* done is to capture a snapshot of the film in the state it was in before it really got interfered with in post-production – before it got taken out of Fincher's hands. What I'm hoping, with this new version that's going to be on the DVD, is that you're going to see the film before it got completely corrupted in the editorial process.

Alien Encounters

DVD producer Charles de Lauzirika went to great lengths to try to convince *Alien³* director David Fincher to participate in the *Alien Quadrilogy* project. Unfortunately, Fincher declined. Still, as fate would have it, Lauzirika had a couple of amusing brushes with the director over a two-week period in April of 2003. We'll let him tell the story in his own words:

"My editor, David Crowther, had just finished his rough-cut restoration of the special edition version of *Alien³*, and we were planning to have a private screening of the cut that evening—myself, David, and a few other people from the office. Given that it's a two and a half hour cut, we figured we should get dinner. Most of the guys wanted pizza, but I wanted a burrito, so I drove over to this Mexican place near our office called Poquito Mas.

So I'm standing there in line, ordering my ahi burrito, when out of the corner of my eye, I see something that sets off alarms in the back of my head. I look over, and there's David Fincher, sitting there with someone else eating his dinner. Immediately, I seized up like I'd just seen Jesus. And I'm thinking, what do I do? Do I interrupt him? Do I introduce myself? Do I invite him to check out the screening of the film with us?

I immediately call back to the office on my cell phone, and I'm telling the guys, "Fincher's here at Poquito Mas! What do I do?" In those moments, for some reason, I totally geek out. Do I dare talk to him about this film he obviously hates so much? Of course, the guys all said, "You've *got* to get him. You've got to go talk with him." Naturally, as I get off the phone and I'm about to do just that . . . Fincher gets in his car and drives off.

I didn't really feel bad about missing the opportunity, because he seemed to be having a pretty intense discussion. It didn't seem like he was in a very approachable mood. And I figured, what could be worse than going up to him when he's in a bad mood and saying, "Hey, do you want to come and see the long cut of *Alien³*?"

Then, about a week later, Fincher actually called my office. Mark Romanek [who directed *One Hour Photo*—another DVD Lauzirika produced] had talked to him about me and put in the good word . . . which, coming from Romanek, is a major deal to me. I mean, I worship both of these guys. Mark gave Fincher my number, which was incredibly nice of Mark to do. Unfortunately, I wasn't in the office when Fincher called. But he left this really cryptic voicemail: "Yeah, Mark Romanek told me to call you about *Alien³* . . ." and about halfway through the message, he just kind of drifted off. It was almost like he lost the heart to even talk about *Alien³* right then in the middle of this message he was leaving me. We played phone tag for a while and never actually spoke directly. So I've saved this voicemail. For a while, I was toying with the idea of putting it on the DVD as an Easter egg until my better judgment kicked in. I doubt Fox's lawyers would have cleared it anyway.

So those are my two brief *non*-encounters with Fincher on this project."

DB: *Alien³* was basically a flawed film from the very beginning, wasn't it?

CL: It was seriously compromised before the shooting even began. I don't know if it's an actual quote, but I seem to remember at one point hearing Fincher say something to the effect that, "The only way to do a director's cut of *Alien³* is to burn the negative and start over." I know he mentioned once that, during the L.A. riots in '92, when some of the fires and vandalism were getting pretty close to one of the labs where the negative for *Alien³* was stored, he kept hoping that it would get burned to the ground. [laughs]

DB: So the point of this new cut is to show people where the film was in post-production, and where Fincher was going with it?

CL: Basically, the best way to look at this cut is, this is pretty much everything they shot originally — before the reshoots, before the test screenings. This is kind of like the first assembly of the film. Now, it's a bit more polished than the actual first assembly . . .

DB: Because you're finishing the effects and the audio mix . . .

CL: Right. What you're seeing is a reconstruction of the direction the film was going. After this point, it started getting cut down and cut down, and then there were reshoots. So this is the first cut of the film after development hell and after production hell, but *before* post-production hell. As such, it's a very unusual piece of work. When you see this cut, you'll really understand how *Alien³* ended up the way it did, because the film was literally being rewritten as they were shooting. And it shows. The film really feels cobbled together. It doesn't make for a very entertaining experience, but it's fascinating, if you're a fan of the film, to be able to *see* how it got so badly screwed up.

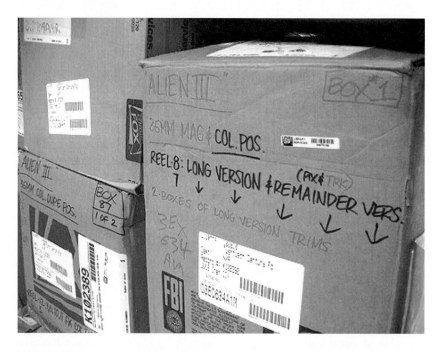

Boxes of film that have been pulled from storage to be cataloged for the *Quadrilogy* production. There are close to 200 boxes of just outtake footage from *Alien³* alone.

More boxes of film that have been pulled from storage to be cataloged for the *Quadrilogy* production. The room you see here is stacked, floor to ceiling, on all sides.

DB: How do you approach trying to put that cut together? Fincher says, "Go ahead," and you know there's a longer cut out there to use as a blueprint. Are you also looking at Fincher's notes, at editorial notes; are you looking at original scripts? What are you using to guide your efforts in this?

CL: Among the things we dug up early on were various drafts of the script, the shooting script, and what's called the lined script, which is what the script supervisor actually had on set and was using to make notes about which takes were going to be used. We also found some alternate cuts of certain sequences that were in these boxes, to use as a reference to see how things had evolved and where they had come from. We went through all the storyboards, all the call sheets we could find. Basically, we took advantage of anything we could use to get a sense of how things were coming together and what the plan was for the way things would be put together.

That said, we really had to be careful, because we're *not* the filmmakers. The one thing I was always adamant about was that we're *not* in the business of revisionism. We're not going to make a cut that we think is a *better* cut. We're not going to tinker and play and have fun with someone else's movie. All we're going to do is to take it as close as we can to what we've ascertained, via all the documents we have and the research we've done, is the original vision of *Alien³* before all the interference occurred in post. I don't know if we're a hundred percent in line with that, but it's not because we didn't try. It's because we didn't have Fincher's guidance, or we don't have the materials to do it more accurately. That's been particularly an issue with the effects shots that were abandoned back in the day before they could be finished.

DB: How many new effects shots for the films are being done?

CL: In total, there are about 30 effects shots. For *Alien³*, there are about 8 or 10 in all. These are a combination of CGI creature design for what's called the "Bambi burster" that comes out of the ox, and a couple of digital matte paintings for the opening scenes on the surface of Fiorina. Then there are a couple little bits here and there along the way.

DB: Can you tell us more about the major scenes in *Alien³* that are being restored for this new cut?

CL: The first big sequence involves Ripley crashing on Fiorina in the EEV. Clemens finds the EEV floating offshore, and Ripley's washed up on the beach. That's a sequence that was alluded to in the early trailers for the film, which show Clemens walking around on the surface. You get to see him carrying Ripley into the facility. All that is the first big chunk. Then there are a couple of subplots that were pretty much gutted from the theatrical version, the biggest one involving the prisoner Golic. He basically ends up worshiping the alien, calling it "the Dragon." He's a very simple-minded person, who starts killing his fellow prisoners so he can get closer to the alien.

DB: He's sort of protecting it, isn't he?

CL: That's right. And I get the sense that this is the stuff Fincher was really interested in, because there's a difference in the direction and the direction of the performance. It's much different than just seeing a guy in a monster suit chasing a bunch of bald guys around in the dark, you know? It's not typical of what you'd normally expect to see in a film like this. It would be the equivalent of watching *Alien*, and following Brett around for a day — it's an interesting little offshoot, but the rest of the story doesn't rely on its inclusion.

There are also some more moments of Clemens and his relationship with Ripley. Then there's an extended action sequence that was heavily abridged in the theatrical version, in which Ripley comes up with a plan to scare the alien into a toxic waste dump. In the final version of the film, they try to do this, but they fail and the place blows up. Several of the prisoners end up getting killed, and the alien gets away. In this version, they actually capture the alien. For all intents and purposes, the alien is defeated and the prisoners go on about their business waiting for the transport to arrive and take Ripley away. Then, reentering the story is Golic, who escapes and frees the alien, which leads to a whole set of other problems. It's mostly Golic's story that's being restored. The character was played by Paul McGann. He must have been crushed when he saw the final version of the film. He had such an interesting role. He was still in the final cut, but like 90 percent of his work was cut.

DB: You also mentioned that the "Bambi burster" scene is being restored?

CL: Yeah. The ox caravan that carries the EEV off the beach and into the facility — originally, one of the oxen was impregnated by a "super face-hugger," which is also a creature you don't see in the theatrical cut. A super facehugger is basically a normal facehugger, but with extra armor, because it's carrying the seed for a queen alien. It's only been seen in a few photos. I think *Cinefex* magazine had some shots of it. You only see it in a long shot in the new cut, but it's there. That leads to the funeral scene for Hicks and Newt. In the final cut, it's basically a montage between the funeral and the dog giving birth to a "dog burster." In this cut, we cut to this dead ox instead of the dog, and what Fincher nicknamed a "Bambi burster" emerges. It's basically the same idea, just with a different animal. And it really doesn't make much sense when you think about it, because the ox is dead, so how is the alien gestating in the body of a dead animal? That's probably one of the reasons why it was cut. Also, there are more dog lovers out there than ox lovers, so seeing the dog go through this pain instead is more emotionally powerful.

DB: Before we move on to *Alien Resurrection* and Jeunet's participation, is the new cut of the original *Alien* basically going to be just the theatrical cut with the deleted scenes added back in?

CL: Here's the thing. The laserdisc and the previous DVD included deleted scenes that Ridley didn't participate in — he didn't get to edit them. This is the first time that Ridley is actually getting hands-on involved with the deleted scenes. So most of them are going to be scenes you've seen before, but they're cut in a different way. They'll be more polished. A few will be seamlessly integrated into the new, alternate cut of the film — the "2003 Director's Cut" — and the rest will be included on the *Alien* supplement disc.

We *had* created a fully expanded version for the DVD, with all the deleted scenes integrated back into the film. But very late in the game, once Ridley saw the whole thing and the prospect of a big theatrical re-release was looming large, he had a serious change of heart. He felt that this expanded DVD version was way too long, and decided that he wanted to create a more streamlined experience, especially if a whole new generation of moviegoers would be seeing it in theaters. So he brought Dody Dorn in, who edited *Matchstick Men*, as well as *Memento* and *Insomnia*, to cut the whole thing down. This was a bit of a puzzle for me, because it was so late in the process and so much of the DVD had already been approved and finalized—this was a *major* curve ball to be thrown. As a longtime fan of the film, and as someone who is well aware of not only *Alien*'s background, but also of both its place in film history and with its fans, I have to admit I expressed serious reservations about this—especially having it called "The Director's Cut," which incorrectly implies that he wasn't happy with the original version. But it's Ridley's movie. It's his call.

So this is the first time you're going to see extra scenes not only back in the film, but approved by the filmmaker. And I will say that there are a few things you've *never* seen before. It's nothing big, but there are some cool little moments that haven't been released previous to this new DVD.

You'll also notice that some of the deleted scenes on Disc Two of *Alien* include a 5.1 track. Those were the scenes that were originally in the expanded version we had been working on, but were later dumped when Ridley and Dody recut the film. The deleted scenes in 2.0 were never under consideration for the expanded version. They were somehow incomplete, either because of missing elements or because some sequences were never shot in their entirety.

DB: We understand that Ridley's new "2003 Director's Cut" of *Alien* is actually *shorter* than the original theatrical version, even with the restored scenes. Can you tell us, specifically, what's been cut from the original version and what's been added back in?

CL: The whole is not really shorter. It's just a hybrid of the original 1979 version and the expanded DVD version we had originally restored. And as a result, some scenes from both versions have been cut down a bit. Additions include the cocoon scene, the scene when Lambert slaps Ripley outside of the infirmary, and the alien transmission scene, which now includes Ridley's preferred version of the transmission sound effect. It kind of reminds me of the sound effect that Luke and Ben trigger as they cross the entrance threshold of the Mos Eisley cantina in *Star Wars*, which comes as no surprise since Ben Burtt designed both sound effects. Omissions include Dallas' final session with Mother, Ripley's conversation with Ash about running the transmission through ECIU, and several little trims throughout. As a longtime fan of the film, I was sorry to see these scenes removed, but I just had to console myself with the knowledge that the original cut would still be there on DVD for purists like myself who love and cherish the original version.

A Quick Chat with Ridley Scott

During a break in the recording of his first audio commentary for the *Alien Quadrilogy*, we spoke briefly with *Alien* director Ridley Scott about DVD and the digital future of film.

The Digital Bits (DB): As someone who's been directing films for many years, how do you feel about the opportunity that DVD gives you to go back and revisit your work?

Ridley Scott (RS): It's nice, because DVDs help you to recapture the quality—the way you originally meant for the film to be seen. DVD really lets you hit that standard you want in terms of the picture and sound.

DB: Obviously, it all started with VHS, then laserdisc, and now DVD is the thing. How willing are you to come back and revisit your films for new formats as they come along?

RS: I'm always willing to be involved in whatever next-phase format they want to take it to. Because it's all basically a device to carve the film in stone as much as you can—to get it just the way you want it.

DB: You and Charlie [de Lauzirika] have worked together on the DVD versions of a number of your films now. Do you find that long working relationship helpful when you approach a new DVD project? Is there a shorthand that develops between you?

RS: Oh, yeah. I think that's absolutely true. We started off on the first *Alien,* and we've worked on pretty well everything since then. It's worthwhile to do, you know? Film is worth preserving 'cause that's what DVD is really doing.

DB: With this new trend toward digital production, is there any thought in the back of your mind that the work being done for these DVDs is ultimately going to be the lasting testament to the film?

(continued)

RS: Well, I don't know how long a DVD disc will last, and the studios are still pulling out the original negatives. I'm surprised—Paramount just did that for *Duellists*, and it's perfect. But will the negative still be around in another 20 years? Martin Scorsese's been complaining about that for years—that we need to spend the extra bit of money on the original negative to preserve it. So, thank God for the digital record, I suppose.

Digital seems to be a faster, better way of doing things. But I think it's all about the legacy of the film, digital or not. Pretty soon, you're going to have a screen that you hang on the wall like wallpaper, and you'll get a better image from it than you'll ever get from projection. So whatever the technology is, as long as you preserve the film the way it was meant to be seen, it's fine by me.

DB: We've already covered Cameron and we know *Aliens* is the same extended cut that was on the previous release. So tell us about the new scenes being completed for *Alien: Resurrection*, and Jeunet's involvement in the production of the new DVD.

CL: I met with Jeunet in Paris last October to interview him and to record his audio commentary. I told him what we had in mind and showed him the deleted scenes we'd found. He seemed on the fence about a lot of it. He was very nice and supportive, but at the same time, he's very happy with the theatrical cut. So I don't know if he was completely sold on the idea of doing a special edition cut of the film. It was really when we got into the idea of restoring the footage, and doing a branched version, that he seemed a little hesitant.

The two biggest additions to *Resurrection* are an alternate opening and an alternate ending. The alternate opening is an incredibly complicated effects shot that Jeunet really wanted to do, but that Fox wouldn't pay for back in the day. When I told Jeunet that Fox wanted to finish it for the DVD, he was shocked. The alternate ending was something Jeunet didn't want to do, but Fox convinced him to shoot it. It shows Ripley and Call back on Earth after they've landed. It existed as just the actors in front of a green screen, which has now been replaced by an amalgam of the various concepts for the ending that were considered back in 1997 – a spaceship graveyard, post-apocalyptic Paris, and so on.

So Jeunet wasn't a hundred percent behind it. But at one point during our first meeting, he said, "Okay, I trust you. Just let me see the result."

So a few months later, as the effects shots were gearing up, I sent Jeunet an e-mail with all the details, letting him know who was doing the effects work and that we'd show him the finished shots for approval. Suddenly, I don't know if it was the language barrier or what, but he didn't understanding what we were doing. There was one day where a flurry of e-mails went back and forth between myself, Jeunet, and Fox. Finally, Fox brought in a French interpreter and we explained exactly what we wanted to do. The result is that Jeunet basically said the same thing to us that Fincher did. "Go ahead and do whatever you want. But you can't call it a director's cut." So then it's my job to make sure we stay loyal to their original vision – to do all the research, to dig up as much conceptual art and pre-viz material as we can, so that we can get as close as possible to how they originally visualized the shots. But it needs to be stressed that these special editions of *Alien³* and *Alien Resurrection* are merely alternate and partially enhanced versions of what once existed as a rough cut in the post-production sequence. These versions are just a treat for fans, and in no way represent a filmic document of a director's compromised vision.

DB: Can you describe the alternate opening for us? What made it interesting or different?

CL: It was an incredible idea for a shot. It starts out with a closeup on these viscous-looking teeth. You think it's an alien warrior. Then you pull out and you see that it's actually a tiny little insect crawling around on a joystick. A finger comes in and squishes it, and you see that it's some guy sitting in a cockpit drinking this Big Gulp. He pulls the straw out and sticks the bug guts on the end of the straw, then blows them right at the camera. There's a rack focus, so we see where the guts have presumably hit the cockpit glass; then we pull out and see that he's in this giant "node" they called it, hanging underneath the command tower of the Auriga, which is the big scientific research vessel in *Alien Resurrection*. And the shot just keeps pulling out and pulling out, for like two minutes. I think it was meant to be used for the opening titles. But if you take the opening shot from *Star Wars*, with the Star Destroyer, it's like 10 times that long. It's a fantastic shot, but I know why they didn't do it back then. It would have cost a bundle. Even with CG, it's difficult to pull off today. The final result, for the DVD, combines an original live action plate and some of the original pre-viz with all-new animation.

DB: So who's doing the new effects work?

CL: It's a handful of companies. We met with several effects houses and got bids from them all. There was a whole casting process for the effects work. The first one we spoke with was Svengali. One of the partners of Svengali is Rocco Gioffre, who is a pretty famous matte painter. He worked on *Blade Runner*. They were extremely enthusiastic about doing the work from the start. Then there's a company called Riot, which is handling a few of the composite shots. Riot also has a sister company called Encore, which is handling some of the CG character creation, and they've done a fantastic job. Then there's a fourth company called Frantic Films, up in Canada, which is handling the opening shot from *Alien Resurrection*. And then there's Modern Video Film, who've done some significant clean-up work and final effects tweaks. They've been our trusty safety net throughout.

DB: Can we assume that new high-definition transfers of the films are being done?

CL: The theatrical versions have been done, but they'll require grading and color timing. The new footage will have to be done. That's more Fox's domain, but I'll be involved in *Alien* and *Alien³* for sure. In fact, I just went in with Ridley to color time the theatrical version of *Alien*. Along the way, he looked at the deleted scenes and gave us some direction. It's a bit darker and more moody now. As far as the other films, we've delivered our cut lists of what needs to be pulled and how it should be edited together, so then Fox goes through boxes and pulls out the neg. They take the original negative and create an I.P. to use, and that gets transferred to video. Then that goes to Modern Video Film, which is where they're doing the tape-to-tape transfers and final color timing. The new edit is assembled from that based on the cut lists we've given them. At this point, just about everything has been onlined and delivered. There are just a few fixes and tweaks left to do on *Alien*.

DB: Is there anything else you'd like share with regard to your work on the special editions?

CL: I think the thing I most want DVD fans to understand, particularly with regard to the interviews and commentary, is that if we *don't* have someone on the disc, it's not because we didn't ask. We asked everyone we could find—we went right down the list. But if they're not there, chances are it's because they said no.

DB: That doesn't just apply to interviews and commentary, does it? It could be any feature. A piece of footage or a documentary or whatever. You start out wanting all of that stuff, but for whatever reason, you just can't include it.

CL: There's no reputable DVD producer I know that would try to put together a disc and intentionally leave stuff off, or not try to find material that they know exists. If it's not on the DVD, it's either because it got rejected for legal reasons, or the director doesn't want it on, or it simply can't be found, or it physically doesn't fit on the disc. There's any number of reasons why things might not be included. That's just part of the challenge of DVD production. And it's getting tougher and tougher all the time.

DB: A lot of *Aliens* fans are probably hoping to see the infamous "Burke cocooned" scene that was shot but not included in the final cut of the film. What's the status of that?

CL: Jim Cameron never wants that scene to see the light of day. Having now seen it myself, I sort of understand why. But I tried to get it on the DVD, I swear! [laughs]

Seriously, sometimes you just get lucky and the stars line up and everything works out. Filmmakers dig into their boxes and give you tapes of amazing stuff you never knew about. And sometimes that doesn't happen. The reality is you do everything you can as a DVD producer to create the definitive disc. I'm pleased with much of the *Alien Quadrilogy* in that respect. In terms of supplements, everything on the first eight discs will be pretty much all new material.

Someday, when HD-DVD or whatever's next comes along, and you can fit 50 hours of content on a single disc, will some future supplement producer top what we've done with this project? Maybe. Probably. Someone, someday, will probably include all the dailies ever shot, a photo gallery of a hundred thousand still images — you name it. At some point, there will be probably entire virtual libraries of literally everything that was ever produced for a film. But given the technology and limitations and the resources available today, we're giving you everything we possibly can on this nine-disc set. I'm not kidding when I say this — we've completely maxed-out the technical capabilities of the DVD format with this set.

DB: We should also point out, knowing you as we have now for so long, that you're as much a fan of this material as anyone.

CL: Absolutely. And it's fans of these movies who I'm really doing all this for. As both a DVD producer and a fan, I want to create a disc that *I'll* want to own and have at home and watch myself. Did we get absolutely everything and everyone into this box set? No. Did we blow away everything that's come before? Yes, I think we did. And hopefully the next supplement producer to tackle the *Alien* films down the road will do an even better job. The bottom line is it's about the film. It's about the filmmakers. And it's about the fans. So long as the technology and the presentation and quality of the material keeps improving . . . everybody wins.

Ridley Scott enjoys a light moment during the recording of his first audio commentary for the *Alien Quadrilogy*.

Lauzirika looks on as *Alien* star Sigourney Weaver chats with Ridley Scott during another audio commentary session.

There you have it — a glimpse behind the scenes at one of the biggest special edition DVD projects to date, Fox's nine-disc *Alien Quadrilogy*, straight from the producer himself. As you can see, realities of DVD production and the technical limitations of the format make for an incredibly complicated process. Not everything turns out the way it's planned, but, as we think you'll see by the details that follow here, the result *should* be worthy of the effort.

As it stands now, each film is THX certified and is presented in anamorphic widescreen video, with Dolby Digital 5.1 audio. *Alien* and *Alien Resurrection* will also include DTS 5.1 audio (disc space limitations prevented the inclusion of DTS on *Aliens* and *Alien³*). When you start the movie discs, you're given the choice of viewing two different versions of each film via seamless branching. In the case of *Alien*, you'll find the 117-minute theatrical version and Ridley Scott's new 115-minute director's cut (which will be shown in theaters later in 2003). *Aliens* will include the 137-minute theatrical version and James Cameron's 154-minute special edition cut. *Alien³* features the 114-minute theatrical version and the new 155-minute restored work print version created just for this DVD release. And *Alien Resurrection* will include the 109-minute theatrical version and a new, 119-minute extended cut, again created just for this DVD release.

New video introductions have been created to place in proper context the alternate cuts of *Alien*, *Aliens,* and *Alien Resurrection*. In addition, each movie disc will include a newly recorded audio commentary track featuring members of the cast and crew for each film. The participants include (but are not limited to) Ridley Scott, James Cameron, Jean-Pierre Jeunet, Sigourney Weaver, Terry Rawlings, Tom Skerritt, Harry Dean Stanton, Veronica Cartwright, Bill Paxton, Michael Biehn, Lance Henriksen, Jenette Goldstein, Carrie Henn, Gale Anne Hurd, Stan Winston, Alec Gillis, Tom Woodruff, Jr., Richard Edlund, Paul McGann, Ron Perlman, Dominique Pinon, Leland Orser, and Pitof.

Each film also has a second disc of newly created supplemental material. The supplemental disc for *Alien* includes Dan O'Bannon's first draft of the screenplay; galleries of conceptual art (from Cobb, Foss, Giger, Moebius, and others); cast, set, and production photos; creature design art; poster art; visual effects stills; premiere photos; a storyboard archive; a gallery of Ridleygrams; Sigourney Weaver's screen test (with optional director's commentary); a multi-angle breakdown of the Chestburster scene (viewable with production audio or director's commentary); seven deleted scenes; and a three-hour documentary entitled *The Beast Within: The Making of Alien*, composed of several shorter featurettes with a "play all" option (these include *Star Beast: Developing the Story*, *The Visualists:*

Direction and Design, Truckers in Space: Casting, Fear of the Unknown: Shepperton Studios, 1978, The Darkest Reaches: Nostromo and Alien Planet, The Eighth Passenger: Creature Design, Future Tense: Music and Editing, Outward Bound: Visual Effects, and *A Nightmare Fulfilled: Reaction to the Film*).

The *Aliens* supplemental disc includes James Cameron's original treatment for the film; galleries of conceptual artwork; cast, set, and production photos; continuity Polaroids; photos of the vehicles and weapons; creature design stills from Stan Winston's Workshop; visual effects photos; photos of the music recording and the premiere; multi-angle pre-visualizations; and another lengthy documentary (well over three hours long) entitled *Superior Firepower: The Making of Aliens*, composed of several shorter featurettes with a "play all" option (these include *57 Years Later: Continuing the Story, Building Better Worlds: From Concept to Construction, Preparing for Battle: Casting and Characterization, This Time It's War: Pinewood Studios, 1985, The Risk Always Lives: Weapons and Action, Bug Hunt: Creature Design, Beauty and the Bitch: Power Loader vs. Queen Alien, Two Orphans: Sigourney Weaver and Carrie Henn, The Final Countdown: Music, Editing, and Sound, The Power of Real Tech: Visual Effects*, and *Aliens Unleashed: Reaction to the Film*).

The supplemental disc for *Alien³* will include a storyboard archive, time-lapse set construction footage, a multi-angle look at E.E.V. bioscan video, galleries of conceptual artwork, set and production photos, ADI workshop photos, visual effects photos of the vehicles and creatures, creature design stills, and another documentary entitled *The Making of Alien³*, composed of several shorter featurettes with a "play all" option (*Development: Concluding the Story, Tales of the Wooden Planet: Vincent Ward's Vision, Pre-Production: Part III, Xeno-Erotic: H.R. Giger's Redesign, Production: Part I, Adaptive Organism: Creature Design, Production: Part II, Post-Production: Part I, Optical Fury: Visual Effects, Music, Editing, and Sound*, and *Post-Mortem: Reaction to the Film*).

The *Alien Resurrection* supplemental disc includes Joss Whedon's first draft of the screenplay, galleries of conceptual artwork, production and visual effects photos, Mark Caro's artwork, ADI workshop photos, multi-angle pre-visualizations and rehearsals, makeup test footage and a storyboard archive, along with yet another lengthy documentary, entitled *One Step Beyond: The Making of Alien Resurrection*, composed of several shorter featurettes with a "play all" option (including *From the Ashes: Reviving the Story, French Twist: Direction and Design, Under the Skin: Casting and Characterization, Adaptive Organism: Creature Design, Death from Below: Underwater Photography, In the Zone: The Basketball*

Scene, *Unnatural Mutation: Creature Design*, *Genetic Composition: Music*, *Virtual Aliens: Computer-Generated Imagery*, *A Matter of Scale: Miniature Photography*, and *Critical Juncture: Reaction to the Film*).

Finally, the ninth bonus disc encapsulates both new and previously released material. Among other things, this includes complete archives of the supplemental content from the *Alien* and *Aliens* special edition laserdisc editions, *The Alien Evolution* documentary (64 minutes), a 1979 promotional featurette called *Experience in Terror*, an all-new featurette called *Aliens in the Basement: The Bob Burns Collection*, a question and answer session with Ridley Scott, an advance preview featurette for

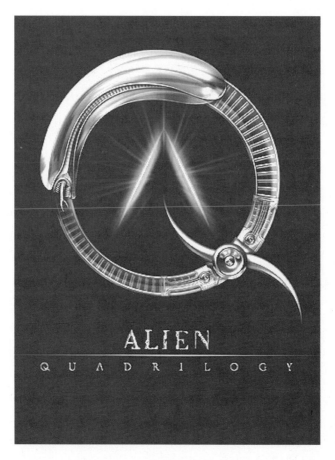

Preliminary cover art for Twentieth Century Fox's *Alien Quadrilogy* DVD box set

Alien³, theatrical trailers, teaser trailers, and TV spots for all four films in the series, a gallery of Dark Horse Comics artwork, and DVD-ROM script-to-screen comparisons.

So now we've gotten you all worked up about the *Alien Quadrilogy*, which won't be available at your local video store until sometime in mid-December 2003. If you're still eagerly awaiting its release as you read this book, never fear—the box set should be worth the wait. And there are *lots* of other great DVDs to keep you busy in the meantime. Rest assured, your friends at *The Digital Bits* have plenty of recommendations for you (well over a hundred in fact).

So turn the page and let's start checking them out, shall we?

Part IV

The Best of DVD (Great Discs Reviewed Here!)

What Makes a Great DVD?

There are really three parts to any DVD, and no one is more important than another. There's the film, the presentation, and the extras. When all of these things come together in the best way, the result is a truly great DVD experience.

The Film

This would seem obvious. Without a great film, how can you have a great DVD? But there's more to it than that. The film has to connect with more than one person in the world. We all have our favorite films . . . and some of them are pretty lame. One of Bill's favorites is *The Final Countdown*, while Todd loves *Killer Klowns from Outer Space*. But don't look for reviews of either of these DVDs within the pages of this book, because for various reasons, they can't be deemed "must own" for everyone.

The reason so many of us have off-beat favorites is because movies define who we are. They make us laugh and cry for personal reasons, and trying to explain that to other people is a very hard thing to do. Film critics try to show why films work or don't work for aesthetic reasons, how there are flaws in the storytelling, or things that shouldn't be too obvious actually are, sometimes to the point of distraction. But what no one can account for is how a film affects us personally. Movies speak to us on many different levels and when they work for a person, you can't ask for anything better than that.

The movies listed in this book have stood the test of time. We picked these films to be on our list, because they're *good* films, plain and simple.

These films define the genres they live in and have repeated viewing value. While some of these films could also be found on our own lists of favorites, not one of them was chosen to be included here for that reason. To be fair, we can't promise that *you* will love every one of these films — everyone has their own unique opinion. But if you give these movies a chance, they just might grow on you as much as they've grown on us.

The Presentation: Video and Sound

Presentation's the thing, so we also have to look at the overall production quality of the disc — the way the film looks and sounds. Is the video and audio quality the best it can be? Is the picture clear and free of artifacts or any other visual flaws? Does the sound draw you into the film or distract?

There are many factors that go into presentation. Many films we love are older and, because of that, they don't look as good as they should before making the journey to DVD. Many of these films need to be restored, which is a lengthy and expensive process that studios don't always want to do. Once a film *does* look great, it goes through the process of telecine and compression for DVD. This process is very sensitive and many things can go wrong. Older DVDs, released in the first year or two of the format, just don't look as good as current releases do, because the technology still wasn't quite up to par. We tend to be hard on the video and audio quality of DVD these days, because *every* DVD should look good. The technology has improved to the point that there really is no excuse anymore for artifacting or shoddy presentations on disc. Bottom line, if a disc looks and sounds like crap, even if the film is a work of utter genius, you won't find it in this book.

The Extras

The last thing that can make or break a great DVD is also the most superficial. When you look at any great DVD, you *have* to take a serious look at the supplemental features on the disc. These days, even terrible DVDs often have lots of extras.

The question you have to ask is, are the extras worth your time? For example, listening to audio commentary tracks can be an amazing experience, but some filmmakers are better at it than others. There is nothing worse than a commentary with participants who have little enthusiasm and nothing of interest to say. What about those short video featurettes studios are fond of putting on DVDs? Were they produced simply to promote the film, or do they actually inform and entertain? Does the packaging oversell the special features to the point that when you actually pop in the disc, you find that the studio considers interactive menus, subtitles, and trailers promoting *other* films "special"? Basically, for anyone to say that a DVD is great, you have to look at the complete package. Great extras must have real value to fans of the film and, more importantly, must serve the *film* instead of the studio.

All of these things together — the film, the presentation quality, and the extras — are what make a DVD great. There are plenty of good DVDs out there that are worth adding to your collection, but the titles reviewed in this section of the book represent what we feel are some of the very best titles available on the format. For a variety of reasons, we're confident about recommending them to you, and we know your collection will be all the better for having them.

So, for all it's worth, we present *The Digital Bits' Best of DVD* reviews section. We hope you enjoy it. And remember — you can always visit our web site for recommendations on more recent releases.

Editor's Note: There are a *few* titles in this list that are movie-only editions. In rare cases, when both the film and presentation quality were considered superior, *and no special edition version of the film was otherwise available on DVD*, we deemed them worthy of inclusion here. Also, sadly, some of the titles reviewed here are no longer available through retailers, either because they're now out of print, or the studios decided to release movie-only editions of the films and have placed the special edition versions on moratorium. (Disney often does this with their DVDs: "Buy it now, because *The Little Mermaid* won't be around after July third!") You should know, though, that if a disc is on this list, it's worth having, even if you have to hunt it down. Many places carry out-of-print discs, including used disc shops and online retailers who over-stocked. Even eBay can be a good source, if the price is right. The point is a great disc is a great disc, so even if we discovered that a title is no longer widely available, it still deserves to be included here.

Must-Have DVD Special Editions

Everyone has advice on the movies you "must have" in your DVD library. You've gotta have this, you've gotta have that. We're no exception to that at *The Bits*, but then we know a little more than the Average Joe about movies. So we're gonna tell it to you like it is. The titles that follow should absolutely be the foundation of any good DVD collection. It's that simple. These are, simply put, the best special editions that have ever been released on the DVD format, bar none.

As we said, if you can't find any of these discs in your local retail stores, they're still worth hunting down. And we suggest you do so soon, whether you seek them out on eBay, at used disc shops, or from friends who need cash. It doesn't matter. Just find these discs and you'll have the great fortune (and pleasure) of enjoying the best that DVD has to offer.

So find these discs! Got that? Don't make us stop this car . . .

The Lord of the Rings: The Fellowship of the Ring—Special Extended DVD Version

New Line

"In the lands of Middle Earth, legend tells of a ring . . ."

For years, people said *The Lord of the Rings* could never be brought to the screen. It was too big, too vast, too expensive. Director Peter Jackson and his team have proven the doubters wrong.

The Fellowship of the Ring is the opening act of Jackson's three-part adaptation of the beloved J.R.R. Tolkein literary epic. It's many years after the events told in the book *The Hobbit*. Bilbo Baggins (Ian Holm) has grown old in the Shire and now longs to retire in peace. But Bilbo has a secret – he's been keeping a ring that he found on his adventures. And it's no ordinary ring. This is the One Ring, created by the dark lord Sauron many thousands of years ago to enslave the world. Sauron was defeated then, and the Ring was thought lost. But Bilbo passes it on to his nephew, Frodo (Elijah Wood), without realizing that the dark lord has risen again and is now scouring all of Middle-Earth for it. When he learns what's at stake, the wizard Gandalf (Ian McKellen) instructs Frodo to leave the Shire for his own safety and take the ring with him. Though Sauron's forces are hot on his trail, a band of loyal companions joins Frodo on his journey, a Fellowship tasked with the seemingly impossible goal of destroying the Ring once and for all. But to do so, they'll have to take it back to Mount Doom where it was originally forged . . . straight into the very heart of evil itself.

The Fellowship of the Ring manages to stay almost perfectly true to the spirit of the original novel. Jackson has cut out all of the unfilmable literary texture — the limericks, the irrelevant characters, the slow build-up of detail — so this film gets right to the story and keeps the action moving all the way through. But lest fans get too upset, he's managed to replace much of that literary texture with its equivalent in visual, production design texture. So this film *feels* like the world we pictured in our heads as we read the novels. Better still, the casting here is magnificent. Ian McKellen simply *is* the wizard Gandalf. While Elijah Wood might have seemed an unlikely choice to play Frodo Baggins at first, he proves in this film that he's more than up to the task, infusing the Hobbit with the perfect measure of pathos and humanity. And the supporting cast delivers in spades as well, including the likes of Viggo Mortensen, Sean Astin, Cate Blanchette, John Rhys-Davies, Ian Holm, and Christopher Lee. Even Liv Tyler manages to hold her own here, and that's saying something.

The original theatrical cut of *Fellowship* is available in a two-disc set, but for this four-disc, extended DVD version, an additional 30 minutes of footage has been restored to the film. Several major new scenes have been added throughout the entire length of the film, as well as numerous smaller scenes, scene extensions, and additional brief character moments. The cumulative effect is to make this film seem vastly more epic in scope. There's a greater sense of distance to the Fellowship's journey, with many more points of interest along the way. We get to learn much more about Hobbits in a longer opening. You see that Gollum has continued to follow the Fellowship after leaving Moria. Lothlorien is depicted in much greater detail. The battle scenes are all much more intense now, and several characters are given added moments that make them feel more rounded, particularly Boromir (his last stand is now much more heroic and emotional). There's more humor in this cut, and the new footage adds significant texture and depth. The result is a much more satisfying viewing experience.

The anamorphic widescreen video quality is excellent. The film is split over the first two discs in the set, allowing for a maxed-out video bit rate. As a result, the video exhibits tremendous clarity. There's more depth to the image, colors are more vibrant, and more detail is discernible. Those of you viewing the film on large widescreen displays (the best way to watch an epic film like this) will appreciate the quality here. The audio is also excellent, available in both Dolby Digital 5.1 and DTS 6.1 ES. Both tracks are highly active, with tremendous spaciousness in the imaging and thunderous bass. The DTS is clearly the better of the two options, with notably greater subtlety and clarity. While the differences between the new Dolby Digital and DTS tracks aren't huge, high-end users with quality equipment will certainly appreciate them.

Discs One and Two include no less than four full-length audio commentaries. As you're watching the tracks, subtitle text appears at the top of the screen when different participants speak, identifying not only the speaker, but also their role in the production (or their character in the case of the actors).

The most immediately engaging of the commentaries is the actors' track. You can tell that these actors really enjoy both their involvement with the film and each other's company, and there are many thoughtful observations, particularly from Ian McKellen and Christopher Lee (who each make frequent appearances in addition to nearly all of the rest of the cast). Peter Jackson's track with the writers provides wonderful insights into the story of the film and the adaptation process. There's also a great deal of discussion about the new scenes — why they were cut from the theatrical version and the value of adding them back in here. The production design commentary addresses the extensive detail that went into the design of every onscreen element, no matter how trivial it may seem, and how it all can be traced back to the original Tolkien books. And the production/post-production track deals with the more practical filmmaking issues, including the sheer effort required to shoot the three films back to back.

That's impressive enough, but Discs Three and Four, also known as *The Appendices*, contain the lion's share of the supplemental material. Disc Three specifically deals with the effort to adapt the story and to formulate a vision for the film that would remain true to Tolkien's books. Disc Four looks at the process of taking that vision and crafting a film from it. Each features a brief video introduction (by Peter Jackson and Elijah Wood respectively) explaining what you'll find on the disc and how to access it. There's also additional help text on how to access the material, along with a complete index of the disc's contents, and a "play all" mode that will allow you to view all of the featurettes and documentaries. Virtually all of the materials are presented in full anamorphic widescreen, and the animated menu pages on all of the discs feature full Dolby Digital 5.1 sound — very nice touches.

The first major piece on Disc Three (*The Appendices, Part I — From Book to Vision*) is an in-depth look at the historical background of the man behind the original books, called *J.R.R. Tolkien: Creator of Middle-Earth*. It's a good starting point for the supplements, because you'll learn how these stories came to be, as well as what Tolkien himself intended them to mean (and, as importantly, what he didn't intend). In *From Book to Script*, Jackson and others associated with the production talk about the process of "cracking the code" of the books, and their effort to craft a workable script based on them (first with Miramax, who wanted it done as a single film, and then with New Line, who thankfully pushed for a trilogy). The disc then takes you into the process of "visualizing" the story—in a featurette specifically on this subject—*Storyboards and Pre-Viz: Making Words into Images*. This section is illustrated with several terrific storyboard and animatic videos for major scenes, as well as multi-angle comparisons that illustrate the development process and give you a peek at roads not taken. For each multi-angle piece, you can switch back and forth on the fly between one angle, the other, and a split-screen comparison of the two.

The next major section of Disc Three is on designing and building Middle Earth, and it contains the real meat of this disc. There's a fantastic, 41-minute documentary, *Designing Middle-Earth*, addressing the effort to conceptualize the look and feel of each race and character, and to add a sense of history for every item as well. Next, Richard Taylor takes you on a 43-minute tour of the Weta Workshop, where an army of hundreds of craftsmen and artisans designed and created nearly every visual element of the film, including the props, sets, armor, weapons, creatures, miniatures, and special makeup effects. The *Costume Design* featurette hints at the massive task of creating the wardrobe elements for the films, which often included dozens of versions of each of the hero costumes (the Hobbits, for example) in various scales. Finally, this section features some 19 separate design galleries packed with sketches, paintings, and photographs that illustrate both the peoples and realms of Middle-Earth. You can view these as a slideshow, or you can page through a scrapbook and view them one at a time. There are literally hundreds of images to see.

Disc Three is rounded out with a pair of interactive maps that help you understand the geography of the film. The *Middle-Earth Atlas* allows you to follow, step by step, the journey that the Fellowship takes in the film. *New Zealand as Middle-Earth*, on the other hand, allows you to see where in the "real world" each film location was shot and includes viewable location scouting video for each place.

Moving on to Disc Four (*The Appendices, Part II—From Vision to Reality*), you're immediately provided with a trio of interesting, day-to-day looks behind the scenes at the production. In *The Fellowship of the Cast* documentary, each cast member recalls funny moments and memories about their fellow actors. *A Day in the Life of the Hobbit* is just what it sounds like—a look at a typical day of filming, from getting feet glued on early in the morning to getting them taken off late at night (and everything in between). *Cameras in Middle-Earth* is the longest documentary piece on this DVD set, clocking in at nearly an hour. It's *the* major look behind the scenes, following the production from location to location (and back through the soundstages and sets). It provides a taste of the massive effort required to capture the story on film. This section also includes a gallery of behind-the-scenes production photos.

The next section on Disc Four relates to the visual effects of the film. There's a featurette on *Scale*, in which you see how the filmmakers developed the various tricks that allowed them to make Hobbits look like Hobbits . . . and everyone else look much taller and bigger. There's a subsection here on the *Big-atures* created for the film (so called because there wasn't anything "miniature" about them), as well as galleries of close-up photos of each model. There's also a featurette on the amazing CGI effects work of Weta Digital.

The post-production section of the disc begins with a featurette on the editing process, *Editorial: Assembling an Epic*. There's also a multi-angle demonstration of the Council of Elrond scene, showing how it was assembled from all of the footage shot on set. And the *Digital Grading* featurette shows how nearly all of the location and live-action footage was enhanced using color-timing and adding a variety of lighting effects to change the weather, make the footage match, and create a more ethereal, other-worldly look to the final film.

The final major section of the disc focuses on the sound and music work done in post-production. *The Soundscapes of Middle-Earth* featurette takes you behind the scenes on the creation of various sound effects and the mixing process, while *Music for Middle-Earth* highlights the work of composer Howard Shore.

Finally, the entire set is capped with a short featurette called *The Road Goes Ever On . . .* , in which Jackson looks back at the first film and briefly ahead at the next two. We also see the premiere of the film and get a taste of how the overwhelmingly positive reaction to it affected the cast and crew. It's a nice way to close out the set. It's worth noting that all of these amazing extras come enclosed in a gorgeous slipcase, designed to look like an old leather-bound book, which holds a fold-out Digipak that contains the discs and a 12-page insert booklet.

We should also tell you that this impressive set is but the first of three. When all is said and done, the two sequel films in this series, *The Two*

Towers and *The Return of the King*, will get matching, four-disc treatment on DVD as well, each with significantly extended footage and expansive extras.

Simply put, *The Lord of the Rings: The Fellowship of the Ring — Four-Disc Special Extended Version* is the most impressive special edition DVD release we've ever seen. Its overall presentation quality, breadth and depth of content, and thoughtful attention to virtually every detail are unsurpassed. There are certainly titles that are better in individual areas (*Star Wars: Episode II* features better video quality, for example), but no other single title can match this set, blow for blow. It's absolutely worthy of the incredible effort that was mounted to make this film, and it is, hands down, *the* single, must-have cornerstone of any good DVD enthusiast's movie collection.

Toy Story: The Ultimate Toy Box

Disney/Pixar (Buena Vista)

In this day of mega-DVD special editions, the bar for quality has been raised awfully high. But Buena Vista and Pixar's *The Ultimate Toy Box* definitely hits the mark and then some. Part of the reason we like *The Ultimate Toy Box* so much, completely apart from how great these films are, is that the folks who produced these DVDs (namely the staff at Pixar) seemed to have had a blast doing so. The creators of these films (including director John Lasseter, writer Andrew Stanton, codirectors Lee Unkrich and Ash Brannon, and others) are everywhere on this set's three discs, explaining how the story developed, talking about the animation process and generally guiding us through what it's like to work on a film at Pixar. The result is truly impressive — an in-depth look at computer-animated filmmaking, with the filmmakers themselves as your tour guides.

For the few of you out there who may have been living under a rock for the last six years and missed these glorious films, *Toy Story* was the first feature-length film to be generated entirely by computer, and it tells the story of a cowboy doll named Woody (voiced by Tom Hanks). Woody is the favorite toy of his owner, a young boy named Andy . . . that is until Andy gets a brand new Buzz Lightyear action figure. When Buzz (voiced by Tim Allen) shows up, Woody's world falls apart. Facing the ultimate worst fate for any toy (no longer being loved), Woody "accidentally" pushes Buzz out Andy's bedroom window. Moments later, Andy's family goes out for pizza and takes Woody with them. But Woody's conscience (and a little pressure from his fellow toys) is starting to get the better of him. Little does he know that Buzz has hitched a ride in the family car . . . in the hopes of getting back at Woody. In a series of accidents and misadventures, Buzz and Woody find themselves lost in the big wide world. Then they fall into the hands of Andy's vicious neighbor Sid, who tortures toys. Can Buzz and Woody resolve their differences and find their way back to Andy? Did we mention that Andy's family is moving to another neighborhood so time's running out? Are you ready to laugh for about 80 minutes? Toy Story is a very funny film, made all the better for its terrific supporting cast, which includes the voices of Don Rickles, John Ratzenberger, Wallace Shawn, Jim Varney, and many more that you'll recognize. Better still, each is playing a toy straight from your fondest childhood memories — think Mr. Potato Head and others like him.

Things obviously turn out well for the toys . . . because there's a sequel (and it's even better than the first film). *Toy Story 2* takes place after the move to Andy's new home. Andy's mom decides to hold a rummage sale to get rid of stuff they no longer want. Among those things is one of Andy's older toys . . . one of Woody's friends. Woody immediately leaps into action to save the toy, but he inadvertently finds himself sold to a rare toy collector named Al (the infamous owner of Al's Toy Barn). It turns out that Woody was based on a 1950s TV show called *Woody's Roundup*, and Woody is extremely valuable. Al wants to clean him up and sell him to the highest bidder, as part of a set of the Roundup toys, which include cowgirl Jessie (Joan Cusack) and Stinky Pete the Prospector (Kelsey Grammer). But Buzz and Woody's other friends just aren't about to let that happen. They band together and journey out into the world to bring Woody back home. Naturally, the going won't be easy, and our heroes will face both long odds and sinister forces . . . including the dreaded Emperor Zurg! *Toy Story 2* is that rare movie sequel that actually manages to top the original, while retaining all the humor and heart you liked about the first film.

In *The Ultimate Toy Box*, both films are included separately on their own DVD disc. Each is presented in anamorphic widescreen (aspect ratio 1.77:1) and features a straight-to-digital transfer, meaning the picture you're seeing is virtually identical to what the animators were watching at Pixar — no film was involved in the making of these DVDs. The result may very well be the most stunning DVD transfer you can get. The colors are rich and accurate, with startlingly clear and crisp detail, tremendous contrast, and deep blacks. Because of the all-digital nature of the image, there's not a speck of dust or dirt to be seen, nor is there any kind of print artifacts for that matter. There's no edge enhancement and virtually no digital artifacting.

The sound is also excellent. *Toy Story* is presented in full Dolby Digital 5.1 surround sound. *Toy Story 2* goes a step further, adding Dolby's 5.1 EX sound scheme. The clarity of the dialogue and crispness of the sound effects are remarkable. The sound mixes for each film are completely encompassing. When Buzz climbs into the claw game (thinking it's a spaceship) in the original film, you're going to hear "inside a fishbowl"-style ambience. Then suddenly you're surrounded by a sea of little green aliens and you'll feel like you're right in the middle of them. And when Buzz takes on Zurg in the opening of *Toy Story 2*, you're going to be awash in creative channel-to-channel panning, nifty directional sound effects, and thunderous bass. The soundstage is deep, wide, and natural. And Randy Newman's music is very well represented in the mix. This is great DVD audio that perfectly matches and supports the quality of the video presentation. Note that each disc also includes a sound-effects-only track.

Each movie disc also includes its own set of supplemental materials. *Toy Story* features the Academy Award-winning short *Tin Toy*, a 27 minute "making of" featurette (*The Story Behind Toy Story*), two "on set" interviews with Buzz and Woody, a four-minute multi-language reel, the original Buzz Lightyear TV commercial seen in the film (with introduction), 52 *Toy Story Treats* (interstitials produced for ABC Saturday morning TV – 10 to 30 seconds each), and an entertaining audio commentary track with writer/director John Lasseter, writers Andrew Stanton and Pete Docter, producers Ralph Guggenheim and Bonnie Arnold, art director Ralph Eggleston, and technical director Bill Reeves. *Toy Story 2* includes the very first Pixar short, *Luxo, Jr.*, along with five minutes of funny outtake footage from the film, preview trailers *for Buzz Lightyear of Star Command: The Adventure Begins, Monsters, Inc.*, and an ever better audio commentary track with writer/director John Lasseter, writer Andrew Stanton, and codirectors Lee Unkrich and Ash Brannon.

But none of this even comes close to topping the bonus material included on the third disc of this set. The folks at Pixar have created a *ton* of new material supplementing the two films, including new video introductions and interviews, and they've made it easy and fun via witty film-themed animated menu screens. To give you an idea of the sheer scope of this third disc, it took us more than an hour just to *scan* through the extras for the first film alone. We're not talking about looking at it all – just simply scanning through it to see what was there! It's outrageous what you get on this disc. The Pixar staff introduces the disc themselves and takes you from story development and character design all the way through the animation process to the publicity and marketing materials created for the film's theatrical release. It's a great beginning-to-end look at the process.

There's just no way we can list everything this set includes in the way of bonus material — this set would need its own chapter to do it justice. But we can give you some of the highlights. You get galleries of some two thousand still images — design sketches, production artwork, and the like. You get video "once-arounds" of virtually every character in the film, along with more artwork and animation tests to illustrate the development of the characters themselves. You get in-depth bios on the voice cast and text essays on the history of Pixar. You get video clips of abandoned and deleted scenes, storyboard pitch sessions, and development concepts. You get video explanations of the steps involved in editing, sound design, music recording, and on and on and on. You get storyboard-to-film comparisons of key scenes in both films. You get animation production progression videos that allow you to use the angle button on your remote to step though and compare the various stages in the animation process. You get original song demos by Randy Newman and even a pair of videos by Riders in the Sky for the *Woody's Roundup* theme song. You get an interactive sound mixing tutorial with Gary Rydstrom, in which you can listen to the Buzz/Zurg battle at the end of *TS2* with music only, dialogue only, sound effects only, or any combination thereof. You get trailers, TV spots, and poster artwork. You even get a guide to the "in" jokes in *Toy Story 2* and a look at some of the fake-but-ultra-cool *Woody's Roundup* collectibles that the animators developed for the film. There's something like six hours' worth of great audio and video material alone . . . and that's just on the third disc. If you have any questions about the process of making a CGI animated film, they're probably answered somewhere on this disc. Best of all, the set includes a foldout guide to all the supplements, which also features comments by Lasseter and the Pixar staff on how they used the capabilities of DVD to really give you a look at their world. These guys clearly love what they do and have a ton of fun with it. And thankfully that love just pours out *of The Ultimate Toy Box*'s three discs.

This is one of the most impressive convergences of superior audio and video quality and truly outstanding supplemental materials on the DVD format. Add to that the fact that these two films are so universally loved, and it's very easy to argue that *The Ultimate Toy Box* is far and away the best family DVD ever created.

Brazil

Universal (The Criterion Collection)

When Criterion finally (after a three-year production schedule) released the ultimate version
of Terry Gilliam's incredible, retro-futurist film
Brazil on laserdisc, fans were in heaven. It's
what made many people go out and buy
laserdisc players in the first place (which in
effect allowed many of us to go spiraling into
the world of DVD). The fact that the laserdisc
even saw the light of day was considerable, especially when you know
about the political battles waged over the production of the disc – including MCA's refusal to let Criterion use added footage, and some not very
nice comments about MCA made by Gilliam on the commentary track. All
of this, by the way, is on top of the drama surrounding the production of
the film itself. *Brazil* is funny that way. No matter what, somehow, somewhere, someone wants to step in the way of people seeing it.

Thankfully, all the problems were ironed out. Criterion got everything
they wanted, and the laserdisc was released to incredible critical praise.
The *Brazil* laserdisc proved to be one of the most important special editions ever released to that point . . . and when it was ported over to DVD
by Criterion, it was just as well received.

On the surface, *Brazil* is about one man's fight against the big corporate machine. In the years since its release, Terry Gilliam has turned the
meaning of the film into something a little more autobiographical – himself
against Hollywood. There's also an argument that the film is only about
love. Any way you look at it, *Brazil* is simply a brilliant film. It takes place
in a retro-futurist world, which means that we are seeing a society that
looks like it's set in the past, except it's obviously the future time. The
world is run like a bureaucratic corporation and it doesn't take any bull.
Anyone who's against the government is quickly swooped up, tortured for
information, and then killed. It's all done very quickly and, most important,
efficiently. As we start the film, we find that the government is after one
Archibald Tuttle, but a literal bug in the system causes one of the arrest
sheets to read Archibald Buttle . . . and we think you can guess what happens next.

Enter into the story Sam Lowry, an average salary man who finds an
overcharge (very nudge-nudge – the government charges those who are
tortured a fee for their services). He tries to remedy the situation but

discovers problems at every turn, until he is eventually caught so far up into the complex machination of the system, that he eventually finds himself in the torturer's chair. Along with that story summary, add a mother fighting the ravages of age, a series of fairy tale dreams (symbolizing the on-going story) featuring baby-faced monsters, and Sam dressed as angel in armor battling a samurai made from technological bits. Those who know Gilliam are used to his visually visceral style. He knows how to conjure up images, and he doesn't hold back an ounce. *Brazil* is not necessarily Gilliam's best work, but it *is* his personal masterpiece. That is to say, it sums his art up in one brilliant work.

Brazil was released early in the days of DVD, and you can tell in terms of video and audio quality. The video transfer is rich, with bold colors and deep blacks. Unfortunately, it's non-anamorphic, and grain and edge enhancement create a layer of noise that really makes itself known. You'll notice a haze of grain, and with all the detail and neon in the film, the little dose of extra edge enhancement stands out quite a bit. Still, the overall quality of this film, and this special edition, make it very easy to both get used to and forgive the flaws in the video. On the other hand, the Dolby 2.0 Surround soundtrack is very good. There's loads of nifty mixing work going on in this film, and it all comes across very nicely. There's a good low bass growl, explosions are felt, and the dialogue is natural. This isn't dynamic surround sound, but it works for the film.

As a special edition, you'll have to go pretty far to find a more thorough DVD in terms of film history. It includes everything you could ever possibly want to know about this film. Along with the featured (and "Gilliam-Approved") theatrical version of *Brazil*, you get to see the Universal TV version, also known as the "Love Conquers All" version. On top of that, you get two documentaries, which are both as witty and brilliant as they are informative, a stack of storyboards from all of the dream sequences (cut and uncut from the film), and interviews with most every creative person behind the making of the film. There are three discs in this very distinctive package, and you'll wish there were more before you're done (and we mean that in a good way). On the director's cut, Gilliam is featured on the commentary track, and he's always a joy to listen to. He likes himself and his work, and he has no problem discussing either. The "Love Conquers All" disc (the third in the set) features a commentary track by *Brazil* historian David Morgan. He's not as passionate about his commentary as Gilliam is, but he does shed a lot of "film school" light on this cut, pointing out which scenes were alternate takes and illuminating the plot points. It's good to see this cut with a commentary track, because it's amazing how different it is from the original vision. Keep in mind that all the footage was shot by Gilliam, but edited by someone else without his supervision, and you will definitely see how editing can change everything about a film, from character to tone, and even storylines.

Brazil has always been considered one of the most important films ever made. It makes our list here because we think it's important not just as a film, but as a production as well. No matter what you think of it, your reaction to *Brazil* will likely be passionate—it's just that kind of film. Once you learn about all of the problems and fights it took to get the film made (or even seen), you'll simply be amazed. Don't let this DVD's lack of anamorphic video and 5.1 surround sound keep you from checking it out. Simply put, as a special edition, Criterion's version of *Brazil* still stands as a landmark of the medium.

Snow White and the Dwarfs: Platinum Edition

Disney (Buena Vista)

Released in 1937, *Snow White* wasn't the first animated production, but it *was* the first feature-length animation. An adaptation of the Grimm fairy tale, this story of a young woman betrayed by an evil queen and taken in by a group of hard-working dwarves has enthralled for generations. As the story begins, a vain and beautiful queen asks her magic mirror who the most beautiful woman in all the land is. Over and over again, the answer is "You." But one day, a beautiful young girl is born . . . and the queen suddenly has a run for her money. The magic mirror soon reports that "Snow White" is the most beautiful woman in all the land, so the angry queen sends an assassin to kill her. But the assassin doesn't have it in him to kill the girl, and decides to abandon Snow White in the woods instead.

The queen thinks Snow White is dead, but seven little men have found her and taken her in. Each has their own unique personality and a name to match. As the years pass, Snow White and the dwarfs live happily together. But when the queen learns from her mirror that Snow White is *still* the fairest of them all, she decides to take care of business herself. Disguised as a witch and armed with a poisoned apple, the queen sets out to kill her competition . . . and only love stands in her way.

If you want to see how classic animation can, and should, be presented on DVD, look no further than *Snow White*. The transfer given to this film is nothing short of remarkable. It's been lovingly restored with the latest digital technology, so the video is flawless — there isn't an artifact to be found. The colors are lush and accurate, having been color timed to match the original animation cells stored in the Disney archives. Presented in the original 1.33:1 aspect ratio, *Snow White* is absolutely stunning on DVD. The audio is presented in both a remastered Dolby Digital 5.1 and the original mono. As much as we love surround sound, we liked the mono track best. For our money, it fits the film better.

The extras on this disc are surprisingly substantial. For those new to the complex interactivity of DVD, Angela Lansbury hosts a guided tour of the extras available on the set — a nice touch. First up is a feature-length audio commentary by Walt Disney himself. Now, don't cry foul — this isn't an impersonator. It's actually audio of the real Disney, pulled together from taped interviews done over the years. John Canemaker, a well-regarded Disney historian, fills in the gaps with insightful comments and other information. Also available here is the 1934 *Silly Symphonies* short, *The Goddess of Spring*. Next, Michael Eisner introduces a video for Barbra Streisand's rendition of "Some Day My Prince Will Come." There's also a 40-minute documentary about the making of the film, also hosted by Lansbury, entitled *Still the Fairest of Them All: The Making of Snow White and the Seven Dwarfs*. For the kids, there's also a karaoke version of "Heigh Ho," an interactive game, and DVD-ROM materials.

Disc Two contains the lion's share of the extras, broken up into five areas. First there's *Snow White's Wishing Well*, which includes text timelines about Walt Disney (the man), and the production history of this film, as well as the original Brothers Grimm fairy tale. There are also four storyboard-to-film sequences hosted by Canemaker.

Moving on to *The Queen's Castle*, you'll find the *Art and Design* featurette, along with a gallery of conceptual art and the like. The *Layouts and Backgrounds* featurette is hosted by Disney's chief restoration expert, Scott MacQueen, and includes a look at more original art.

The *Camera and Tests* section features two excerpts from episodes of the *Disneyland* TV series: *The Story of Silly Symphonies* and *Tricks of Our Trade*. Scott MacQueen pops in again as the narrator of the featurette *Camera Tests*, which looks at the development of color, animation, and style. There's also a featurette on the voice talent used in the film, a short entitled *Live Action Reference*, another *Disneyland Tricks of Our Trade* episode, and more art galleries and concept art. *The Queen's Dungeon* section of the disc showcases storyboards and rough pencil tests for three unfinished sequences from the film. You'll also find information about the beautiful work done to save this film and prepare it for DVD in *The Restoration* featurette.

Next, we journey into *The Dwarfs'* Mine, where you'll find some fully animated deleted scenes and a documentary entitled *Disney Through the Decades*, which features a variety of hosts who guide you through each decade of Disney history (including Roy Disney, Angela Lansbury, Fess Parker, Robby Benson, Dean Jones, Jodi Benson, Ming-Na, and D.B. Sweeney (all of whom have been featured in Disney films, either on camera or as voice talent). Also available here are trailers for *Snow White*'s theatrical re-releases.

The final section is *The Dwarfs' Cottage*. If you're looking for audio-based extras, you've come to the right place. There's a radio broadcast from the 1937 Hollywood premiere, a couple of interviews from the Lux Radio Theater, a *Mickey Mouse Theater of the Air* broadcast, seven radio ads for *Snow White,* and even audio clips from a couple of recording sessions done for the film.

When it's over and done with, you'll have spent about *16* hours going through all the material this DVD has to offer . . . and you should enjoy every minute of it. The experience of *Snow White* on DVD is definitely time well spent.

Fight Club: Special Edition

Twentieth Century Fox

"The first rule of Fight Club is . . . you do not talk about Fight Club."

Based on a novel by Chuck Palahniuk, *Fight Club* follows an unnamed narrator played by Edward Norton. He has an office job, a great apartment filled with many beautiful things, and what most of us would call a good life. But he's having problems sleeping and goes to the doctor about it. He learns that this has nothing to do with a physical problem, but it's recommended that he attend a support group meeting. He does, and soon becomes addicted to it. If there's a disease for something, he'll be at the support group. It's cathartic for him to see others whose suffering is worse than his own. He can feel better about himself by holding these dying people in his arms. At one of these meetings, he meets Marla (Helena Bonham Carter), a chain-smoking chick with a similar addiction to human suffering. Like him, she's a tourist at these meetings, and seeing her is making him feel guilty. So he confronts her and they agree to attend alternate nights and groups. This also invites her into his life.

About this same time, Norton bumps into Tyler Durden (Brad Pitt), a messiah of sorts who works as a hotel waiter, film projectionist, and all-around entrepreneur. When Norton's apartment is mysteriously blown up, he calls his new friend Tyler for a place to crash. It's there that the two of them start Fight Club. Together they create an underground movement for men who have lost their way in life . . . and figure they can literally beat it back into each other. With Tyler as ringleader, the club soon becomes an organization, and the one-on-one fights turn into full-scale war on the world. They form a terrorist group known as Space Monkeys (because the first astronauts were chimps taught to pull levers and buttons) to wage that war. When Norton finally realizes what's going on, it's up to him to stop Tyler and the Space Monkeys before it's too late. But first he has to *find* the elusive Tyler, and that may prove harder than he thinks.

Fight Club is a very easy target for critics. The point can validly be made that director David Fincher and company are glorifying violence. But there's a greater point to be made too, and you'd be cheating yourself out of the film's deeper meanings if you didn't take the time to try and understand it. The film (and Palahniuk's novel) present to us a world (*our* world) in which men are bored and have gone looking for the caveman within. The logic is that human males need to hunt and fight . . . to do just about anything but thumb through Abercrombie and Fitch catalogs. That's both a frightening concept and a refreshing perspective. Whether you agree with that idea or not, there's no denying the power (and yes . . . even value) of this film.

Fight Club looks fantastic on DVD in full anamorphic widescreen (at a 2.40:1 aspect ratio). *Fight Club* is a dark, gritty, color-washed out, under-exposed head-trip . . . and you see every bit of that on this disc. You'll find deep blacks befitting a Fincher picture, excellent detail (particularly in shadows, where most of this film takes place) and nicely accurate (if muted) color, which makes all that badly bruised flesh look like . . . well, badly bruised flesh. The English Dolby Digital 5.1 surround sound mix is terrifically atmospheric. This is a film jam-packed with subtle little sound cues, audio transitions, and distant wild sounds . . . and you'll hear every one. Dialogue is nicely clear and the soundstage is very deep and wide, with rich and substantial bass. The mix goes from blissfully subtle to aggressive with ease, without ever sounding forced or imbalanced. And the Dust Brothers' music never sounded so good.

Although it's currently only available as a movie-only disc, *Fight Club* was originally released on DVD as a two-disc special edition. What set it apart from the DVD pack was the quality of its interactivity. Everything here is true to the theme of the movie. This is a true multimedia experience perfectly tailored to the strengths of the DVD format. Starting on Disc One, you'll find no less than *four* individual audio commentary tracks. The first features director David Fincher solo, and it's a very fun and easy listen. What will strike you most about Fincher's comments here is his sense of humor, his attention to detail, and his perceptiveness. On track two, we have Fincher, Brad Pitt, and Ed Norton all in the same room. Helena Bonham Carter is also on this track, but was recorded separately. This is definitely the most entertaining of the commentaries—you can tell the boys were having a good time looking back at the film, and their easy banter is extremely funny. A third track features novelist Chuck Palahniuk along with screenwriter Jim Uhls, who are fascinating to listen to as they delve into the characters and their motivations. The final track features other members of the production crew, including the director of photography, the costume designer, and the effects supervisor, among others.

Disc Two contains the meat of this set, divided into five basic sections: *Crew*, *Work*, *Missing*, *Advertising,* and *Art*. *Crew* is a pretty straightforward section of cast and crew bios and filmographies. There are some 18 detailed listings, including everyone from the actors and director to the writers and even the Dust Brothers. These are some of the most comprehensive bios on any set to date. The *Work* section is where you'll find lots of behind-the-scenes video clips, also divided into sections: *Production*, *Visual Effects,* and *On Location*. But these aren't your ordinary featurettes—just about every one of them has multiple video and audio tracks for you to explore. For example, you can watch a segment on the *Alternate Main Titles* without text, as the preview version, or with two alternate font styles (you can change between them on-the-fly using your remote's "angle" button). You can also listen to one of two audio tracks— the original main title theme or an alternate, unused theme. Finally, you can access production art serving as a map for the sequence. And this is just one of *14* such video featurettes, some of which even have optional Fincher audio commentary.

In the *Missing* section, you get seven deleted or alternate versions of scenes, including Marla's infamous "I wanna have your abortion" line. The video quality of these is very good—not anamorphic, but very good. When you select each scene from the menu, a bit of text appears to tell you why the scene was cut or changed. Moving on, the *Advertising* area gives you access to three theatrical trailers (including one that was unused but was finished just for this DVD); 17 TV spots used in the U.S., international, and spanish markets; a Dust Brothers music video for the theme; five Internet spots; two hilarious PSAs done by Norton and Pitt ("Did you know that urine is sterile? You can drink it."); a packed gallery of promotional artwork (featuring lobby cards, one sheet art, production stills, and the film's press kit); and a transcribed interview that Norton did at his alma mater, Yale University, about the film.

If that's still not enough for you, the *Art* section features loads of production artwork, storyboards, more production photos — you name it. There are literally hundreds of still images to sift through on this disc. Just to top it all off, there's a cool liner notes booklet included with this two-disc set as well.

Bottom line — if you wanna check out everything on this set, you'd better quit your office job, stock up the fridge, and set aside a few days to do so. *Fight Club* has to be one of the most visceral and head-twisting film experiences in a long time. You'll either love it or hate it. But *Fight Club* on DVD is one of the best special editions ever created, and it absolutely demands the attention of any serious fan of the format.

Se7en: Platinum Series

New Line

In a rainy, unnamed city, a serial killer starts a grisly cycle of murders over a seven-day period, handing out death in the pattern of the Seven Deadly Sins. For those without a Bible, those sins are Gluttony, Greed, Sloth, Lust, Pride, Envy, and Wrath. The killer seeks out people that embody their particular sin . . . making everyone a possible target.

Detective William Somerset (Morgan Freeman) is set for retirement in seven days, ironically enough, when this cycle begins. He's breaking in a new partner, David Mills (Brad Pitt), who's fresh from the country and eager to tackle big-city crime. When the two detectives accidentally stumble onto the case, they find themselves in the middle of a moralistic master plan that looks, right from the start, like it won't be easy on anyone. No one walks away from this unscathed, and the detectives find themselves playing right into the killer's master plan. If there's any hope to be found, it isn't apparent in the final, startling minutes of this well-crafted thriller. *Se7en* is a stark and horrifying look at the corruption of evil . . . as well as a study of sin.

Director David Fincher makes it a point to push our faces into the grime here . . . yet we see almost nothing. People complain about how violent and disgusting this film is, but there are really only two actual moments of violence in the film — the rest is subjective. We're given glimpses of it, yet nothing more than a taste — the aftermath, rather than the actual violence. Our imagination fills in the blanks, making it that much more effective.

Se7en is presented on DVD in truly amazing anamorphic widescreen video. The transfer was mastered directly from the film's original edited negative, which (according to New Line) was a first for DVD. The high-definition master was then digitally cleaned of dust, dirt, and other print defects. Then, under Fincher's supervision, it was completely recolor timed shot by shot. In many cases, Fincher was able to finally achieve the look he'd wanted originally, but had never gotten before, even theatrically. The result is truly something to behold. You'll see stunning blacks, which still retain incredible detail. The colors on this transfer are gorgeous, with amazing subtleties to skin tones and shadings. This has always been a dark, gritty, and grainy film . . . but it's never looked better than it does here.

The audio options are equally impressive. New Line went back to the original audio stems for the film and completely digitally remixed *Se7en*'s soundtrack for home theater, again under the supervision of Fincher and sound designer Ren Klyce. Included is a Dolby Digital 5.1 soundtrack with Surround EX compatibility, which features a nicely wide soundstage, very subtle directional cues and panning, good low frequency when necessary, and delicious creation of ambient sound environments. Dialogue is crisp and clear, and Howard Shore's score comes alive. This isn't the most aggressive soundtrack, but it's very atmospheric and it's never sounded better . . . aside from the DTS 6.1 ES soundtrack also available here. Listening to the opening credit sequence in DTS is an almost sublime experience. It's as if the sound goes straight to the pleasure center of your brain. This mix is extraordinarily cool, all the more so for its subtlety. It's easy to blow people away with the kind of attackingly aggressive surround sound you get in a big action film, but this is much more impressive.

Included on Disc One are a series of four audio commentary tracks. All feature director David Fincher, with a variety of other cast and crew members. First there's a newly recorded "Stars" track, where Fincher is joined by actors Brad Pitt and Morgan Freeman. The easy interplay between

Fincher and Pitt is entertaining and really shows how intelligent these guys are. There are plenty of interesting insights into the film, the story, and the angle these guys approached it from . . . plus it's occasionally very funny. Freeman also adds significantly to the track, although it sounds like he was recorded separately. Also available is a "Story" commentary with Fincher, professor of film studies and author Richard Dyer, screenwriter Andrew Kevin Walker, editor Richard Francis-Bruce, and New Line President of Production Michael De Luca; a "Picture" commentary with Fincher, director of photography Darius Khondji, production designer Arthur Max, editor Richard Francis-Bruce, and author Richard Dyer; and a "Sound" commentary with Fincher, sound designer Ren Klyce, composer Howard Shore, and author Richard Dyer. This last one is cool because it includes isolated 5.1 music and effects cues.

Disc Two adds an amazing quality and quantity of supplemental materials on this film. This isn't your usual lame-duck mix of cross-promotional trailers for other films, studio-produced fluff pieces, and other uninteresting crap. This is a hardcore, film buff experience — a serious exploration into the minds of the creative individuals who contributed to this film.

You start off with an in-depth look at the film's opening credit sequence that uses both multiple angles *and* multiple audio tracks. You can choose to view the original storyboarded sequence, the rough edit, or the final, completed version. And you can listen to the sound in Dolby Digital 2.0, 5.1 EX, DTS 6.1 ES, or PCM stereo sound. You get commentary on this as well. Next, you've got a section of six extended scenes, all with optional Fincher commentary. Included in this section are the film's original opening, the original test ending, and a storyboard video of an unshot ending. There's a video of production design artwork with commentary, a packed section of still photographs compiled to video with commentary, a video on the creation of John Doe's notebooks with commentary, a promotional featurette, the film's theatrical trailer, and extensive cast and crew filmographies. Via DVD-ROM, you can watch the film while following along in the screenplay, and access a web site New Line created with even more material.

Our favorite section of the extras looks at the process of mastering this film for home theater. You get video on the picture and sound mastering, with commentary, along with a video demonstration on the color correction process. It's as if you're in a post-production suite, watching as the colorist corrects skin tones and matches sky color from shot to shot using directions given him by Fincher. As he's working, he's explaining the process to us — very cool. Finally, there are three scenes from the film here where you can use your angle and audio controls to switch on-the-fly between the film's original and remastered video and audio. The difference is startling and it really illustrates the whole process perfectly.

Everything that makes *Se7en* a great movie is examined closely on this DVD. You get optimal picture and sound, a firm sense of the film's history and design with the supplements, and even a great look at the DVD production process. *Se7en* is the kind of film that, the more you study it, the more you appreciate it. There's no better way to do that than with this fantastic DVD release.

Star Wars: Episode I—The Phantom Menace

Lucasfilm (Twentieth Century Fox)

A long time ago, in a galaxy far, far away, the Old Republic is beginning a steady decline. The greedy Trade Federation is putting the economic pinch on the peaceful planet of Naboo with a massive military blockade. Queen Amidala (Natalie Portman) is none too pleased with this arrangement and has appealed to the Supreme Chancellor of the Republic for help. As the movie starts, the Chancellor has dispatched a pair of Jedi knights (Qui-Gon Jinn and his apprentice Obi-Wan Kenobi, played by Liam Neeson and Ewan McGregor) to settle the dispute. When the negotiations go south, the Jedi must find a way to protect Amidala and save the people of Naboo. Naturally, their task will not be easy. There's plenty of evil afoot, courtesy of the vile Sith Lord and all-around galactic rabble-rouser Darth Sidious, not to mention his rather nasty saber-wielding apprentice, Darth Maul. And, along the way, our heroes will meet characters new and old, who will play a significant part in the rest of the Star Wars saga—C-3P0, R2-D2, Yoda, Jedi master Mace Windu, and, of course, the young Anakin Skywalker destined to be a villain named Darth Vader.

Given all the years of waiting, and more hype than has ever been seen for one film in the history of cinema, there is almost no way that *Episode I* could have met fan expectations. It doesn't help that there are *plenty* of flaws to the film. The dialogue is tin-ear flat. The film moves slower than a Jawa Sandcrawler at times because there's just so much groundwork that needs to be laid for the later (and earlier) films. We barely see Darth

Maul the whole film, and he doesn't get to be really *bad* until the end. And you don't really get emotionally invested in any of these characters, so that when the big climaxes happen, they have little impact.

Still, the film mostly works as an entertainment if you give it a fair shake. Liam Neeson is solid and likable as Qui-Gon, and Ewan McGregor absolutely nails Alec Guiness's mannerisms as Ben Kenobi. And the lightsaber fighting in this film is first-rate — the duels in the original films look geriatric compared to this. Probably the most interesting thing about this film is that you watch it knowing that none of these characters are going to have happy endings. That's not giving away the plot here — anyone who knows anything about *Star Wars* already knows that things will ultimately turn out badly for almost *every* character in this film. Given that the fate of these people is already set in stone, the very fact that Lucas manages to tell their stories in a way that seems at least somewhat fresh and new, and gives us plenty of interesting things to see along the way, is a feat in itself.

The first thing you should know about this two-disc DVD set is that it features a longer cut of the film, done just for this DVD release. One complete scene (the air taxi sequence) has been restored to the film, and two other major scenes (the pod race starting grid and the race's second lap) feature extended footage. It's about two minutes of new material in all. The presentation quality on the disc is excellent. On the video side, we're given a great anamorphic widescreen transfer, marred only by the use of a little too much edge-enhancement. The contrast is excellent, with deep and detailed blacks. Color is rich and natural at all times, without bleed. Flesh tones, in particular, look wonderful. If not quite perfect, this is great DVD video.

On the audio side of things, however, this disc is absolutely reference quality. This is one of the best Dolby Digital 5.1 sound fields ever heard on DVD. The soundtrack can go from explosively loud to whisper quiet in a heartbeat, and the track captures that perfectly. The sound design by Ben Burtt is perfectly translated into the home theater environment. Best of all, John Williams's amazing score is blended nicely throughout the mix. You'll be hard pressed to find more active surround sound on DVD.

Disc One of this set only has one major extra, which is audio commentary by director George Lucas, producer Rick McCallum, sound designer Ben Burtt, animation director Rob Coleman, and visual effects supervisors Dennis Muren, John Knoll, and Scott Squires. Most of them were recorded separately, with the track edited together later. It sounds a little choppy at first, but stick with it because there's a lot of interesting information conveyed in the track, particularly when Lucas gets going on issues relating to the story of *Episode I* in the context of the eventual six-film saga. As you're listening, electronic "subtitles" appear on the top of your screen to identify the person talking.

Disc Two features one of the best behind-the-scenes documentaries we've ever seen. Directed by filmmaker Jon Shenk, it's called *The Beginning* and it runs a little over an hour. There's no annoying studio narrator and no talking head interviews anywhere to be seen. Instead, this documentary presents key events in the making of the film, from beginning to end, in classic cinema verité style. You are simply a fly on the wall, watching as Jake Lloyd is cast as Anakin, as Ewan McGregor gets to pick his very own lightsaber, and as the ILM effects supervisors struggle to complete shots in time for the film's release. You're there when a desert storm destroys the sets in Tunisia, when Lucas himself questions the structure of the film, and when he takes buddy Steven Spielberg on a tour of the production. There are some great candid and emotionally honest moments (including more than one instance of swearing). *The Beginning* is one of the best looks at the making of a film you'll ever see.

Next up are a series of seven deleted scenes, all in full anamorphic widescreen and Dolby Digital 5.1 EX sound that were finished by ILM just for this DVD release. You can view them separately or in the context of a documentary about the scenes that explains why each was deleted, what purpose it served in the story, and what was done to complete it. Be sure to stay for the end credits, which feature funny outtakes and special effects gags.

Also available are all 12 of the web documentaries that were created to preview various aspects of the film online. There's also five additional featurettes on the production, the film's infamous teaser and theatrical trailers, a series of TV spots, "Duel of the Fates", two multi-angle "animatic" segments, three still galleries (of production photos, poster artwork, and the film's print campaign), a brief look at the making of a LucasArts video, and a web link to an exclusive online site (that can only be accessed by people who have the DVD), which is occasionally updated with additional bonus material. Finally, you get a trio of fun Easter Eggs on the set as well (check the review for this DVD at *The Digital Bits* web site for instructions on how to find them).

This is a great two-disc special edition, which immediately takes its rightful place among the best DVD has to offer. The deleted scenes and longer cut of the film alone are worth buying the disc if you're a *Star Wars* fan. The sheer dynamic power of the surround sound also makes purchasing the disc a no-brainer for anyone who wants to show off their home theater gear. Add to that the excellent documentary, and this should be an easy decision.

Star Wars: Episode II—Attack of the Clones

Lucasfilm (Twentieth Century Fox)

It's 10 years after the blockade of Naboo and the events of *Episode I*. Padmé Amidala (Natalie Portman) is no longer queen of her people, but now a senator and one of the key figures in a bid to oppose the creation of an army for the Republic. The apparent need for such an army is mounting, because a growing separatist movement of thousands of star systems is threatening to plunge the galaxy into civil war—a threat the limited number of Jedi seem helpless to counteract.

As the film opens, Padmé narrowly escapes an attempt on her life designed to keep her from casting a vote against the army in the Senate. Shortly thereafter, Chancellor Palpatine (Ian McDiarmid) convinces the Jedi Council to assign Obi-Wan Kenobi (Ewan McGregor) and his now grown apprentice, Anakin Skywalker (Hayden Christensen), to protect her. The assassins make a second attempt to kill Padmé, but thanks to the Jedi, the attempt fails and instead results in a frantic chase through the streets of Coruscant. When the dust settles, Anakin is ordered to take the young senator back to Naboo to keep her in hiding, while Obi-Wan follows the trail of evidence back to her would-be assassins. The clues soon lead Kenobi to a distant and hidden water planet, where a massive clone army is being secretly created from the DNA of a rough-edged bounty hunter named Jango Fett (Temuera Morrison).

While his master struggles to fit these new pieces of the puzzle together, back on Naboo, Anakin finds that his long-simmering love for Padmé is beginning to overwhelm his commitment to the Jedi Order. Before all is said and done, a series of tragic events will forge their love—events orchestrated by the mysterious Sith that will unleash the legendary Clone Wars, begin the transformation of Republic to Empire, and turn young Anakin inevitably down the path to the Dark Side of the Force.

Attack of the Clones is a significantly better film than *The Phantom Menace*, but it still isn't really a great *Star Wars* film. Once again, Lucas is a slave to the complex plot, which undermines the drama. As with *Phantom*, it seems difficult for *Clones* to build a real sense of jeopardy. There's never those "edge of your seat" moments you got with *Star Wars* and *Empire*. Though it starts strong, it nearly grinds to a halt midway thanks to a few painfully bad scenes of romance between Anakin and Padmé. Not only is the dialogue flat as ever, George's direction makes these two actors look stiff as boards.

But the things that do work in *Episode II* tend to work well. Ewan McGregor's mannerisms as Obi-Wan are once again perfect. There are lots of moments fans will appreciate, where events resonate with the original *Star Wars* films. Jango Fett gets to mix it up in the rain with Obi-Wan — a fight scene that makes the whole subplot worthwhile. We're also briefly introduced here to characters that we know will play a major part in the later films — Owen Lars, Beru Whitesun, and Senator Bail Organa. Best of all, Yoda gets to kick some *serious* ass before this film is over. We've all wanted to see the small green one in action, in his fighting prime, ever since he uttered his first circular sentence in *The Empire Strikes Back*. It's everything we could have hoped for.

We should note that, as with *Episode I*, the film on this disc includes a few additional moments newly restored by Lucas just for the DVD. Among these are a slightly extended version of Anakin's breakdown on Tatooine and the shot of Anakin's mechanical hand at the end of the film. *Attack of the Clones* is also the first full-length, live action film shot on high-definition video to arrive on DVD with a straight-digital transfer. In other words, the anamorphic widescreen DVD image was created *directly* from the final digital master of the film — no film was involved. The result is absolutely spectacular. From the most lush, vibrant hues to the subtle tones of skin and mist, this is breathtaking color. There's a smooth, natural character to the image that we don't recall seeing theatrically, even in digital projection. There's almost zero compression artifacting visible and edge-enhancement is nonexistent. Being a digitally originated and mastered image, there's not a spec, nick, or fleck anywhere — it's absolutely rock solid all the way through. Simply put, this is probably *the* best looking video you'll ever see on DVD. It absolutely demands to be viewed on a big, widescreen anamorphic display. If you've been looking for an excuse to buy that new digital TV . . . here it is.

As with the *Episode I* DVD, the audio is also top-notch. The sonic wizardry of Ben Burtt and Gary Rydstrom is fully evident in this aggressive

5.1 EX mix. The soundstage is nicely wide up front, deep and enveloping front to back, and smooth all around. The surround channels deliver terrific ambiance to create the various sonic environments of the film and are also extremely active with panning and directional effects. Dialogue is always clean, John Williams's score sounds wonderful, and there's substantial low-frequency reinforcement. Our favorite sound effects from the film were those seismic charges Jango fires at Obi-Wan's starfighter. The "audio black hole" (as Burtt describes the effect) is immediately striking . . . and then the blast wave races towards you, envelops you, and then goes groaning away behind you. Amazing.

The extras are not as impressive here as they were on the *Episode I* DVD, but that shouldn't be taken as a strike against the disc. On Disc One, you get another full-length audio commentary track with George Lucas, edited together with several other members of the production team. It's a good track with lots of interesting bits of information, some good story points, and fun little pieces of trivia. We do wish George had a track all his own to talk more about the story and his thematic ideas. Once again, when you're listening to the track, subtitle text appears at the top of your screen to identify each speaker. There's also an Easter Egg on Disc One and a web link to the exclusive *Star Wars* DVD web site (visit our review for this film on *The Digital Bits* web site for Easter Egg access instructions).

The extras on Disc Two are patterned very similarly to the *Episode I* DVD. There are eight deleted scenes here that were completed just for this DVD release. They're presented in anamorphic widescreen video (as are most of the extras on this set) with full Dolby Digital 5.1 EX surround sound. Most of these scenes feature additional character background and political detail with Padmé. Unfortunately, a disappointment here is that a number of other more interesting deleted scenes (that are known to exist by fans) aren't included here, probably because of the time that would have been required to finish them.

Also on Disc Two are a trio of in-depth, behind-the-scenes documentaries. The most extensive of these is *From Puppets to Pixels: Digital Characters in Episode II*. It runs some 50 minutes and focuses on the elaborate effort to realistically create the characters of Yoda, Dexter Jetster, and others with CGI. The second documentary, *State of the Art: The Previsualization of Episode II*, runs 23 minutes and shows you in detail the process by which Lucas and company imagine, evaluate, and refine the various scenes and sequences in these films. And the final documentary is a 25-minute look at the work of Ben Burtt and his crew of sound designers, called *Films Are Not Released: They Escape*. These just aren't in the same league as *The Beginning* from the *Episode I* disc. That said, they're worth checking out, with lots of fun little moments that you should experience fresh yourself. A favorite of the three is *State of the Art*, for several reasons. First, it includes several shots from the original *Star Wars* in anamorphic widescreen, whetting our appetite for eventually having that film on DVD. It also has a funny CGI blooper reel on the end during the credits. But the best thing about this piece is that it features extensive footage from the original animatics created for the Clone War battle sequence, much of it extremely cool action that was not included in the final film.

The rest of the extras include all 4 theatrical trailers for the film; 12 TV spots (including the *Spider-Man* spoof with Yoda); the "Across the Stars" music video; 3 more short featurettes; the complete 12-part web documentary series; the *R2-D2: Beneath the Dome* trailer; galleries of production photos, posters, and print campaign art; and an ILM visual effects breakdown montage showing effects shots in various stages of completion. Finally, there's also another DVD-ROM web link and an additional Easter Egg. It's all interesting stuff and well worth a look, particularly for you die-hard fans.

Okay, so it's not the original *Star Wars*. Sadly, we'll have to wait until 2005 to get that film on DVD (that comes straight from Lucas himself). Still, *Attack of the Clones* is better than *Episode I*, and this disc boasts the best video and audio quality you'll probably ever experience on DVD. You want to show off your home theater to friends with something that will blow them away? Just spin this film on disc and watch their jaws drop in amazement. That alone should justify adding this disc to your collection.

Moulin Rouge: Special Edition

Twentieth Century Fox

The American musical format, from vaudeville to Broadway and on to the silver screen, has always been a way to exploit pop songs. Even the cartoon format has been exploited as a means to promote a studio's library of pop songs and film tunes — it shouldn't strike anyone as odd that the most famous brands of classic cartoons are *Looney Tunes*, *Merrie Melodies,* and *Silly Symphonies*. So to rave about *Moulin Rouge* as a wholly original idea is a bit silly. It just hasn't been done in a while. Still,

Baz Luhrmann's approach is pretty impressive, and nowhere is that fact made more clear than on this DVD.

The plot, admittedly formulaic, involves forbidden romance, tragedy, and a hint of danger. It's got everything you need to keep you on the edge of your seat, revolving around a young English writer who's just arrived in Paris to take part in the Bohemian revolution around the end of the nineteenth century. He finds himself drawn to the Moulin Rouge, and one irresistible woman in Satine. Of course, she's a courtesan and he's a penniless poet, and if that's not enough to come between them, there's a smitten Duke who wants Satine for himself . . . the same Duke who's funding our young writer's play. True love becomes complicated by reality, but if you couldn't guess that love conquers all, you haven't been watching movies, friends.

Moulin Rouge plays all the clichés, but it works. Stars Nicole Kidman (as Satine) and Ewan MacGregor (as the penniless poet) don't have Tom Hanks/Meg Ryan chemistry, but they're actually good actors, and they muster more than enough energy to keep the film alive. And let's be honest . . . this is a *musical* romance. As such, the romance needs only to be serviceable, and let the musical numbers and dynamic cinematography carry the day. Carry it they do, and admirably. *Moulin Rouge* uses ultra-lavish production design, inventive camerawork, and hyper-stylized editing to convey the emotion the music brings to the piece. In so many ways the music is like the third actor in this romantic melodrama, complemented by the actors and an absolutely opulent milieu.

This set, by the way, is even more decadent than the source material it's based on. By the true definition of envelope pushing, this DVD simply rocks. If you're like us, left slightly unimpressed with the film overall yet strangely attracted to it at the same time, you'll be quite impressed with this top-notch special edition. It's worth noting that the disc and many of these materials were conceived and crafted by Luhrmann himself, along with the very same production team that worked on the film. It's tough to beat that.

The film itself is presented in incredibly luscious anamorphic widescreen video. The colors are sharp, the detail is razor fine, and blacks are hard and clean. This is an incredible transfer. The sound (in Dolby Digital and DTS 5.1) is equally stunning. Most of the mix's activity is front and center, but when the first huge musical production kicks in, so do the surround channels. The music will swirl around you as fast as one of the dancers onscreen, and if you don't feel your toes tapping along with these reconstructed pop songs, you'd better turn up your hearing aid.

If you're up for an inside look into the magical world behind this film, this two-disc set will open all the doors for you. First, there are dual audio commentary tracks, both featuring director Baz Luhrmann. On one, he's joined by different members of his crew, who talk about how the production was mounted, the effort it took, and how things could have looked differently. The other track takes a decidedly different tone. It's more story based, with Baz and co-writer Craig Pearce talking like the old friends they are about their creative process, different story lines, and working with the cast. Both tracks are pretty fascinating. Neither features any members of the cast, but they don't suffer for it.

After you've viewed the film three times (once to just watch the film and twice more with the commentaries), set yourself up for one more run. Using a feature called "Green Fairy," you can watch the film interactively. Occasionally, whenever a green fairy icon pops up on the left-hand side of your screen, you can access behind-the-scenes footage and conceptual art for the major scenes. This function will suck you in, because it's all pretty amazing stuff and gives a grand historic overview of the story and its source.

Disc Two is even more jam-packed. There's a "making of" documentary entitled *The Nightclub of Your Dreams: The Making of Moulin Rouge*, and heaps of video interviews with Nicole Kidman, Ewan McGregor, John Leguizamo, Jim Broadbent, and Richard Roxburgh. A section called *The Story is About* features more video interviews, this time with Luhrmann and Pearce discussing early story ideas and script comparisons. *The Cutting Room* section has interviews with Luhrmann and editor Jill Bilcock and about five abandoned edits and some mock pre-visualizations used to show the effects group what Baz was trying to accomplish. *The Dance* section contains an intro from Baz and focuses on four extended and multi-angle features on each of the major dance sequences, along with a video interview with choreographer John "Cha Cha" O'Connell and some pre-production dance rehearsals. *The Music* section contains *The Musical Journey* featurette, as well as an interview with Fatboy Slim, the unforgettable *MTV Movie Awards* performance of "Lady Marmalade," and the "Lady Marmalade" music video. *The Design* section has video interviews with production designer/co-costume designer Catherine Martin and co-costume designer Angus Strathie, nine galleries of set design photos and art, and four galleries of costume design photos and art. *The Graphic Design* area is a video gallery of poster art used in film set to music. Rounding out the disc are the *Smoke and Mirrors: The Evolution of the Intro* and *The Green Fairy* featurettes, as well as a *Marketing* section containing the international "sizzle reel," photo galleries, marketing materials, two theatrical trailers, a *Red Curtain Boxed Set* DVD trailer, and some Easter Eggs. Whew!

If you have a question about anything involved in the making of this film, it's answered here. Baz and company kept video records of just about everything they did, and much of it is here for us to enjoy. Some of it is standard, behind the scenes stuff (like the glossy made-for-TV documentary about the film), but the rest is pretty cool (like video of Pearce and Luhrmann acting out the script to get a feel for the characters, or Baz using stills from the film to stage reshoots, thus saving the time and money of rebuilding the sets). There's tons of interesting stuff here. Believe us when we say that this is an incredible DVD package . . . and once you see it, it's pretty tough not to fall in love with the film.

The Rock

Hollywood Pictures (The Criterion Collection)

By now, you're no doubt familiar with the Michael Bay/Jerry Bruckheimer drill — establish the plot in about 10 minutes and then let the shooting begin, progressively blowing up more stuff until the end. The good guy wins, the bad loses, and the body count is usually well into double digits. *The Rock* is probably the best example of this formula.

Here's the story. A highly decorated but embittered Marine Corps general (Ed Harris) steals a handful of rockets loaded with deadly VX gas and leads a rogue military assault on Alcatraz Island. His plan is to hold the city of San Francisco (and 81 civilians on the island) hostage until the U.S. government pays reparations to the families of men that died under his command while on covert operations. Of course, the government will have none of this and assembles a SEAL team to retake the island. Enter Agent Stanley Goodspeed (Nicholas Cage), an FBI chemical weapons expert (with zero field experience) who must accompany the team to deal with the VX gas. Pretty straightforward, right? Well, there's a catch — the general's got the island locked down tight. And the only person who knows the way in has been rotting in prison without trial for 30 years. He's a former British spy (Sean Connery), who knows just about every secret the U.S. has. And he's the only man that ever escaped Alcatraz alive.

Connery, Cage, and Harris all do a commendable job of bringing life to largely cardboard characters. The script is gimmicky, but hip (ya gotta dig that "alien landings at Roswell" reference). Still, it's the action that's the real star here, and director Michael Bay keeps it coming fast and furious, right to the very end.

Originally released as a laserdisc box set, this Criterion DVD version of *The Rock* gives a darned good 2.35:1 anamorphic widescreen picture, with nice detail and bright color representation. The sound is available in both Dolby Digital and DTS 5.1 formats. Both tracks are very active with plenty of work in your surrounds and subwoofer. This DVD presentation does the film proud.

Considering that this is a two-disc Criterion edition, you know there are some kick-ass extras. Want a commentary with your film? You got it. Director Michael Bay, producer Jerry Bruckheimer, technical advisor Harry Humphries, and actors Nicolas Cage and Ed Harris discuss the making of the film. They're recorded separately and edited together, with chatterboxes Bay and Cage chiming in the most. It's actually pretty interesting stuff, considering they're talking about a "big, dumb" action flick.

Disc Two has volumes of supplemental material. The *Production Secret* section has a slew of featurettes on the technical advisor, how to shoot a gun right, and the special effects. In *Publicity and Promotion*, we can watch the trailer and TV spots. There's also a stills gallery, outtakes, a historical view of the island in *Secrets of Alcatraz,* and a video interview with Jerry Bruckheimer.

If you dig *The Rock*, really want to put your surround speakers to the test, or just like action flicks, this is the special edition DVD for you. It's a must have for any action fan.

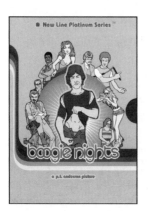

Boogie Nights: Platinum Series

New Line

Cast your mind back a few years to 1997. Burt Reynolds was generally thought to be washed up, having just failed at a Travolta-style comeback in the '96 Demi Moore disaster *Striptease*. Mark Wahlberg was an underwear model, better known as Marky Mark of the Funky Bunch fame. Only a handful of hardcore movie fans had ever heard of Paul Thomas Anderson, having caught his debut film, *Hard Eight* (originally titled *Sidney*) during the five minutes it played theatrically. And nobody thought that making a mainstream movie about the porno industry was a particularly viable idea.

Boogie Nights changed all that. Porn has since become . . . well, maybe not universally accepted, but certainly chic (witness the bizarre media appearances by Ron Jeremy in support of the documentary *Porn Star* — surely the Hedgehog's guest spot on the ladies yak-fest *The View* is a sign that the world has turned itself inside out). P.T. Anderson is now one of those filmmakers whose work is analyzed frame by frame by rabid fans. His most recent film, *Punch-Drunk Love*, premiered in competition at the Cannes Film Festival and it starred Adam Sandler . . . who actually proved he can act dramatically. See what we were saying about the world turning itself inside out? Mark Wahlberg has pretty much wiped out all memories of *Good Vibrations* and established himself as a movie star (and, like all movie stars, proceeded to squander his talent on junk like *Planet of the Apes*). As for Burt Reynolds, *Boogie Nights* turned out to be the comeback vehicle he'd been looking for. Of course, he hasn't really capitalized on its success, and today he's generally thought to be washed up again. The more things change . . .

Regardless of who was made or broken by it, *Boogie Nights* is an extraordinary achievement for Anderson, both as a writer and director. His screenplay expertly juggles a huge ensemble of fascinating and distinct characters. Often with a movie this episodic, certain characters tend to stand out, and you find yourself wishing the story would get back to those folks whenever the focus shifts away from them. *Boogie Nights* is the rare exception where each character is as interesting as the rest. This is, in no small part, due to the consistently outstanding performances here, including Julianne Moore, John C. Reilly, Philip Seymour Hoffman, William H. Macy, and Melora Walters (all of whom appear in Anderson's subsequent picture, *Magnolia*), along with Don Cheadle, Heather Graham, and, of course, Wahlberg and Reynolds.

Boogie Nights has been released twice by New Line in their Platinum Series. The first release was a single disc affair that's virtually impossible to find anymore, which is fine — you can get this spiffy two-disc set instead. The movie looks and sounds spectacular here. Apparently, Anderson's dissatisfaction with the transfer of the movie on the original DVD was the primary reason for the re-release. When Anderson decides to do a new transfer, he doesn't mess around. This is a vibrant, gorgeous-looking anamorphic widescreen video, with solid colors and very little detectable edge enhancement. In his note to the viewer on the slipcase, Anderson asks that you turn the volume up loud while watching. That's good advice. The Dolby Digital 5.1 mix fills the room with the groovy hits of '70s and '80s hits. The "Rahad Jackson" scene (better known as the "Sister Christian" scene) plunges you directly into the middle of that dizzying soundscape with spectacular results.

The supplements are pretty decent, with two commentary tracks, a bunch of deleted scenes (with commentaries of their own), the music video for "Try" by Michael Penn (with yet another commentary that explains why it's included since the song and video have nothing to do with *Boogie Nights*), and the first installment of *The John C. Reilly Files* (a feature also found on the *Magnolia* DVD—more on that disc next) spotlighting Anderson's favorite actor.

The commentary tracks are both well done. Anderson's solo commentary is a little better, because it's more focused than the actors' commentary. The actors' track too often devolves into a mutual admiration society between Anderson and the cast about what a privilege it was to work together. Still, if you'd rather have fun than learn something, the actors' track is much looser and more entertaining. We should note that there are a number of references made in the various commentaries to *Exhausted*, a documentary about porn star John Holmes that inspired much of *Boogie Nights*. It was supposed to appear on this DVD, but legal issues prevented its inclusion. That's unfortunate, because it would have provided context and a basis for comparison with the film itself. Finally, Easter Egg hunters (and/or fans of large, prosthetic penises) will definitely want to seek out the Dirk Diggler makeup test hidden on Disc One.

Boogie Nights is a sprawling, ambitious movie that succeeds on almost every note. It is, perhaps, a little longer than it absolutely needed to be, but this is a minor complaint. The movie holds up to repeated viewings very well and, as such, is a welcome addition to any DVD library. And, unlike some movies that have seen multiple releases on DVD (like any random *Evil Dead* flick), we actually believe Anderson's promise that this double-disc Platinum Series release is the *definitive* version of the movie. You can plunk down your hard-earned cash for this and be reasonably secure in the knowledge that New Line isn't going to sneak out yet another version any time soon. Even if they do, it's unlikely that they'll be able to improve on this.

Magnolia: Platinum Series

New Line

Magnolia is a nice step away from the world of *Boogie Nights*, in that it's less about reference and more about the inner workings of Anderson himself. It's about death and life, love and forgiveness. It's about the one moment we've all have had, where whatever was going on in the world seemed to pale in comparison to what was going on in our own lives. In *Magnolia*, Anderson seems to be asking the big questions of life. Whether any answers are found, or you get anything out of the film, well . . . that's for you to explore.

Magnolia follows the lives of people who all seem to be connected somehow—six degrees of separation, if you will. Except here, there only seems to be one or two degrees difference. We have game show host Jimmy Gator (Philip Baker Hall), his daughter Claudia (Melora Waters), and his wife Rose (Melinda Dillion). There's also Jim Kurring (John C. Reilly), a lonely cop who talks to himself while on the beat (à la *Cops*). There's Earl Partridge (Jason Robards), Jimmy's producer who is dying of cancer, his wife Linda (Julianne Moore), his caretaker Phil (Philip Seymour Hoffman), and his son Frank "T.J." Mackey, an infomercial host (played by Golden Globe-winner Tom Cruise). Rounding out the main cast is Donnie Smith (William H. Macy), a former quiz-kid genius and child star of Gator's game show (who's now a pale shadow of himself), and the newest quiz-kid champ Stanley (Jeremy Blackman). Also making appearances are Anderson mainstays Ricky Jay, as a TV producer, and Alfred Molina, as Donnie's boss. What this film does is follow the lives of these characters as they bump into, around, and off of one another.

Whether you like *Magnolia* or not will depend largely on your attention span. The film definitely sets its own deliberate pace. The good thing about that is you get to know these thoroughly drawn characters very well. You understand their motivations perfectly, and you can understand where they're coming from. The bad thing is it's easy to see why people complain about the film being overly long. There's a definite payoff here, but it takes quite a while to get there.

The anamorphic widescreen video on this DVD edition is incredibly good. Given that this is such a long film, it's amazing that they fit the whole thing onto one side of one disc in any kind of quality. Somehow they managed — the film looks beautiful. The colors are well rendered and the blacks are deep. The sound is also decent, but for a Dolby Digital 5.1 mix, it's awfully centered up front. There's not much action in the surround speakers. We do hear some things come out here and there, but only when Anderson feels playful. In any case, dialogue is clear and the music is full and vibrant sounding.

Here's where we get into a tricky arena. Over the years, some have complained about the lack of extras on this disc — most notably, the lack of a commentary or "real" deleted scenes. We say, "Ridiculous." What we do get here, producer Mark Rance's *That Moment*, is a beautiful portrait of the making of a film. Anderson invited Rance to document the production in diary form, from day to day. If you look between the lines on this documentary, you know that Anderson says everything he needs to on this film here. Watch his body language and his comments about people and situations. Listen to the cast's stories in the interviews. Within this documentary, we also see some deleted scenes in production and get a hint as to why they didn't work. There are even candid moments of humor and frustration. It's all there . . . you just have to pay attention. So as far as we're concerned, this might as well be a visual commentary track . . . and it's one of the best on any DVD.

In terms of other supplements, you also get Mackey's infomercial (the one that plays through the film) isolated by itself, along with some excised flashback material from the "Search and Destroy" seminar. There's Aimee Mann's haunted "Save Me" video (which was directed by Anderson), a teaser trailer, the theatrical trailer, TV spots (including one never released) and a loop of outtakes hidden in the color bar section (which is the same on both discs). Right there, with all that, this two-disc set has everything we would ever want or expect for this film.

Magnolia is going to be something different for everyone. For Anderson, it might have been a way of dealing with his father's death. For you, it might be a creative way of hearing the F-word repeated in new and exciting ways. Whatever it is, *Magnolia* is a good film that means something more than it is. Do yourself a favor and check it out. You just might be surprised.

The Godfather DVD Collection

Paramount

Ah . . . the sins of the father. They're pretty hard to escape, aren't they? Some of us spend our entire lives running from worlds we never created and have no control over. Poor Michael Corleone never even had a chance. *The Godfather Trilogy* tells the story of Michael and his family. And Paramount has done a great service to film fans everywhere by releasing this most sacred of film trilogies onto the format we all love. So without further ado, let's take a closer look at the family and the films.

The Godfather

"That's my family, Kay. It's not me."

In the 1940s, Don Vito Corleone (played by Marlon Brando) rules organized crime with an iron fist. But rather than exploring the mechanics of his regime, we uncover the true heart of his world — his family. There's Sonny (James Caan), the hotheaded Don-in-training; his younger brother Fredo (John Cazale), a slow-witted go-to guy with a heart as big as his eyes are sad; baby brother Michael (Al Pacino), a returning war hero who wants to go back to college; and their little sister Connie (Talia Shire), whose marriage to a family foot soldier is the reason all have gathered together at the start of film.

A grand wedding is the setting and, as the tradition goes, no man can refuse a request on his daughter's wedding day. When you're a Don, the request line can stretch around the block. The story slowly unfolds from here, when a small time connection from another family comes to ask for the Corleone's blessing and investment in the drug trade. Don Vito doesn't want any part of it, even if Sonny initially thinks it's a good idea. When things escalate, Don Vito is shot down in the middle of a market, so the rest of the family goes after their enemy. And young Michael, who never wanted to be involved in the family business, is thrown front and center into the fray with one calculated move that will affect the rest of his life.

This is epic filmmaking at its best. *The Godfather* is a three-hour film that grabs you with believable and tangible characters. The story is ripe with psychology and emotion, and the artistry at work behind the scenes is enthralling. *The Godfather* will always be at the top of many people's favorite film lists. But this story's about to get even better.

The Godfather: Part II

"Michael, your father loves you very much."

Seven years have past since the ending of the first film. Michael is now the undisputed Don of the family and his intention is to not only protect his family and their fortunes, but to legitimatize the family and its business as well. When we begin here, Don Michael has moved away from the family's New York stronghold to Lake Tahoe, where he is pushing into the gambling arena. Finding it more difficult, but possible with the right tactics, to control politicians, Michael pushes for control over several casinos in Vegas. At the same time, he's setting up shop in Cuba. But realizing the political climate there is a bit off-center he wants to pull out . . . which makes a quick enemy of an age-old friend. All the while this story is unfolding, we also see the rise of young Vito Corleone (played by Robert De Niro) in flashback, from a young boy in Corleone, Sicily to the golden streets of New York City. Young Vito goes from delivery boy to head of a crime syndicate with a few well thought out but very bloody moves.

Coppola shines as a director here, jumping back and forth between parallel stories that not only further the mood and tone from the first film, but also show us the link between father and son. And not only does the film extend the ideas presented in the original, it also goes back and fills in some gaps. For many fans of *The Godfather Trilogy*, this is the better, and more satisfying, film in this series. But we're not done yet.

The Godfather: Part III

"Just when I thought I was out, they pull me back in!"

It's now eight years later, and Don Michael has followed through with his plan to make the family business a legitimate operation . . . at least in spirit. Don Michael is busy taking over a company led by the Vatican. But some of the other families aren't happy with the way the Corleones have treated them in the past, or the family's direction. So when a small timer gets too big for his britches and starts flexing his muscle, all the other families side with the hood, leaving Michael in a lurch. Now he struggles to protect his family from the problems hiding in the shadows . . . not realizing that, while protecting his family, he's also destroying it.

A lot of people don't like *The Godfather, Part III*, and admittedly this isn't where we would have liked to see the Corleone story go. The acting in the film is all very good (even the nonactor parts, such as Sophia Coppola, who stepped in at the last minute for an "ailing" Winona Ryder). But *The Godfather: Part III* doesn't have the same force behind it that the other films had. It's a good film, but not great.

In any case, at long last, all three films have been pulled together in a great five-disc, special edition DVD box set from Paramount. The video quality of all three films is fine. It may not blow you away, but given the notoriety of the films' storage condition, you have to be mildly impressed that they're on DVD looking as good as they do. The transfers are a bit on the dark side, but a lot of the dust, tears, and odd splices that have riddled the picture in the past have been cleaned up. To put things in perspective, the presentation quality of the first film is good, the second film is better, and the third film is about the same. But if they aren't reference quality, this is very possibly the best we'll ever see these films looking in our homes.

The sound, however, is stellar. Each film has been remastered under Coppola's supervision in Dolby Digital 5.1. And the newly remixed audio rocks. The music is well represented, the tone and dialogue are there . . . there's even some fun play in the surrounds. This was never an audiophile's film series, but it certainly sounds better than it ever has before.

As for extras, each film disc features an audio commentary track with Coppola. These tracks blow any commentary you've heard before this away. Coppola walks us through his world with so much personality and clarity that when he's finally done, you'll swear you know the guy personally. There's a lot of respect to be gained for Coppola as an artist of true humility and vision by listening to these tracks. Coppola talks about his collaborations with various actors, writers, and craftsmen; he discusses the parallels between the films and his own family, and even his future film projects. This commentary should be required listening for all film fans, students, and professionals.

We should note that the first and third film in this series each are contained on their own disc, while the second film is split over two more discs — that's four in all.

For our money, we would have been happy with just the commentary tracks. But Paramount gives us an additional fifth disc of nothing but extras. To start with, it features the incredible documentary *The Godfather: A Look Inside*, packed with interviews and behind-the-scenes footage. The rest of the disc is broken up into different areas. In the

Acclaim and Response Gallery, you'll find acceptance speeches for both the 1972 and 1974 Academy Awards. It's interesting to see some of our favorite stars close to 30 years ago and how short the speeches were. In this area, you'll also find a listing of all the awards and nominations the films won, and the 1974 network TV introduction featuring Coppola at work editing *Part II* and pleading with audiences not to find ill will in his portrayal of violence and Italian-Americans in the film.

The Trailer Gallery collects trailers for each film. Also on the disc are several photo galleries. There's one of production and behind-the-scenes photos and a *Rogue's Gallery* with photos of all the thugs and conspirators in the film. An interactive *Family Tree* takes us through the history and major players of the Corleone family, with biographies of the characters and photos of the actors playing them. *The Filmmakers* section is a nicely drawn biography of the major artisans who worked on the film. The *On Location* section features Dean Tavoularis walking us through the original locations used in the filming, along with archive footage of the neighborhoods. One of the more fascinating extras (for future filmmakers at least) is the *Francis Ford Coppola's Notebook* featurette, where Coppola shows off the original notebook that he used to find the right tone, story, and method for shooting the film. His ideas are all there in blue pen notes on the margins. *The Music of The Godfather* section is broken into two parts. One includes audio excepts from meetings between Coppola and composer Nino Rota, and the other is a section devoted to Carmine Coppola. The latter features a short video interview with Carmine, along with comments by Francis Ford Coppola and clips from the 1990 scoring session for *The Godfather: Part III*.

Continuing on, we have a short featurette entitled *Puzo and Coppola on Screenwriting*, which is just that (along with a clip of novelist Mario Puzo essentially pitching a fourth *Godfather* that would have focused on the rise of Sonny). There's the *Gordon Willis on Cinematography* featurette that has Willis explaining his methods and his madness. You'll also find storyboards for *The Godfather: Part II* and some very cool animated storyboards for *The Godfather: Part III*, along with *The Godfather Behind the Scenes* — a 1971 production featurette. There are no less than 34 deleted scenes here (no kidding) with text introductions. These are what made up the various television and *Saga* edits, and help to strengthen the mythos and pathos of the film. They're presented full frame, probably to prevent fans from creating a bootleg *Saga* set, which has yet to appear on DVD. One can always hope it will someday. Oh . . . and for those looking, you'll find a few Easter Eggs scattered hither and yon, which include a foreign language loop (on the setup menu) and a very relevant and humorous clip from *The Sopranos* (at the end of the DVD credits). The disc also has a cool, unadvertised feature. Every so often, as you navigate the menus, you'll be treated to an audio clip (taken from Coppola's own tape recorder) of cast and crew discussing the script, or of Puzo and Coppola unraveling an important sequence. And just to round this review out, know that each of the films comes in a thin cardboard and plastic package (much like *Se7en* or *Boogie Nights*), which all fit snugly together in a handsome, protective slipcase.

What more could possibly be said? If you claim to be *any* kind of serious fan of film, *The Godfather DVD Collection* is an offer you simply can't refuse.

Guy Movies on DVD

In this section, we look at "guy" movies. After reading this far, you've probably come to the conclusion that we're bespectacled intellectual types you can trust. But no . . . even though we're bespectacled and like to think of ourselves as intellectual, like most "guys" we love us some serious shoot 'em up, blow 'em up, and love 'em up flicks. You know what we're talking about — the type where lots of stuff blows up, testosterone flows like jelly, and morals are in the eye of the beholder. Of course, we do have standards. For a guy flick to be included in *this* book, we had to consider a whole different agenda. Like . . . how *much* stuff blows up and do the explosions kick lots of ass? Is the soundtrack good enough to impress all your poker buddies, rattle the windows, and piss off not just everyone else in the house but your neighborhood as well? Then there's the extras. You've gotta have extras. Michael Mann's *Heat* is the perfect guy movie, but the DVD is as bare-bones as they come, so unfortunately you won't find it here. With such lofty criteria in mind, we've picked the best DVDs that we know all you guys out there will love.

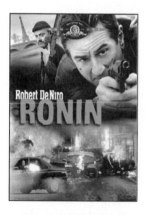

Ronin

MGM

In today's post-Cold War world, many former spies have now become freelance mercenaries, operating in a secret world of uncertain loyalties that's more dangerous than ever before. *Ronin* tells the story of a group of these agents who have been hired by an IRA operative to "retrieve" a mysterious briefcase from an unknown, well-armed party. The pay is premium and there's only one condition — no questions asked. But there are problems. The briefcase is up for sale, and the Russians want it badly. And when you can't trust the other members of your own team, how can you possibly place your life in their hands?

The word "ronin" comes from the lore of feudal Japan, used to describe samurai whose masters had been killed and who wandered the land in shame looking for redemption. The comparison is apt, as *Ronin* is as enigmatic a film as its title implies. Less is definitely more here. In that vein, we just can't bring ourselves to reveal any more of the plot—you simply have to see it.

In *Ronin*, the late John Frankenheimer created a taut, intense, and seductive thriller, placing its characters in harm's way at an absolutely breathtaking pace. As a member of the old school of Hollywood film-making, Frankenheimer eschewed the use of CGI and digital special effects in creating his action scenes. The result is that extra edge—a heightened sense of realism that's lacking in so many of today's thrillers. When you see these car chases (and there are several in the film), you'll know what we mean. The story takes the audience on a high-speed tour of France, with 100-mile-an-hour pursuits through city streets, back alleys, and winding mountain roads. Frankenheimer hired a team of French formula one drivers to do these stunts, and actually placed the actors themselves in the cars with the drivers. So when you see Robert DeNiro inside a car that's doing a high-speed, four-wheel drift around a Paris intersection, that's *really* Robert DeNiro. *Ronin* absolutely raises the bar for this kind of film action—you'll never see better.

The script, as originally written by J.D. Zeik (and doctored by the acclaimed David Mamet under the pseudonym Richard Weisz), is tight and well woven, with sparse dialogue and minimalist characterizations. These characters could easily come across as one-dimensional, but the impressive cast makes them all seem real and well lived. You may not know much about these people, but you know everything you need to. And what a great cast it is—Frankenheimer has assembled some serious talent here. DeNiro is terrific as always, conveying so much information with just a subtle glance, or a slight movement. Jean Reno (*The Professional*) matches him perfectly, step for step. The two play almost effort-lessly against each other—we can't wait to see them in another film together. And the rest of the ensemble, which includes Stellan Skarsgård, Natacha McElhone, Sean Bean, and Jonathan Pryce, performs flawlessly. There're also several other terrific actors seen here in smaller roles, and a nifty bit of stunt casting with Olympic ice skater Katarina Witt.

The video quality on this DVD, in both anamorphic widescreen and full frame, is outstanding. The look of this film is very muted, with subtle blues and grays in every frame. The film is dark and moody, with many scenes set at night, or in dimly lit, grungy environments—warehouses, garages, and alleys. Given the visual darkness of the film, there could easily have been trouble with the MPEG-2 compression, but that isn't the case—it looks fantastic, with great shadow detail. That's also important, as the film uses a great deal of deep focus—things are happening in the background that are often as important as the foreground action. All in all, the print is very clean, and the transfer is excellent. Also, the film was shot using Super 35 with spherical lenses, so you're not losing as much in the full frame as you would in a regular pan-and-scan transfer. That said, anamorphic widescreen is the *only* way to view this film. Just trust us on this.

The Dolby Digital 5.1 audio is also excellent. *Ronin* is a film in which some of the major action sequences take place without background music — you're just hearing the growl of car engines, the screech of tires, and the crack of gunfire. The clarity and resolution are as good in softer passages as during the more explosive action. The bass conveyed by the subwoofer channel is thunderous where appropriate. The panning is also good — great surround effects can be heard in this film, yet it's all entirely natural sounding. You'll hear cars roar up from behind, flash past to one side, and fade into the distance in front of you. Bullets ricochet, crowds scream all around — in short, this is a fully immersive mix. The dialogue is also completely natural, and the score is well presented, with pulsing staccato and haunting dirges that hint at classic samurai films. This is simply Dolby Digital 5.1 audio at its finest.

The extras on the DVD are also generally good, if a bit skimpy. The full-length audio commentary by Frankenheimer is terrific — very engrossing and one of the best that we've listened to. You really get a sense of Frankenheimer's skill as a craftsman when it comes to his work. It's absolutely fascinating to listen as he describes the importance and deliberateness of each of the film's transitions, the composition of action within the frame, the use of deep space. He talks about the subtext present in the script and the actor's performances, why he prefers the violence in his films to be quick, the sense of hyperrealism he strives to achieve. This is a true insight into his thought process while directing, making it that much more of a shame that he's no longer around to stand behind the camera. There is also an alternate ending included on the disc, and it's interesting to hear Frankenheimer describe why it was omitted, despite the fact that he liked it.

Ronin also has a series of web links available via DVD-ROM. On Sunday, March 14, 1999, MGM held a special, online event — a tour of the making of the film with director Frankenheimer as our host. For those who missed the event, it's available for replay at any time. The presentation will cue behind-the-scenes footage on your DVD that is not accessible in any other way. You can also download the film's trailer from the online site as well. Sure, this sort of extra is very cool, but we have always had mixed feelings about web content for DVD titles. While it's a good way to expand the experience of the disc, not everyone can access the material (if you have a Mac, or no Internet access, you're out of luck). And worse, there's no guarantee as to how long this material will continue to be available. So five years from now, it may disappear. Maybe we're old fashioned, but *The Bits* always prefers to have all our extras on the disc itself.

ROM issues aside, *Ronin* is a seriously entertaining flick. Keep in mind that this isn't a film that answers all of its questions. It doesn't even try, which we find refreshing. But if you give it a chance, we think *Ronin* will really get under your skin.

The Evil Dead: Book of the Dead— Limited Edition

Anchor Bay

Whether you're a fan of horror films or not, if you want to be a filmmaker, you simply need to see the first *Evil Dead* movie. It's not written incredibly well, nor is the story such that you'll find yourself unable to stop talking about it. No, it's nothing like that. But *Evil Dead* does illustrate an all-important lesson in film: style over substance sometimes actually works. *Evil Dead* was shot with so much creativity and verve that it simply demands to be loved. You may not be able to stomach the film the whole way through, but you will be captivated by it nonetheless. And no one on this planet can say that it doesn't work as a movie.

Evil Dead's history has been chronicled so much online and in print that it's a waste of time to regurgitate the details again here. All you need to know is that it was made by a handful of twentysomethings, with cameras literally nailed to boards and Kyro syrup by the barrelful standing in for body fluids. Whatever it cost to make, every penny of it shows up on screen in one form or another, from the quirky makeup and art design (courtesy of artist Tom Sullivan) to the incredibly cartoony camerawork (wiggled straight from the mind of director Sam Raimi — late of *Spider-Man* fame).

The minute you see this DVD on the store shelf you'll want it in your collection. *The Evil Dead: Book of the Dead — Limited Edition* looks like the "Necronomicon" featured in the film. The book in the movie is made from human flesh and written in human blood. Simply reading from it will unleash the evil spirits of the dead. Instead of flesh and blood, Anchor Bay went with foam latex and print, but it still works incredibly well.

Sculpted by the designer of the original book, artist Tom Sullivan, this DVD case was a little over a year in the making. It looks and feels just like the real thing . . . and oh, the *smell*. You'll be higher than a kite after you open the protective plastic covering on this thing. And when you open the book itself, you'll find a series of pages that look like they were pulled right from the original book. They might as well *have* been, because these too were designed and drawn by Sullivan, who did the original interior art seen in the film's book. The book includes a fun "thank you" to the fans from Tom (along with 13 pages of his art), a list of web sites (that you can use to learn more about the film), and 24 pages of liner notes written with incredible passion by *Dead* scholar Michael Felsher . . . and a pocket containing your DVD. The whole package is very cool.

So is the disc as good as the packaging? You're damn tootin'! To start with, this is the first time the original *Evil Dead* has been presented in anamorphic widescreen. Originally released as a full-frame transfer from a 35mm print by Elite Entertainment, this DVD was pulled from a recently discovered 16mm negative. Raimi always wanted the film to be seen the way it's being shown here (matted at 1.85:1) and Anchor Bay went one better with an anamorphic transfer. There's some minor faults here, including a few shots that seem to be overmatted and the fact that the transfer almost seems too clean. But the transfer clearly shows a lot of TLC, and this is, without a doubt, the best-looking version of *Dead* you will ever see.

The sound is also pretty arresting. Raimi and company have always been known for their incredible sound design, and both Dolby Digital 5.1 EX and DTS 6.1 ES tracks have been included here. That means that you can now listen to this film in the highest style possible. All these ear-busting thumps, puckers, gushes, and smacks will make your head spin. If you're not knocked on your ass by the presentation of this disc, you're probably dead yourself: Klaatu barada nikto, indeed.

Of course, wicked-as-hell packaging and stellar presentation values aren't all this disc is about. Packed onto this set is everything you'd ever want in relation to the original film. First up, we get the original commentary tracks produced for the Elite Entertainment edition a few years back. No sense messing with perfection, right? The first one is with Sam Raimi and his producer and business partner, Robert Tapert. They talk about everything, half-remembering facts about the shoot and making fun of everyone. It's a great track — always has been. There's also an additional commentary by the one and only Bruce Campbell. We love Bruce, you love Bruce . . . everyone loves Bruce. And after you listen to his hilarious commentary, you'll love him even more.

Bruce fans will also appreciate the inclusion of his short film, *Fanalysis*, in which Bruce guides us through the world of sci-fi conventions and the geeks who live for them. It's a very funny and welldone feature. You'll also find the film's original theatrical trailer, four TV spots, and two Easter Eggs if you want to hunt for them (including a *Special Makeup Effects Test* and video of a panel discussion with producer Robert Tapert and actresses Betsy Baker and Sarah York filmed at Grauman's Egyptian Theater in L.A. during a recent screening). Looking to click the night away with your remote? There's an extensive stills gallery featuring production photos and poster art. Wanna learn more about the guys who made the film? Hop over to the cast and crew biographies and filmographies. The featurette *Discovering Evil Dead: The Palace Boys Meet the Evil Dead* follows the film's adventures in the U.K. in theaters and on video. You also get a nice loop of behind-the-scenes footage and outtakes, mostly of monster work.

One thing worth mentioning: This disc was originally going to have included the short film, *Within the Woods*. This was a project Raimi was involved in while still in college, which became the inspiration for *The Evil Dead*. Unfortunately, while all the music licensing issues with the film were resolved, there were so many people involved in its production that clearing its release with everyone became too expensive. For this reason, the film ultimately had to be dropped. The DVD doesn't suffer from its absence one bit, though it would have been great to see.

The Evil Dead: Book of the Dead – Limited Edition is a must own. It gives one of horror fandom's all-time favorite films the sort of form and format that you just have to chuckle and gush about. Bravo to the Bay for a great DVD!

Big Trouble in Little China: Special Edition

Twentieth Century Fox

Who doesn't love this frickin' movie? *Big Trouble in Little China* follows the adventures of Jack Burton (Kurt Russell, in one of his best roles ever), as he goes block-head to hollow-head with an ancient Chinese wizard that's holed up underneath Chinatown, waiting to take over the world. Burton's an idiot though. He's the type of man that's so self-absorbed he refers to himself in the third person. His days are spent hauling pigs from coast to coast (in his truck "The Porkchop Express") and waxing moronic on his CB radio. His latest trip sets him up with an old friend who, as the story plays out, keeps getting ol' Jack deeper and deeper into trouble. But half the fun of this film is letting its wackiness unfold as you watch it. If you've ever seen *Big Trouble*, you know what we're talking about. If not, pick up this disc and you'll quickly learn what we mean.

Big Trouble was made long before special effects relied on CGI and wire-erasing technologies. This is an old-fashioned, chop-suey serial, and it's a whoppin' fun flick. The pace is quick, the writing is witty and razor sharp, and all of the effects are pretty impressive — even by today's standards.

This two-disc special edition presents the film in a beautiful anamorphic widescreen transfer, with a lot of detail and minus any unwanted blemishes. The sound is amped up to the max in both DTS 5.1 and Dolby Digital 4.1. The DTS sounds a bit better here, because there's much more balance to it. But those of you without DTS aren't going to be disappointed with the 4.1 track.

Extras on Disc One include a wickedly funny audio commentary track with director John Carpenter and Russell himself. These guys love each other and every track they've done is wonderful. Half the time, they simply talk about their lives and it's a nice change of pace (made nicer by the fact that they actually have a lot of interesting points to make about life, Hollywood, the current state of directors, and the like).

Disc Two gives us a plethora of deleted scenes (seven in total) which can be viewed either as a work print (which is of slightly better viewing quality, in full anamorphic widescreen) or from a backup video that shows how the scenes would have fit into the final film. Among them is one that's more storyboard than actual footage and shows how a scene in the final film was originally imagined. It's pretty fascinating and, through alternate angles, you can compare the storyboards to the final scene as you watch. There's also an extended ending, magazine articles about the making of the film, trailers, TV spots, an interview with the visual effects supervisor Richard Edlund, a stills gallery, and a wonderfully kitschy music video for the film's theme. Yes, that's John Carpenter on lead vocals . . . and the guy who played the original Michael Myers on keyboard as well. Scary, huh? Extras aside, one of the things that makes this set most fun is the menu design, which is pretty damn cool. And be sure to look through the credits page for a fun Easter Egg.

Just having this classic film on DVD is reason enough for you to run right out and pick it up. Add to that the extras on this set, and your decision should be a no-brainer. The commentary track alone is worth standing in line for. Carpenter and Russell are magic together and this special edition proves it. Do check it out.

Die Hard: Five Star Collection

Twentieth Century Fox

Die Hard takes a straight formula and turns it on its end. It's really nothing more than a simple heist flick, at least story wise. It's about a group of supposed terrorists, who take over a high-rise office tower with the sole intent of robbing it of $600 million worth of U.S. bonds. If the film focused on the criminals, it would be just like any other heist film of the '60s and '70s. Who are these guys, what's their agenda really, and will they get away with their caper? But *Die Hard* takes that formula and focuses more on cops trying to foil the plan. In this case, one John McClane. As played by Bruce Willis, McClane is a tired man. He's separated from his wife and kids, he's sarcastic and stubborn . . . really he's the perfect New York cop. He's fresh off the plane in Los Angeles, on his way to a reunion with his wife at an office Christmas party at the Nakatomi building, headquarters for the Nakatomi Corporation.

The reunion is less than sweet and turns out to be pretty damn short too, because at that exact moment, the building is taken over by these so called "terrorists," led by the nefarious Hans Gruber (played with scene-stealing charm by Alan Rickman). So McClane (clad only in a pair of pants and a wife-beater tee) must take matters into his own hands to stop these criminals from killing his wife and her coworkers.

Die Hard is an expert action flick from its writing and direction, right on down to each and every performance. Willis is great, Rickman is wonderful, and De'voreaux White and Reginald VelJohnson steal the movie in their supporting roles (as Argyle and Sgt. Al Powell, respectively). You can't ask for much more than this from any "guy" movie.

The picture quality on this DVD is awesome. There's no distracting film grain, no digital artifacts, and no edge enhancement. The widescreen is anamorphic, the colors are rich and accurate, and the blacks are solid and detailed. It's a really great video picture presentation. The sound is also excellent, with both a new DTS 5.1 mix as well as Dolby Digital 5.1 (strangely, a Dolby 2.0 Surround option listed on the box is missing from the disc). Both 5.1 tracks feature ever-present bass. When this film explodes . . . you definitely feel it. Dialogue is tight and centered, there's plenty of action in your surrounds, and the music is nicely blended in the mix.

In terms of extras, this Five Star Collection disc is one damn spicy meatball. Disc One features not one, not two, but *three* commentary tracks. The first stars director John McTiernan and production designer Jackson DeGovia. McT talks about his reservations about the film, its origin, and how it went from being a bad script to maybe one of the most perfect action films ever made. DeGovia sheds light on his profession as production designer—something we don't hear much about on DVD. There's also a "scene-specific" audio commentary by visual effects producer Richard Edlund. The last track is a text commentary focusing on some of the things not talked about in the other two. There's a lot of film history and trivia notations (like the musical odes to Kubrick) and a bunch of other neat references. Reading the track takes some getting used to, but once you get the hang of it, you'll want to watch it all the way through. Finally, Disc One features an "extended branching" option that allows you to watch the film with (or without) an extended scene. The scene is also available separately on Disc Two if you choose to skip it here.

Moving on to Disc Two, the *From the Vault* section includes a reel of deleted sequences, all of the isolated newscasts seen in the film, and text articles on the film from *American Cinematographer* and *Cinefex* magazines. The entire screenplay is here too. You also get an interactive stills gallery with lots of behind-the-scenes video footage hidden within it. As you move through the photos, a little icon will pop up occasionally. Click on it, and you'll find dailies footage, the extended version of the "Twinkie" scene, a few outtakes, reams of building blueprints, behind-the-scenes footage from the production, a deleted dialogue sequence, and alternate corporate logos. It's a little odd at first, but the more we watched, the more we liked the concept. For you folks who like your DVD interactivity, there's *The Cutting Room*, which is essentially a scene editing workshop. Packed in as well are a multi-camera shooting demonstration (with alternate angles from different cameras used during filming), an audio remixing lab, a video lesson on the differences between widescreen and pan-and-scan presentation, and a glossary of filmmaking terms. Rounding out the second disc are three theatrical trailers, seven TV spots, an EPK-style "making-of" featurette, and the usual DVD-ROM materials (web links, games, and so on).

Any way you slice it, the *Die Hard: Five Star Collection* rocks. This is *the* prototype for great action on film, and it's a must have on DVD.

Black Hawk Down: Deluxe Edition

Columbia TriStar

In October of 1993, CNN gave us all a window seat on one of the (then) most disturbing moments of recent history. As we sat transfixed in our homes, the bodies of American military soldiers were dragged through the Somalian streets by an angry group of men, women, and children. Author Mark Bowden saw these events too and wanted to know more about what lead up to them. The story he uncovered became the basis of a bestselling, nonfiction page-turner . . . and eventually this film as well.

Black Hawk Down is the story of the ambush of U.S. military soldiers in Mogadishu. It doesn't try to point fingers, or even turn American soldiers into mythic heroes. It simply tells the truth (although sometimes with Hollywood flourish) about the events of October 1993. It seems that several teams of soldiers were attempting to capture a militia warlord who had been stealing shipments of food aid from the U.N. — food meant to ease widespread starvation deaths and political grief in the region. But through a series of tactical mishaps during the raid, a Black Hawk helicopter was shot down, leading to an unfortunate change of mission and a horrifying night of bullets, blood, and grief.

Groundbreaking director Ridley Scott joined uber-producer Jerry Bruckheimer to bring the story of this remarkable day to life onscreen, and it's thrilling to watch. The film moves with breakneck speed from start to finish. Because it's thankfully not a preachy "America Rocks" type of film, it doesn't get bogged down with politics. This is all about a group of men who were doing their jobs . . . and ended up fighting for their lives.

This three-disc special edition DVD features a gorgeous anamorphic widescreen transfer, replete with great color and sharp detail. Scott is a painter of light and motion, and this is one of his most beautiful films in a long time. Backing up the video is a Dolby Digital 5.1 sound field that pulls you right into the film and never lets go with amazingly active surround sound.

But the extras on this release are what really make it a must-own set. To start, you'll find no less than three full-length audio commentary tracks.

The first is with Scott and Bruckheimer discussing the film in-depth. Recorded separately, each talks about what brought them to the film, what they were attempting by making the film and the effect the filming had on the public, considering the historical event was still so fresh in people's minds. The next track features author Mark Bowden and screenwriter Ken Nolan. They talk about the story like old chums, bringing up the differences between the script and the reality, and what subtle changes were necessary in order to convey the story cinematically. Finally, and perhaps the most interesting of the three, there's a commentary with several of the original Task Force Ranger veterans involved in the *real* events. They dissect the film between fact and fiction, and share their own stories and memories about the events. It's really amazing stuff.

On the second disc, which mainly looks at the production of the film, there's a two and a half hour documentary covering just about every aspect of the making of the film you could fathom. Some of this is more interesting than the film itself. Disc Two also houses a nice selection of deleted scenes, "Ridleygrams" (Scott's own hand-drawn storyboards), and photo galleries (among other things). The second disc alone is packed with great material, and we still have one more disc to go.

The third and final disc tries to bring the true events into perspective. Two documentaries originally produced for television appear here. The first is the History Channel's *The True Story of Black Hawk Down*, which recreates the events through background information from Bowden, interviews with the soldiers, and reenactments. Then we get the PBS documentary *Frontline: Ambush in Mogadishu*, which has a more news-story approach but is just as fascinating. Disc Three also features a text-based *Mission Timeline* of the events, and a multi-angle sequence showing you how Scott shot a complicated scene involving the soldiers dropping from a helicopter. On top of all this, three question and answer sessions (recorded at various special screenings of the film), trailers, TV spots, poster concepts, and a music video round out the bonus material on this last disc.

The *Black Hawk Down: Deluxe Edition* is, without a doubt, one of the most comprehensive explorations of the making of a film, and the real-life events that inspired it, to find its way to DVD in a long time. To be fair, this set could probably be included in our list of Must-Have Special Editions in addition to its status as a great guy movie. Any way you categorize it, this is an absolutely wonderful DVD release that demands the attention of any self-respecting movie fan.

The Matrix
The Matrix Revisited

Warner Bros.

The Matrix focuses on a mysterious man known as Morpheus (Laurence Fishburne), and his search for a person known only as "The One." In his mind, that person is Tom Anderson (Keanu Reeves), a computer hacker that goes by the alias Neo. Neo is beginning to suspect there's more to life than meets the eye. When they finally meet, Morpheus gives Neo a choice — return to his normal life in blissful ignorance . . . or learn the truth. Of course, Neo chooses the truth, so both he and the audience are in for one *serious* ride.

For those few of you who still haven't seen *The Matrix* yet, to reveal anything more would be to blow the film's major twist. Suffice it to say that *The Matrix* is one of those "simple-yet-complicated" films, where nothing is what it seems and everything starts to make sense the longer you watch it.

As for the cast, well . . . believe it or not, the cast is great. Who would have ever thought Keanu Reeves would be perfect in *any* role? But he's perfect here. Carrie-Anne Moss is cool and sexy, as is Fishburne in his own way. And Joe Pantoliano is at his slimy best. The direction is wonderful — the Wachowski Brothers clearly knew what they wanted to do with this film, and they got it. This story could have become *awfully* stupid, but it absolutely works. The film's saving grace are some groundbraking stunts and special effects sequences, which set it apart from most everything that American audiences had seen before.

The film is presented in anamorphic widescreen on DVD and it looks pretty good . . . if not perfect. The brightly lit shots are super, with crisp colors and solid flesh tones. The transfer looks very good in these scenes. But the darker scenes show course grain and some distinct MPEG-2 compression artifacting. It's nothing that would disappoint anyone buying the disc, but it's evident enough that it should be pointed out. It's also worth mentioning that Warner Bros. plans to release all three films in this trilogy (including this first one) as special edition box set later in

2004, so look for these slight problems with the video to be rectified in time for the new release.

The audio on this DVD really stands out. It's not a subtle or particularly well-rounded sound field, but it still packs a tremendous punch. There's some great play with directional effects in the surround mix (the entire soundstage rotates in 360 degrees along with the picture during the "bullet time" moments, for example). All in all, this is pretty sweet DVD sound.

In terms of special edition material, there's plenty of stuff here to keep you satisfied while you wait for the new version on disc. There are three behind-the-scenes documentaries, two of which are hidden in the special features section as "red pills" that you have to highlight. One of them (*What Is the Concept?*) is a look at the production art, paintings, and digital effects put to music, and progressively shows how each scene was done. Another (*What Is Bullet Time?*) focuses on the making of the film's signature effects shot, the infamous "bullet time." It's hosted by visual effects supervisor John Gaeta. You won't understand a single technical word that comes out of his mouth, but hats off to him for coming up with visual effects like this. The third (and most easily found) documentary is a more comprehensive look at the making of the film itself, as originally seen on HBO. It shows the Wachowski Brothers at work, the special effects and stunt work, and the training of the cast by famed Hong Kong director (and master of "wire" fu) Yuen Wu Ping.

There's also a full-length commentary track with Carrie-Anne Moss, John Gaeta, and editor Zach Staenberg, which is nice enough. Still, one can only hope the new DVDs will actually feature the Wachowski Brothers in commentary form. The coolest extra is something Warner calls "Follow the White Rabbit." It allows you to watch the film and break out to special, behind-the-scenes featurettes whenever a little white rabbit logo pops up on your screen. At this point, tons of DVD special editions use this technique, but this is where it all started, kids. There are nine featurettes accessible during the film in all, and they're definitely worth checking out.

Also included on the disc, via DVD-ROM, are the screenplay, storyboards, concept art, theatrical trailers, a trivia game (that, if solved, will allow you into a special web site), and screensavers. There were also links to access special live chats and web events that were held in the months after the disc's release (one of which even featured "virtual" commentary by the Wachowskis).

If all of that still isn't enough for you, and you want to explore *The Matrix* universe even further, we suggest you pick up a second DVD release, called *The Matrix Revisited*. This is basically a collection of additional featurettes, stock footage, and interviews about the making of *The Matrix* and its sequels. There's even tons of audio tracks that were under consideration for use on the film's soundtrack. Think of *The Matrix Revisited* as the second disc of a two-disc set, containing additional bonus material. It's actually pretty loaded — we weren't quite expecting the disc to be as good as it is.

In many ways, *The Matrix* is the perfect film for release on the DVD format. As such, it became the biggest selling DVD release of its day. With two sequels on the way, and more hype than you can escape surrounding them at the moment, this film is sure to continue impressing — and selling — on disc for years to come.

If You Like These Films . . .

. . . don't miss *Léon: The Professional*, *The French Connection*, the original *Rocky*, and *Speed*. As we mentioned earlier, *Heat*'s worth a look as well, if you don't mind bare-bones titles. As a general rule, if the significant female in your life rolls her eyes when you mention the title, the film's probably a deserving guy movie worth a look.

DVDs for Kids of All Ages

Oh, to be a kid today, with the all the great videogames, gadgets, and toys . . . and the cool movies to experience for the first time. Sure, life might be more complicated for kids these days, but we still think they've got it made. So, in honor of the child in all of us, we've pulled together a list of movies we think any kid would love on DVD. We sure do love them anyway.

Editor's Note: *Bits* writer Greg Suarez contributed to this section.

The Wizard of Oz

MGM (Warner Bros.)

Wizard of Oz belongs in every child's movie library. It's our cultural duty to turn every new generation of children onto this film, because as it happens, *Oz* is one of the most perfect children's films ever made. With great effects, beautiful design, and characters everyone can love . . . what's not to like?

Wizard of Oz is based on the series of children's stories by L. Frank Baum, who created a fictional fantasy world of little people, talking lions, and pumpkin-headed people who live under the rule of a wizard, a witch, and, for a period, a scarecrow. In this film, we're treated to an adaptation of the original tale, a story about a disenchanted girl named Dorothy (played by Judy Garland), who travels to a land over the rainbow called Oz. While there, she meets new friends, saves their world, and, most importantly, realizes that there is no place like home.

Generations have already grown to love this film, seeing it first in theaters and later on yearly CBS TV broadcasts. And now, we can enjoy *Oz* on this beautifully crafted DVD. *Oz* is packed with lavish production design, incredible songs, and beautiful Technicolor, all of which gets loving attention in this special edition from Warner Bros.

Warner's done this film proud on DVD. The transfer is clean and the colors are bright. There are a few problems with the video due simply to the age of the film and the condition of the print, but Warner's done a great job in bringing this film to us in the best quality possible. The soundtrack has been remastered in Dolby Digital 5.1. Sadly, the original mono audio track isn't an available option. Still, the 5.1 track is of good quality and supports the film well. You have to be generally impressed by what was done here.

The disc is loaded with extras. In fact, three guides inside the packaging help you navigate through everything. The best of the extras is the original TNT documentary, *The Wonderful Wizard of Oz: The Making of a Classic*, hosted by Angela Lansbury. It's an entertaining and useful primer on the film, and gives foundation to the rest of the extras you will find on this disc. These include a series of outtakes, deleted scenes (in both film and still recreation form), behind-the-scenes sketches, storyboards and still galleries, excerpts from the original books (as well as previous incarnations of *Oz* in film), archival interviews, newsreel footage, and clips from the 1939 Academy Awards. You'll also find a series of "jukeboxes" featuring audio clips of cut songs, rehearsals, alternate soundtrack cues, and radio broadcasts. For *Oz* fans, this set is the treasure at the end of the rainbow.

The Wizard of Oz is one of those classics that everyone can love. Not only is the film timeless, Warner has made sure that it doesn't show its true age. This is a great DVD for both the young . . . and the young at heart.

Tarzan: Collector's Edition

Disney (Buena Vista)

Created by Edgar Rice Burroughs, Tarzan first appeared in the pulp magazine *All-Story* in 1912. The character quickly expanded to books, comic books, and radio serials, and went on to become one of the most well known figures in the history of entertainment. But after a series of successful MGM films, and a handful of attempted "reinventions," Tarzan looked like he was finally going to be just another guy in a loincloth. After all, you have to stay in the mainstream in order to stay cool.

Enter Disney, which decided to release an animated version of the Tarzan legend, combining traditional character animation with CGI backgrounds and effects. The resulting film is fast paced, is full of fun Phil Collins songs, and features a traditional "child loses parents, gorilla raises child, child becomes man, man falls in love, and love conquers all" story. Oh . . . and there's a sneering bad guy in the mix too.

Disney presents *Tarzan* on DVD in anamorphic widescreen video, once again transferred straight from the original digital animation files. The picture quality is gorgeous, with no faults or blemishes to be found. The color is lush and vibrant, and the contrast is excellent, with deep, detailed blacks. The audio is presented in Dolby Digital 5.0. That's not a typo—there is no LFE or bass channel. Low frequency is still coming out of your speakers, just not on an isolated channel. Not to worry—this track is an accurate representation of the original theater audio. Most will never notice that the LFE channel isn't active.

This is a two-disc special edition and, as such, it's packed with enough historical info and behind-the-scenes material to choke an elephant. Running with the film is a pretty standard audio commentary with directors Kevin Lima and Chris Buck, as well as producer Bonnie Arnold. Nothing earth-shattering is mentioned here that doesn't appear elsewhere on the disc, but the track is still nice to have. Also appearing on Disc One is a preview trailer for *Dinosaur*, as well as an interactive trivia game and a "read-along" version of the story. You'll also find some DVD-ROM features including a game demo and web links.

Disc Two has a lot of fun stuff broken up into sections. Addressing everything from the history of the character to the development of the film, we're treated to early animation tests, research video, production notes, and more. If you're interested in the music, you'll find behind-the-scenes footage, music videos, and interviews with Collins and 'N Sync (which also provided a song for the soundtrack). There's also conceptual art, marketing material, storyboard-to-film comparisons, abandoned sequences, and more. All in all, this is a packed special edition that kids and parents alike will have fun exploring.

Tarzan is a modern classic and has become a favorite of many Disney fans. If you didn't give it a chance when it first appeared in theaters, get yourself a copy of this Collector's Edition and swing on in.

The Princess Bride: Special Edition

MGM

The Princess Bride tells a story of adventure, bravery, revenge, and – above all – the power of true love. The plot is very simple. One day, a long time ago, a girl of nobility named Buttercup (played by Robin Wright) falls in love with a poor but honorable stable boy named Westley (Cary Elwes). Westley knows in his heart that their love is true and meant to be, but he has nothing to offer for Buttercup's hand in marriage. So he sails off across the sea to seek his fortune, intending to come back for her one day. But after a time, word comes back to Buttercup that the dread pirate Roberts attacked Westley's ship . . . and Roberts spares the life of no one.

Not long after, the egomaniacal Prince Humperdinck (Chris Sarandon) chooses Buttercup to be his wife. She doesn't love him, but she can't refuse the man who will soon be king. Before the wedding can take place, however, Buttercup is kidnapped by the dastardly Vizzini (Wallace Shawn) and his two henchmen, a Spanish swordsman named Montoya (Mandy Patinkin), and a gentle giant named Fezzik (André the Giant). Who will come to Buttercup's rescue? Will Humperdinck find and marry her? Will Montoya ever find the six-fingered man who killed his father and enact his revenge? And who is the mysterious Man in Black? You're just gonna have to watch the movie to find out now, aren't you?

There are so many things that make *The Princess Bride* an almost perfect movie. First of all, the screenplay is first-rate, adapted from his own novel by the acclaimed William Goldman (who also wrote *All the President's Men* and *Butch Cassidy and the Sundance Kid*). It's wickedly funny and very smart, and has as much (if not more) in it for adults as it does for children. How many movies do you know of that feature a battle of wits . . . to the *death*?

The other thing that makes this movie special is the exceptional ensemble cast. They say that good casting is 70 percent of the work in making a film, and no film proves that to be truer than *The Princess Bride*. Mandy Patinkin is absolutely brilliant as Montoya, a character who starts out as bad guy, but for whom we ultimately end up rooting in the end. How many people would cast André the Giant in a film? He's just wonderful here – sweet and funny. Wallace Shawn almost steals the movie in the aforementioned battle of wits – what a great villain! Cary Elwes is so good here as Westley that it makes you cringe at all the smarmy bad guys he's played on screen since (in *Twister*, *Liar Liar*, *Kiss the Girls*, and so on). Did we mention Christopher Guest and Billy Crystal? Yep . . . you get them too. Add to that a clever wrap-around story that has a wily Grandfather (Peter Falk) actually reading the film's story to his sick grandson (played by a young Fred Savage of *The Wonder Years*) and you've got a surefire winner. Rob Reiner has directed a lot of great movies over the years (among them *A Few Good Men*, *The American President*, *The Sure Thing*, and *When Harry Met Sally*), but this is definitely our favorite.

Right now, without us saying another word, you should just rush out to pick up the new *The Princess Bride: Special Edition* as soon as you can possibly do so. Finally, the film can be seen in the anamorphic widescreen transfer it deserves. The quality of the image is very smooth and natural, with very little dust and dirt on the print. Grain is visible but isn't distracting, the color is natural and accurate, and the blacks are again deep and true. On the audio side, this Dolby Digital 5.1 track is full and rich, with a smooth, wide soundstage and prominent bass. Dialogue is clear, Mark Knopfler's score is well blended in the mix, and the rear channels are actively utilized to provide good ambience.

The extras included are quite nice. First up, you get a newly recorded audio commentary with director Rob Reiner. He's very engaging to listen to and has a lot of great stories to tell. It's obvious that he loves this film as much as the rest of us do. Better still, you get a second audio commentary with the writer of the original book and the screenplay, William Goldman. The real gem of the disc, however, is a new 27-minute documentary on the film, entitled *As You Wish*. It features some really great behind-the-scenes footage and stories, along with newly recorded interviews with the entire cast (save for Wallace Shawn and André the Giant — André is sadly no longer with us). The fondness with which they all recall the experience of making this film, the wonderful stories about working with André, and particularly Mandy Patinkin's recollections about creating his character really make this a gem. The remainder of the extras is mostly filler, but it's good filler. You get a pair of 1987 EPK-style featurettes, a few TV spots, the theatrical and foreign release trailers for the film, and a gallery of some 80 production photos (nicely indexed by subject). Good but frustrating is Cary Elwes's video diary, made up of home video footage the actor shot during the production. It's good because it's a great — and honest — look behind the scenes, particularly at the interactions of the cast off camera. It's frustrating because it's not nearly long enough (only about four minutes in all).

The only thing that's really left wanting on this disc is the lack of deleted scenes or outtake footage. Given that the ownership of the film has transferred hands a few times over the years, it's probably not surprising that this kind of material isn't on the disc. It is probably impossible to find anything of this nature that had survived. But at several points in both the commentary and the documentary, the participants talk about how they kept constantly cracking each other up during filming. You just know that the bloopers and outtakes would have just been a riot. Oh well . . . it's a minor and probably unavoidable weakness on an otherwise solid disc.

The Princess Bride is simply wonderful, any way you slice it. It's the *Wizard of Oz* for Generation X. The film is charming, sharply funny, sweet, romantic, and it is fit to share with the whole family. What more could you want? As the trailer says: "Heroes. Giants. Villains. Wizards. True love. Not just your basic, average, everyday, ordinary, run-of-the-mill fairy tale." Run out and pick this one up quick.

The Nightmare Before Christmas: Special Edition

Touchstone (Buena Vista)

Every once in a great while, the stars and planets are in perfect harmony. The heavens part, the muse sings her enchanting melody, and inspiration is born. Behold *The Nightmare Before Christmas* — an example of the birth of a truly original and groundbreaking epiphany from the fertile and haunted place that is Tim Burton's imagination. Mix a little bit of Dr. Seuss with a dollop of German Expressionism . . . and you get a genuine classic.

Jack Skellington is the Pumpkin King of Halloween Town . . . the place where Halloween is born, and where it lives 365 days a year. Jack becomes bored with the yearly Halloween routine and feels that somehow his life could have more meaning — that there's something else out there for him. In a wooded area far from Halloween Town, he stumbles upon a mysterious doorway to Christmas Town, the place where Christmas comes from. After witnessing the Christmas joy and spirit, Jack becomes enchanted with the jubilant feeling and decides that it's his destiny to become the new King of Christmas. So after training the ghouls, ghosts, witches, and vampires that live in Halloween Town on the Christmas philosophy, Jack kidnaps Santa Claus and takes over the role of the Jolly Fat Guy in Red (even though Jack is, ironically, nothing more than a skeleton). Jack's frightening brand of Christmas sees December 25th twisted into a scary and distorted combination of the two holidays that leaves the children of the world terrified. Jack means well, but let's face it . . . all he's never known is Halloween. And severed heads, snakes, and evil toys don't exactly leave the children of the world singing "It's the Most Wonderful Time of the Year." Will Jack come to his senses and save Christmas, or will the holiday forever be a freakish nightmare?

The Nightmare Before Christmas is filmed in a classic, Rankin-Bass style of stop-motion animation, and it took quite a long time to complete. Never before had such a large-scale stop-motion film been done, which definitely makes this movie unique. The stop-motion technique disconnects the audience from more traditional methods of filmmaking, giving them an otherworldly feel, yet it has much more charming eeriness than live-action or traditional cell animation. This film definitely would not have had the same effect if it were made any other way. The atmosphere and sets are very creepy and perfectly set the mood for the movie. Halloween Town's color palette is dreary and dark, giving it a black and white look, while Christmas town is bright and colorful. Many of the sets have a very textured look, giving them what the filmmakers call a "sketched look" – as if they were drawn from a live-action storybook penned by Edward Gorey.

There's plenty of bold humor and twisted visions here (not too twisted, as this is a PG film, but still very strange). The story is simple, intriguing, and completely original, and the dialog is entertaining. But Danny Elfman's music is the real star of this film – this is probably some of the strongest work of his career. The amazing thing about the music is that it's not quite Halloween and not quite Christmas. Much like Jack's vision, the line between the holiday moods in these songs is blurred. There are many subtleties in the lyrics and music that beg for repeat attention. Songs like "This is Halloween" and "Kidnap the Sandy Claws" are deliciously humorous and just plain fun, while "Jack's Lament" and "Sally's Song" are quite delicate and beautiful. Both kids and adults will find this soundtrack marvelous, as the humor is perfectly balanced for both generations to enjoy.

Want to hear something ridiculous? After releasing this film once on DVD already with a mediocre, non-anamorphic transfer, Buena Vista decided, three years later, to re-release *The Nightmare Before Christmas* as a feature-packed special edition . . . but with the same non-anamorphic transfer. While the video is okay, it has the same soft image and annoying NTSC video noise that plagued the first release . . . which an anamorphic transfer would have cured. The color scheme of this film is very important to the overall mood of the story, so thankfully the colors on this disc do not disappoint. Compression artifacting is only rarely visible and fine picture detail is acceptable but lacking in some of the darker scenes. Any way you slice it, the lack of a new anamorphic widescreen transfer is *really* a missed opportunity by Buena Vista.

The audio on this special edition fares better, with two 5.1 soundtracks – one from Dolby and one from DTS. The Dolby Digital 5.1 track is a high-caliber mix. The front soundstage is gigantic and wide during musical passages, flooding the listening space with rich, full sound. The music is smooth and never harsh or strained. There's a definite sense of airiness and depth to the mix. The low end fills in the bottom octaves for the score and surround usage lends a nice sense of space and envelopment. Several nifty directional sound effects find their way into the rear channels as well. The DTS 5.1 soundtrack sounds very similar to the Dolby track but adds a bit more low-end tightness and overall definition to the mix. These differences are definitely subtle.

Thankfully, there are tons of extras here. There's a commentary track with director Henry Selick and director of photography Pete Kozachik (while this film is based on Tim Burton's ideas, vision, and artwork, he only produced and consulted due to his commitment to Warner Bros. to direct *Batman Returns*). The track is entertaining in a very technical and informative way. There aren't many anecdotes or stories – just a lot of information about how the movie was made and where many of the ideas originated. Next up are two sets of deleted scenes, all introduced by the director. Three of the scenes were never animated but are presented in storyboard form, while the other four were fully completed. A majority of the scenes are extended or alternate scenes, including the unused surprise identity of Oogie Boogie. On the subject of storyboards, you get a number of storyboard-to-film comparisons, with the storyboard on the top half of the screen, while the finished product plays in conjunction below it.

A 25-minute, behind-the-scenes featurette is also included, which focuses on how the puppets were created from start to finish, the painstaking work of animating them frame by frame, and how the sets were designed and lit. There are interviews with Burton, Selick, and many of the animators. The featurette also covers the motion control camera used in filming to give the movie a more contemporary and exciting feel. Traditionally, with stop-motion animation, the camera cannot be moved around the characters as they are animated. But with their new computer-controlled camera, the filmmakers could pan the camera around the sets as the characters moved, just as one would in a live-action film. This is a must-see featurette that's as educational as it is interesting. Next on the list of supplements is a gallery of character and set concept art, including some animation tests for several of the lead characters. You also get the film's theatrical and teaser trailers, as well as a trailer for *James and the Giant Peach*.

Probably the most anticipated of all the supplements are Tim Burton's two short films, *Vincent* (1982) and *Frankenweenie* (1984). Both are black and white in their original full-frame ratio. *Vincent* is Burton's tribute to his long-time idol, Vincent Price, and the featurette is actually narrated by the Master of Horror himself. It's a highly imaginative little tale about a seven-year-old boy, named Vincent, who adores Vincent Price. Shot mostly in stop-motion, the nine-minute short is written in rhyme and is a real treat to watch. *Frankenweenie* is a modern interpretation of Mary Shelley's *Frankenstein*, except the mad scientist here is a little boy (Victor) and the monster is his beloved dog (Sparky) that was killed by a car. Young Victor is heartbroken, but after learning about electricity's effect on the nervous system, he decides to dig up his old friend and bring him back to life. You can imagine where it goes from there. Sparky accidentally gets loose in the neighborhood and terrorizes the people on the block with his freakish appearance. The ending is an amusing homage to the 1931 Universal Studios version of *Frankenstein*, with a slightly different outcome. *Frankenweenie* is about 30 minutes long, with a cast that includes Daniel Stern and Shelly Duvall. It's definitely geared more towards children and lacks the sheer originality and uniqueness of *The Nightmare Before Christmas* or *Vincent*, but it is still fun to watch as part of Burton's roots as a filmmaker.

Words like "brilliant," "inspired," and "masterpiece" can easily be used to describe *The Nightmare Before Christmas*. Watch this film for yourself and you'll see that those descriptions are apt indeed. Without the inclusion of an anamorphic widescreen transfer, this disc falls short of being the "ultimate edition" of this film on DVD. But the extras are fantastic and the film is absolutely deserving of your attention. This is definitely one of our favorites, and we think you'll love it too.

If You Like These Films . . .

. . . also suitable for kids of all ages are *Shrek*, *Chitty Chitty Bang Bang*, *Willy Wonka and the Chocolate Factory*, *Who Framed Roger Rabbit*, *Stuart Little*, *A Bug's Life*, *Ice Age*, the *Spy Kids* series, *Monsters, Inc.*, *The Incredible Adventures of Wallace and Gromit* and pretty much anything with Winnie the Pooh or the Muppets in it.

Aliens and Robots on DVD

Nothing, and we mean nothing, beats watching good ol' aliens and robots on DVD. Imagine it's a rainy Sunday afternoon and you can't get outside to mow, play with the puppy, or toilet paper that annoying neighbor down the street. You dig through your DVD library, through all of the standards you've watched a thousand times already . . . and come upon a good sci-fi flick. Sure, you've probably seen it a thousand times too. But it's sci-fi, so you think, "Hey, it's been a while. Let's give this a spin." And all of a sudden, that rainy afternoon gets a whole lot better.

There's something that's just so comforting about aliens and robots . . . even if the aliens are of the backdoor-probing variety, and the robots are rampaging out of control. With that in mind, here are our picks for some of the very best alien and robot flicks available on DVD—classic, go-to, sci-fi flicks that every self-respecting movie buff should have warming the bench in their collection, just waiting for that rainy day.

Independence Day: Five Star Collection

Twentieth Century Fox

So what happens when a race of war-mongering aliens drops by unannounced for some of your famous barbecue ribs one Fourth of July weekend? They lay waste to your planet's biggest cities without so much as a "thank you very much, but we're taking over, earthlings." That's the plot of Dean Devlin and Roland Emmerich's *Independence Day*, forever after known in this review as *ID4*.

ID4 is one of *those* movies. You know . . . the big summer explosion epics. In fact, at the time it came out in the summer of 1996, there had been few films that even approached its sheer scale and audacity. After a bunch of aliens show up in a mothership a forth the size of the moon, and launch dozens of little (so to speak—they're 13 miles wide each) flying saucer ships to blast the biggest cities in the world, the human race finds their collective tails thoroughly whipped. But good old mankind isn't down for the count—no, sir. They decide to fight back, lead by a yuppie president of the United States (Bill Pullman), a scrappy fighter pilot (Will Smith), his stripper girlfriend (Vivica Fox), an over-educated cable TV technician (Jeff Goldblum), and his dear old dad (Judd Hirsch).

The beauty of this film is that it knows exactly what it is — a massive, testosterone-pumped, B-movie hype-fest — and it doesn't try to be anything else. You can find plenty of faults with this flick, but after a while you'll simply stop counting them. You'll be too busy being dazzled by all those cool special effects. This film is almost like *Fight Club*, in that it just walks right up to you and smacks you square in the face . . . but it's a good sort of pain.

Fox's DVD of *ID4* contains no less than two discs' worth of alien-zappin' fun. First of all, Disc One contains two different versions of the film, thanks to the wonders of seamless branching. There's the original theatrical version, as well as a nine-minute-longer special edition. Both are in very good looking anamorphic widescreen video, transferred from high-quality film elements. The color and contrast exhibited here are outstanding, with very solid blacks and rich, vibrant (and accurate) tones. There's very little artifacting to be found, even in the kinds of images that often give MPEG-2 compression trouble, like exploding fireballs (and there are plenty here). There's also very little dust on the print, even in the newly added footage. The special edition's additional footage restores several scene trims and at least two entire scenes, which actually help the believability of the film's ending somewhat.

The Dolby Digital 5.1 sound is also very good. The sound field is very wide, with tons of directional effects and panning and extremely active rear channels. You'll hear plenty of F-18s and alien attackers screaming by over your head and away into the distance behind you. And the low frequency is ever present and often thunderous. It isn't particularly natural or ambient, but this mix will definitely give your sound system a workout.

Also included on Disc One are a pair of full-length commentary tracks, one with writer/director Roland Emmerich and writer/producer Dean Devlin, and one with SFX supervisors Volker Engel and Doug Smith. The Devlin and Emmerich commentary was done for the previous laserdisc release of the film and is only okay. They often talk about "this laserdisc," and their comments as they watch the film are very production specific — "this shot was . . ." or "and here we have . . ." You get the idea. At one point, after a long period of silence, Emmerich even says, "We're sitting here watching our own movie. Maybe we should say something . . ." That's not to say the commentary is bad — there are some good stories here and plenty of funny little anecdotes. But given that these two were also the writers, we would have enjoyed more discussion of the concepts and ideas behind the film (what little there are). The effects commentary is even more production specific, as one would expect. It's also not great, but is worth a listen for fans.

Disc Two has plenty of goodies as well. To start with, you've got the film's original ending, which featured Russell (Randy Quaid) flying in a biplane rather than an F-18. The sequence includes non-optional commentary, in which we learn that the ending was changed because it was thought that having the biplane (rather than a fighter jet) would stretch believability. That's funny. The biplane ending is actually much more fun, especially given that believability is already stretched well past the legal limit in this flick. There's a pair of 30-minute documentaries on the making of the film, *Creating Reality* and *The Making of ID4* (the latter was hosted by Goldblum for HBO). Our favorite is *The ID4 Invasion*, which is a 22-minute "mockumentary" that was produced for actual use in the film. You'll see bits of it throughout the film, as the characters watch news reports on the invasion on TV. It's very well done (those are all real TV news people) and is actually extremely clever ("The LAPD is asking Los Angeles residents *not* to shoot their guns at the alien spacecraft. You could start an interstellar war . . .").

Also included on Disc Two is a whole mess of trailers and TV spots, including the Apple Computer and Superbowl commercials. There's also a gallery of hundreds of production photographs, storyboards (from three sequences in the film), and cool conceptual artwork. Finally, you get DVD-ROM features, which are (as far as such things go) particularly lame. It's nothing more than a whole lot of web links. You also get a link to an online space combat game called *ID4 Online* and a *Get Off My Planet* interactive trivia game, which to this day, we can't find at all.

ID4 doesn't push the boundaries of DVD, but it serves the film well and that's what counts. In any case, this is definitely a DVD that'll satisfy your craving for aliens . . . especially if you like your aliens served with stuff blowing up.

The Adventures of Buckaroo Banzai Across the Eighth Dimension: Special Edition

MGM

You know . . . there just aren't many films like *Buckaroo Banzai*. In fact, about the only other film that even comes close is *Big Trouble in Little China*. Put yourself smack dab in the middle of a 1980s campy, quasi-sci-fi mind frame, and you're in the right ballpark. Well, sort of.

Our story starts as the infamous Buckaroo Banzai (played by Peter Weller) is preparing to test his suped-up, high-powered Jet Car (but not before consulting on an intricate brain surgery—Buck's a man of many talents). Buckaroo climbs into the cockpit of the car and installs a strange device—an Oscillation Overthruster. Once prepared, the car blasts down range on a faster-than-sound test run. But something goes wrong . . . or so it seems. The car careens off the test track and heads right for a mountain. But just as it looks as if Buckaroo's about to bite it, he suddenly engages his Overthruster . . . and the car drives right through solid rock, blasting through the mountain and into the mysterious Eighth Dimension. News of this scientific breakthrough is quick to spread, and it soon reaches the maniacal Doctor Emilio Lizardo (John Lithgow), who has been confined to a mental hospital in New Jersey since one of his experiments went horribly wrong years ago and drove him crazy. But it seems that he isn't really crazy . . . just possessed by the spirit of an alien Red Lectroid named Lord John Whorfin. Whorfin wants to free his Evil Red Lectroid comrades, who are trapped in the Eighth Dimension, and he needs Buckaroo's Overthruster to do it. So with his loyal henchmen John Bigboote (Christopher Lloyd) and John Gomez (Dan Hedaya), he sets out to steal the Overthruster . . . and hopefully destroy Buckaroo in the process. But Buckaroo never stands alone—he's got the hard-rockin', atom-crackin' Hong Kong Cavaliers on his side. And if things really get desperate, he can always call upon his worldwide network of Blue Blazer Regulars. When they're not busy fighting the World Crime League, of course.

Buckaroo Banzai is one of those films that you either already love, don't get, or have just never seen. If you first caught it during its theatrical run back in the '80s, then its off-kilter brand of humor probably hit you right square between the eyes. John Lithgow and Christopher Lloyd absolutely shine in their over-the-top performances here. Other cast highlights include Jeff Goldblum as New Jersey (in a role that's very much hinted at in his later appearances in *ID4* and *Jurassic Park*) and a very young Ellen Barkin as Penny Priddy, Buckaroo's hair-teased femme fatale. You even get Yakov Smirnoff in a bit part.

Still, *Buckaroo Banzai* is not a great film by any stretch. Its major flaw lies in the direction and editing. This is not a well-paced film. It doesn't build on the humor (there are some *very* funny throwaway gags here) and it doesn't build much tension either. It also doesn't help that the film's soundtrack is a lot hokey—a cheesy, early '80s brand of synthesizers and drum machines. You'll either love *Buckaroo Banzai* or hate it . . . but we dig it big time at *The Bits*.

The video on this DVD looks surprisingly good, given the film's new anamorphic widescreen transfer. The contrast is very nice, with deep, detailed shadows. Most impressive is the color, which is surprisingly rich, accurate, and vibrant, while never bleeding or otherwise falling short. The print itself is occasionally soft, and there's moderate grain visible at times, but it's very clean and free of dirt and other blemishes, once you get past the opening credit sequence. You've certainly never seen *Buckaroo* looking this good before.

The disc's audio is also generally good, available here in a newly re-mixed Dolby Digital 5.1. The track sports a surprisingly wide and smooth front soundstage. The film isn't really too active in terms of surround sound effects, but when they're needed, the mix handles them just fine. Mostly, you get ambience from the rear channels. Dialogue can occasionally sound a little flat, and the low frequency is a little wanting at times. But, like the video, the audio is much improved here and should please most fans of the film.

As for the extras, here's the shtick with this DVD, and we think it's pretty funny. This disc was assembled as if the actual Banzai Institute produced it. That means that the material here is rife with in-jokes. Only fans are going to get all of them, but if you *are* a fan, this disc is a real treat. To start with, you get a tongue-in-cheek audio commentary with the film's director W.D. Richter and the real Reno from the Banzai Institute (it's actually writer Earl MacRauch, but play along here). The two dissect the film and talk about how the "real" Buckaroo Banzai liked the film. You see . . . the further schtick here is that the film itself is a dramatization of "real" events, so the people you're watching onscreen are only actors playing the "real" Buckaroo Banzai and the Hong Kong Cavaliers. If you buy into it, it's pretty funny. There's also a subtitle track of *Pinky Carruther's Unknown Facts* (Carruthers being a character in the film played by musician Billy Vera, although again, it's Rauch who wrote all this stuff). It's filled with funny bits of trivia, all in keeping with the gag. Fourteen deleted scenes are available here too, taken from the film's work print (so they're not anamorphic and the quality is poor, but hey—at least they're here). And there's even the film's alternate extended opening, which features Jamie Lee Curtis as Buckaroo Banzai's mother. You can view this separately, or you can choose to view it not quite seamlessly restored to the film itself (when you go to play the film, you choose between the theatrical cut and the extended). We say not quite seamlessly, because the switch from the alternate opening to the rest of the film triggers a long pause—a major complaint.

Hang on there, Blue Blazers! We're just getting started. This disc also includes the film's original teaser trailer (in anamorphic widescreen) and a new Jet Car promo trailer that was created by the folks at Foundation Imaging to sell the idea of a *Buckaroo Banzai* TV series. The idea never flew, but it's cool to have the trailer here (it's otherwise only been seen at sci-fi conventions). You get a featurette on the making of the film from the 1980s, *Buckaroo Banzai Declassified*, along with a detailed profile of the "real" Buckaroo Banzai. Then there are profiles of most of the characters in the film, a look under the hood at the Jet Car, an extensive gallery of photos (broken down by subject) and some Nuon features no one cares about (because only like three people have a Nuon-enhanced player to access them). It's doubtful that you're missing anything if you can't view them. Then there's a fun section called the *Banzai Institute Archives*. This is filled with schematics, photos of movie tie-ins, badges and CD covers by the Hong Kong Cavaliers (gotta love "Your Place or Mayan?"), a list of the film locations, the text of a pair of reviews of the film, a text interview with the "real" Buckaroo Banzai, a history of the Institute, entries from Hikita's diary, and even a 10-minute *Banzai Radio* interview with Terry Erdman (who was the Fox publicist for the film back in 1984). Finally, several Easter Eggs are hidden throughout the disc's menu screens, including quotes, alternate DVD menu designs, and alternate DVD cover designs (oh, how we wish they'd been used). There's even a funny bit about the watermelon—'nuff said. They're funny and pretty easy to find if you search around.

A lot of care has gone into this disc, and if you're a fan, you're really going to love it. Other people are going to completely miss the gags and be left scratching their heads. Screw 'em! This baby's aimed at you die-hards anyway, and it's got plenty of mojo. So fire up your Oscillation Over-thrusters and enjoy.

By the way . . . when we tried to reach Buckaroo Banzai himself, to see what he thinks about the DVD, turns out he was still working on that *World Crime League* sequel and so was unavailable for comment. His publicist at the Banzai Institute did, however, send over this statement: "Nothing is ever what it seems, but everything is exactly what it is." We're still not sure if he meant the DVD or the sequel, but there it is.

Close Encounters of the Third Kind: Collector's Edition

Columbia TriStar

In the middle of the desert, a squadron of planes that disappeared in the 1940s off the coast of Florida mysteriously shows up in prime condition. A team of government scientists floods in and the questions begin: How did they get here? What brought them? Is this important? Does this mean something? Meanwhile, American air traffic controllers are getting reports of mysterious aircraft playing chicken with commercial airliners. Are these two events somehow related? Enter Roy Neary, played by Richard Dreyfus. He's an Indiana power station worker who gets called to the location of several power outages one night. While lost on a country road, he sees something he can't explain . . . something that changes his life forever.

Roy is now plagued by a pounding pain in his head, a pain that only seems to give way when he's sketching or sculpting a dark shape out of whatever he can find at hand. His family doesn't understand what's going on and, after he loses his job, he hits the road to find the truth. It's on this quest that Roy meets another dangling plot thread, a woman named Jillian (Melinda Dillon). Jillian's little boy, Barry, has a connection to the ever-growing mystery and had literally become caught up in it. So, together, Roy and Jillian must work to uncover the answers to their own questions, as well as the larger questions posed by the film.

When it's all said and done, whether you like the current crop of Spielberg movies, *Close Encounters* remains a wonderful film and it's aged very well. Aside from some of the dated pop cultural references and clothing styles, this is a film that could have been made yesterday. The acting is superb, the story is well woven, and everything about it works. Spielberg presents *CE3K* with a very natural honesty—nothing seems forced. Any emotion you feel here is all about great filmmaking. *Close Encounters* has long been a favorite of sci-fi fans everywhere, and deservedly so.

CE3K has traveled a long, strange road over the years. With all of the alternate versions of this film that have been in circulation, since its initial release in 1977, you'd need a blueprint just to figure out what's "new" footage, what's original footage, and what was seen only in the syndicated TV version. Honestly, at this point . . . most of us can't tell anymore. But that's okay—for all intents and purposes, this is Spielberg and Columbia's ultimate, final director's version of *CE3K*, a "definitive" approach that incorporates everything Spielberg was trying to accomplish.

The big question is, "How does *Close Encounters* look on DVD?" While the transfer quality is right up there with most of Columbia's other DVD work, a lot of the night shots in this film look slightly weak. The contrast is fine, but darker areas of the image are very much lacking in detail. There's likely a handful of reasons for this, all of which are both fair to assume and couldn't be helped. First of all, the original negative has lost important density over the years. Also, a number of flawed film stocks were being used for special effects shots back in the late '70s and early '80s, and they've become corrupted over time. Both of these factors have left their mark on *CE3K*, and it's a shame. But, when all is said and done, *CE3K* looks pretty remarkable despite these issues, and it's still very watchable on DVD. Just look at the brighter scenes in the film—colors are bright and striking. You can't help but be impressed with this transfer for the most part.

Even more impressive here is the sound. The best thing on this disc is the DTS 5.1 track. It's a tight, fluttery mix that spins with panning and surround activity. You won't believe how great this film sounds in DTS—it really draws you in. The Dolby Digital 5.1 also does a fine job, but it's nothing compared to the DTS. If you don't have DTS yet, and you're a fan of this film, this disc is a great reason to upgrade your receiver.

The special features here are just candles in the icing of the cake. Since it's a two-disc set, you'll find all the extras on Disc Two. First up, there's the original theatrical trailer, which has an almost documentary feel (à la the trailer for *The Abyss*). There's more behind-the-scenes footage in it than actual scenes from the film. It's pretty neat. There's also the so-called "special edition" theatrical trailer, which heralded the 1979 re-release—the one that featured the awful ending with Roy inside the mothership (thankfully, that's been excised from this DVD version of the film). Next up is the very long, but thoroughly enjoyable, *The Making of Close Encounters of the Third Kind* documentary. It's from Spielberg mainstay Laurent Bouzereau, and it's mostly the same documentary that was featured on the 1998 laserdisc release. It covers pretty much everything you want, aside from the fact that other writers, including Paul Schrader (*Taxi Driver*), wrote the original draft of the script. It's also pretty heavy with Spielberg interviews from the set of *Saving Private Ryan*, along with interviews with other stars from the film. Then there's *Watching the*

Skies, the film's 1977 featurette (where a lot of the documentary footage from the trailer first appeared). It's cool, but feels slightly fluffy. Finally, the most important feature of this special edition is the inclusion of some 11 deleted scenes. Everything you remember having heard about, or seen, in the previous editions of *CE3K* can be found here (including the afore-mentioned alternate ending). A lot of this footage actually helps to flesh out the film a little bit. You don't need to actually see it in the context of the film, but viewing it separately helps round out character motivations and define the plots a bit better. They're definitely worth checking out.

Close Encounters is an important film — we think it's pretty close to per-fect — and it remains one of the most enjoyable sci-fi films ever made. Time hasn't been as kind to *CE3K* as we would have hoped, but that's just the way the nitrate crumbles. That said, this DVD does great honor to the film and preserves it wonderfully.

E.T. The Extra-Terrestrial: Limited Collector's Edition

Universal

Who in the world hasn't seen *E.T.: The Extra Terrestrial*? Who doesn't love the film? E.T. him-self is the most adorable alien ever created — shot, with a long neck, light-up heart, and the biggest eyes this side of an anime vixen. *E.T.* holds a very special place in all of our hearts.

Young Elliott lives with his newly separated mom, older brother, and younger sister in the rolling California suburbs. One night, while getting a pizza from the delivery guy, he happens upon something quite curious in his garage . . . something puzzling that prompts him to go out again later that night for a closer look. What he dis-covers is an alien scientist, who was left behind by his ship when a team of government investigators got too close. Now Elliott has to try and help the little guy get home, while hiding him from his mom and dealing with an odd connection the alien has made with him — one that lets them share each other's emotions, as well as health and physical states.

E.T. is fun for the whole family. Now, thanks to the new digital world we live in, director Steven Spielberg has gone back to not only CG-enhance *E.T.* (fixing some stiff effects from 1982), but also remove all of the guns from the hands of the government agents (he replaced them with walkie-talkies). Is it obvious? No. It's actually pretty clean and, in a few years, no one will probably even care. Spielberg also relooped a voice cue (when Mom says her older son can't go out on Halloween as a "terrorist"), replaced the word "terrorist" with "hippie." But who says a film god doesn't also giveth when he taketh away? Thanks to the same tinkering that helped Spielberg fix some moral issues he's obviously had with the film over the years, he's also added back a scene of E.T. taking a bath. It didn't quite work with the original animatronic puppet, but it does now, thanks to CGI. It's actually a cute scene and fits into the film quite nicely. All these changes aside, those of you concerned that the original version of your favorite film has been lost to time and digital tinkering, fear not. The original theatrical version is alive and well on Disc Two of this special edition, so shotguns and terrorists alike can be found in all of their remastered glory.

Both versions of this film look remarkable. Blacks are deep and hard, colors are bright and shiny, and there's not a moment of edge enhancement to be found. It's really a joy to see a film we so fondly remember looking as good as it does here on DVD. The audio is also pretty remarkable. Both versions feature 5.1 DTS and Dolby Digital tracks that really sound great. The sound fields are playful and wide, with lots of surround activity and incredible replication of John Williams's score.

The extras on this set are pretty cool. A word of warning though – the back of the packaging on many copies lists a 50-minute documentary in the extras portion that isn't on this disc (it was dropped at the last second to allow for the inclusion of the original 1982 theatrical cut). Conversely, some copies don't let you know that the original version of the film is included – but it's there.

Disc One features a very nice featurette about the event of a lifetime: John Williams and orchestra performing the score live while the premiere audience watched the film. Although not as cool as actually sitting in the audience, Laurent Bouzereau brings the experience to us in a very appealing form. That entire performance is also presented as an isolated Dolby Digital track during the film. After a short time listening to it, you'll forget it was recorded live. For the kids, there's a very lame look at our solar system through the eyes and voice of E.T., which gets old fast. E.T. sounds like one of those guys who can recite the entire alphabet through one everlasting burp.

Disc Two, along with the standard *Spotlight on Location* featurette, holds a very cute family reunion of sorts with all the major players in front of (and behind) the camera. It's fluffy and full of major backslapping, but what family reunion isn't? There's also plenty of other fluff: uninspiring galleries of production design, publicity photographs and marketing materials, the 2002 re-release theatrical trailer, a *Back to the Future Trilogy* DVD promo, cast and filmmakers bios, production notes, an archival 1982 Special Olympics commercial (featuring E.T.), a new Dave Thomas Foundation for Adoption commercial, a promo for Universal Studios Theme Parks, and a DVD-ROM feature called *Universal Studios Total Axess*, which promises to add (weekly) original material such as storyboards, deleted scenes, and additional behind-the-scenes footage via an online web site.

If all you want is both versions of the film, this two-disc set is the way to go. On the other hand, if you want everything you can get your greedy paws on, there's a 3-disc collector's box set available, which *does* include the 50-minute documentary missing from this set. Both sets will be pulled from distribution over time, so if you're a fan of *E.T.*, be sure to pick one of them up while you can.

Futurama on DVD

Twentieth Century Fox

You want aliens and robots? Well, for a plentiful helping of both, look no further than cult fave *Futurama*—that *other* animated sitcom from the creator of *The Simpsons*, Matt Groening. Hopefully, you've heard of it . . . and seen it. *Futurama* was the show that was stuck in that no-man's-land of television program scheduling on Fox, right after NFL football and before their primetime lineup. Since many football games run long, *Futurama* was bumped so many times that most of us believe it was cancelled. It wasn't, but the damage was done. After years of such treatment, Fox seems to have given up on the show, leaving nearly *two seasons worth of episodes* unseen by fans. Thankfully, *Futurama* has found rejuvenated life on Cartoon Network's *Adult Swim*—and on DVD—so maybe there's some life left in the old boy. Let's cross our fingers.

The show follows on the adventures of 1990s uber-slacker Fry, a pizza boy who is "accidentally" frozen on December 31, 1999. He's awoken in the year 3000, where he essentially picks up with life right where he left off. In this future, Fry hunts down his last known living relative, Professor Farnsworth, who is close to being the oldest living human on the planet and a very wealthy man. Farnsworth runs a delivery service, so along with his newfound friends, an alien cyclops named Leela (Katey Sagal), and a sociopathic robot named Bender, Fry starts working as a delivery boy.

The writing, animation, and voice acting are what make this show so funny. *Futurama* is one of the smartest shows on TV. Pop culture references abound as Fry and company do battle with angry aliens, soda-spewing slugs, and various interplanetary faux pas. The animation, a beautiful combination of CGI and traditional hand-drawn art, is some of the best we've seen outside of Disney and Pixar features. And the voices are genius. Leela, as played by Katey Sagal (formerly of *Married with Children*), is tough, no nonsense, and full of love for her friends. Billy West (best known as Stimpy on the *Ren and Stimpy* show), brings to life the Professor, Fry, and the crazy doctor/lobster/thing Dr. Zoidberg (along with other minor characters). West is one of the preeminent voice talents working today, and he shows his stuff in every episode. Finally, there's John Di Maggio, who plays Bender. His superb delivery of the most sarcastic (and utterly true) lines is some of the best stuff in any cartoon on TV.

Futurama is presented on DVD in the show's original full-frame aspect ratio. These episodes look flawless, with beautiful colors, nice detail, and no signs of artifacting anywhere. The audio is standard Dolby Digital 2.0, which, while it sounds better than the original TV broadcasts, is still just stereo. It's solid, but there's no frills.

The extras on these *Futurama* complete season sets, although not piled on, are still pretty sweet. Every episode has a commentary track. We hear from writer/creator Matt Groening, writer/executive producer David X. Cohen, co-director/supervising director Rich Moore, co-director/supervising director Gregg Vanzo, Billy West, and John DiMaggio. You won't get everyone on every commentary, but these tracks are terrific, touching on the voice casting, trivia about the productions and characters, inside jokes, Internet fans, and future plot developments. There are spots where they run out of stuff to say, but for the most part, the tracks are super. Also on these discs, you'll find some deleted scenes, featurettes, concept art, animatics, scripts, storyboards, and, of course, Easter eggs.

Futurama is full of aliens and robots—two great tastes that go great together—especially in an animated comedy. We never miss an episode when it comes on, so having them on DVD is the best. If you don't already have this set in your collection, you're missing out.

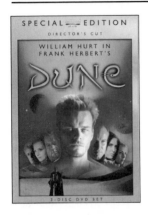

Frank Herbert's Dune: Special Edition Director's Cut

Artisan Entertainment

In the history of science fiction literature, there have been few truly great works – works of high concept that cut through the trappings of pulp sci-fi to describe timeless stories of the human condition. Among the best works of the genre are Arthur C. Clarke's *2001: A Space Odyssey*, Isaac Asimov's *Foundation Trilogy*, and Ray Bradbury's *The Martian Chronicles*. As good as any of these, however, is Frank Herbert's *Dune*. Set many thousands of years in the future, it's a simple tale of two great royal houses engaged in a massive struggle to control the most valuable planet in the Universe – Dune (also known as Arrakis). Rich in political, cultural, and ecological detail, Dune won both the Hugo and Nebula awards for science fiction and remains one of the most highly regarded novels of all time.

For many years, filmmakers struggled to bring Frank Herbert's novel to the big screen. At one point, even Ridley Scott was involved in such an effort (without success). In 1984, David Lynch delivered his own vision of the story to theaters with very mixed results. Recently, a TV producer named Richard Rubinstein discovered that the TV miniseries rights to the Herbert novel were still available. He interested a TV director friend (John Harrison) in the project, and the two acquired the rights with the help of the SciFi Channel. Harrison set out to adapt the novel as a miniseries and to recreate the novel's three-act structure – each act would be told in a separate night of the miniseries. Once the ball got rolling, the production attracted a terrific pool of talent, including actors William Hurt (*The Big Chill*) and Giancarlo Giannini (*Hannibal*), and cinematographer Vittorio Stararo (*Apocalypse Now*). The result is an epic and lavish, 266-minute production of *Dune* that finally does justice to Frank Herbert's original vision. When shown on the SciFi Channel, it was well received and garnered the network's highest ratings to date.

And that was just the original broadcast version. The new director's cut found on this 3-disc DVD set has been expanded by some 30 minutes, with roughly 10 minutes added to each part of the series. This new, longer *Frank Herbert's Dune* is a richer, more layered experience. Certain characters are more fleshed out in the new cut, and the battle scenes are more intense. In addition, more adult scenes and themes have also been added to the story—including numerous (but brief) instances of violence and nudity—which give the overall story a more unapologetically mature, edgy feel. You wouldn't have thought this miniseries could get much better, but this new cut is truly outstanding.

Artisan's special edition DVD release of *Frank Herbert's Dune* features fully remastered anamorphic widescreen video. It looks amazing—this is *definitely* the best way to view this miniseries. Each episode is presented on its own disc in this set, which means that the video bit rates have been maxed out. This video features rich detail, incredibly vibrant color, and satisfyingly deep contrast. Some of the effects shots—okay, most of the effects shots—have a very "digital" look to them, along with some aliasing and other artifacts resulting from budget-grade CGI work. This minor complaint aside, the video here is very pleasing to watch. And the miniseries' extremely lavish and colorful production design really pops off the screen.

This release also features your choice of remastered Dolby Digital 5.1 or DTS 5.1 sound. The DTS track is a slightly preferred option, featuring a smoother, wider soundstage, greater clarity (particularly when conveying ambiance), very active surrounds, and rich low-frequency reinforcement. The Dolby Digital track sounds a little crisper and more directional in its use of surround effects. Both tracks are excellent in their own right —whichever option you choose, you should be plenty happy.

The extras on this DVD release are split over the three discs in the set, but the best of them (and the one constant on all the discs) is the full-length audio commentary by writer/director John Harrison. His commentary is thoughtful and considered, covering such topics as the adaptation of the story, the character development and themes, the production, the special effects, and even some of his thoughts on filmmaking and science fiction in general. Harrison is joined on the track by various production personnel—a different group for each episode—making the commentary an entertaining listen.

A behind-the-scenes documentary, *The Lure of Spice*, is also included on the DVD, along with production notes, cast and crew bios, a gallery of artwork, and cinematographer Vittorio Storaro's *Frank Herbert's Dune: A Cinematographic Treatment* text essay. The gallery artwork covers the visual effects, storyboards, and character and costume sketches. You even get a sneak peek at design art for the sequel miniseries, *Children of Dune*, which is also available on DVD from Artisan.

The DVD also includes a number of featurettes, among them an interview with Harrison (on the film and his work); an interview with composer Graeme Revell (on the score); an interview with cinematographer Vittorio Storaro (on his philosophy/visual approach); an interview with *Dune Encyclopedia* author Willis McNelley (on Frank Herbert and the novel); a roundtable discussion with Harrison, Harlan Ellison, and Ray Kurzweil (among others, talking about science fiction); and interviews with religious scholars (on the meaning of "messiah"). They're all decent in terms of the material they present, but the production quality is pretty bad compared to most other DVD featurettes we've seen.

That complaint aside, if this DVD isn't quite a home run, it definitely satisfies. Despite being a TV miniseries rather than a feature film, *Frank Herbert's Dune* is a first-rate adaptation of the original novel. Better still, the new cut of the miniseries on this DVD is absolutely outstanding. It's not to be missed.

2001: A Space Odyssey—Stanley Kubrick Collection

Warner Bros.

There are a very small handful of landmark science fiction films that can truly be said to have influenced almost every film that followed. Among these are such classics as *Metropolis*, *Forbidden Planet*, *Star Wars*, and *Blade Runner*. But *2001: A Space Odyssey* stands alone in these ranks, as that rarest and most amazing of achievements in science-fiction: a work of unparalleled vision, grounded firmly in the realm of science, yet presented with breath-taking cinematic style and artistry.

To call *2001* high concept is a major understatement. The basic plot is as follows: Millions of years ago, at the very Dawn of Man, the appearance of a mysterious, black monolith inspires a small band of primitive ape-like humanoids to begin using tools, thus triggering the evolution of modern man. Suddenly, it's the year 2001. Humans, by now, space travelers-a-go-go, have just discovered a duplicate monolith that's been buried under the surface of the Moon for four million years. Little can be learned about it, except that it was clearly placed there by an extraterrestrial intelligence. Then the monolith sends a single radio signal towards Jupiter. Within months, the gigantic spaceship *Discovery* has been dispatched on a top-secret mission to determine who (or what) may have received the signal. What follows is an astounding, even metaphysical series of events — perhaps nothing short of the next step in human evolution. This film will definitely keep you thinking for a while.

2001 is a signature piece by director Stanley Kubrick. As you may know, it's based on a short story by writer Arthur C. Clarke. One of the goals these two men set for themselves when making *2001* was to accurately portray, for the first time on film, what it would really be like to travel in space. Keep in mind that at the time *2001* premiered, mankind had only just begun to travel in space and had yet to reach the moon. The fact that the film *still* holds up amazingly well today is an impressive testament to the efforts of Kubrick, Clarke, and an Academy Award-winning visual effects team led by Douglas Trumbull. In fact, the only things that really date this film are the scenes with Dr. Floyd (William Sylvester). He and his fellow scientists are all stiff, Ward Cleaver look-alikes. But *Discovery* astronauts Frank Poole and David Bowman (Gary Lockwood and Keir Dullea) seem very contemporary. One wouldn't be surprised to find either of them on a present-day NASA space shuttle crew.

2001 is a visual feast and features an impressive classical soundtrack as well. Johann Strauss' "The Blue Danube" is widely recognized, in no small part due to its use here. Richard Strauss' "Also Sprach Zarathustra" has become virtually synonymous with this film. *2001* also boasts the most infamous and paranoid computer in all of science fiction . . . HAL 9000. HAL presented the public with perhaps the first accurate representation of what artificial intelligence may look and act like. HAL is still decades ahead of his time, even by today's standards, but the potential of legitimately sentient computers is real . . . and, based on HAL at least, somewhat disturbing.

2001 has been released on DVD in a few different versions. The best of them is Warner's remastered disc, so that's what we'll focus on here. The film on this disc has been given beautiful anamorphic widescreen treatment, and the quality is nothing short of astonishing. The film simply looks amazing, in all its 2.20:1 widescreen glory. Remastered from a brand-new print taken from the original negative and color-corrected to Kubrick's own standards, the video on this DVD features rock-solid color, terrific contrast, and crisp detail. There's absolutely no digital artifacting or edge enhancement to be seen, and shadow delineation is excellent. Best of all, the image has been digitally cleansed to remove dirt, scratches, and other unwanted print artifacts. There's not a speck, pop, or blemish to be seen anywhere.

The audio is presented in newly remastered Dolby Digital 5.1, and once again it's terrific. There's tremendous dynamic range in the mix, such that the most quiet passages are almost reverent, while the thundering drums of "Also Sprach Zarathustra" will crash over you like a sonic tidal wave. There's also a great sense of atmosphere created in the mix with active use of the rear channels — note the quiet wildlife sounds in the Dawn of Man sequence, and the ever-present machine noise aboard the *Discovery* on its way to Jupiter.

Unfortunately, the extras include a theatrical trailer for the film . . . and that's it. This is as obvious a candidate for a special edition re-release as they come (and we've heard through the grapevine that Warner *is* working on one). Still, *2001: A Space Odyssey* is an amazing film, and this remastered DVD is absolutely reference grade in terms of picture and sound. This is a good disc to amaze your friends with in your home theater.

Fritz Lang's Metropolis: Restored Authorized Edition

Kino Video

Originally released in 1927, *Metropolis* tells the story of a utopian city of the future, designed and guided by the genius of its founder, Joh Fredersen. The children of *Metropolis* live in idyllic splendor . . . at least the children of the city's elite class. But deep underground, the workers of *Metropolis* toil endlessly to keep the city running smoothly for those above.

Fredersen's own son, Freder, never gives the plight of the working class a moment's thought . . . until he meets Maria one day in the Eternal Gardens. Freder follows Maria deep into the bowels of the city and discovers a world of suffering that he never dreamed existed. When he confronts his father with what he's seen, he's quickly dismissed. And so Freder journeys back into the underbelly of the great city, actually switching places with one of the workers in order to better understand the kind of lives they lead.

Freder once again meets Maria and soon learns that she's trying to keep the disenfranchised workers hopeful that one day a "mediator" will come to champion their cause with the elite class. But when Joh Fredersen discovers what Maria is up to, he has a scientist named Rotwang replace her with a robot duplicate. Fredersen's goal is to ferment the workers into acts of civil disobedience so that he can clamp down on them once and for all. But he fails to account for the determination of his son to see that justice is done. And it seems that Rotwang has set his own sinister plan in motion.

Fritz Lang's epic tale is not just a fevered and dizzying vision of the future—it also represents a landmark moment in the history of both the German and world cinema. It can truly be said that *Metropolis* was the first great science fiction film ever made, and its mark is seen in nearly every film of the genre that came after it. Heavily influenced by the German expressionist movement, and benefiting from an economic climate that encouraged an explosion of German film production, *Metropolis* featured impressive use of state-of-the-art special effects and innovative cinematography that were highly unusual for the period. The reaction, from both audiences and critics at the time, was at once enthusiastic and extreme. Even by today's standards, few films have inspired such an extensive degree of commentary and analysis.

This version is the most complete *Metropolis* ever seen since the film's debut in 1927. Taking advantage of the latest in digital restoration technology, and employing a massive archeological effort to understand the construction of the original version of the film, *Metropolis* was meticulously restored in 2000/2001 from the best available elements. Unfortunately, some scenes have been lost to time—as much as 20 percent of the original version. For this release, title cards have been added to describe missing scenes, and small black sections of film have been edited in to designate individual missing shots. The quality of the image presented on this DVD is impressive. Presented in its original full-frame aspect ratio, the black and white image is faithfully reproduced. The picture is remarkably free of dirt, dust, and other print damage, which has been removed as much as possible without distorting the original image. The image quality does vary somewhat depending on the condition of the available source material for each scene, but the quality has been generally brought into line whenever possible. You certainly can't call this reference quality, but it's as good as this film has ever looked since 1927, so it's very hard to complain about what defects are visible.

Every bit as important to the emotional impact of the film's visuals is the accompanying musical score by Gottfried Huppertz. We can only imagine how incredible it must have been to experience the film projected with a live orchestra present to render the original music. This DVD represents the first time that the film has been available on any home video format set to the original Huppertz score. Using notations taken directly from Huppertz's manuscripts, as well as the film's original script, the restoration team has been able to reconstruct the timing and musical tempos as closely as possible to the way they were originally intended. Performed by the Rundfunk-Sinfonteorchester Saarbrücken and recorded in full Dolby Digital 5.1, the resulting track is a supremely welcome addition to the DVD. The music adds tremendous impact to the viewing experience.

Arguably the best of the extras available on this DVD is the 43-minute documentary on the history of the film and its production. Not only are we given a great deal of information about the film itself, historian Enno Patalas also places the film in the context of its time period and its place in both art and cinema history. There are excerpts from vintage interviews with those involved in the production, behind-the-scenes photographs, and more. The documentary feels somewhat stilted in its construction and presentation, but it works if you hang with it. Unfortunately, the audio commentary, also by Patalas, doesn't fare as well. Patalas's approach to the track is to describe the emotional and psychological impact of what we're seeing on-screen, scene by scene, as we watch and listen. While much of the information he offers is fascinating, the presentation is a little too theatrical and is very distancing. There are better approaches to audio commentary tracks, and we really wish this disc had taken one of them.

Other extras here include a featurette on the film's restoration, multiple galleries of production photographs, design sketches and poster artwork, cast and crew biographies, and several pages of facts about the film. There's also an excellent insert booklet, which is packed with liner notes on the history of the film and the elaborate restoration process, written in exacting detail by Martin Koerber, who supervised the process.

There can be no doubt that this is the definitive presentation of this film available on any format, both in terms of video and audio quality and the "completeness" of the film itself. For that reason alone, this disc is a must-have for any serious student of cinema. *Metropolis* is a wondrous marvel, even with more than 20 percent of the original cut lost to the ravages of time. And it's a film experience that enthralls more with each new viewing.

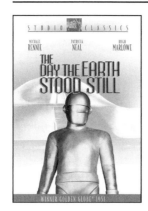

The Day the Earth Stood Still: Studio Classics

Twentieth Century Fox

"If anything should happen to me, you must go to Gort. You must say these words: Klaatu barada nikto."

In 1951, director Robert Wise gave the world a classic science fiction film that would also come to define a socio-political era. The film poses a simple question: "What would happen if a spaceship from another world set down in Washington, D.C.?"

From out of this particular spaceship steps a human-looking and well-mannered alien, Klaatu (Michael Rennie), and his badass robot Gort. He greets the crowd that's begun surrounding the ship and, as he's about to take out a gift for the world, he's instantly shot by a nervous military. So much for diplomacy. While recuperating in a hospital under heavy guard, Klaatu explains that he has a very important message for the people of the earth and asks to have a special meeting with the leader of every country. Klaatu is told that his request is impossible to grant, because quite simply, the people of the earth just don't get along. They would much rather point weapons of mass destruction at each other than listen to some well-spoken alien. So Klaatu deals with that reality the best way he knows how — he sneaks out to see the situation for himself. Adopting the name Carpenter, Klaatu hooks up with a single mother (played by Patricia Neal), her smart kid, and the only man willing to listen to him, a pre-hippy era scientist. All play very important roles in showing Klaatu that there's still hope for we earthlings — that we're worthy of our world.

The Day the Earth Stood Still makes you think. It's a richly layered film, with religious, social, and political allegories woven into the story. Sure, the ending is a bit heavy handed and the effects haven't aged all that well, but this is still a classic. Without it, *Evil Dead*'s Necronomicon wouldn't have a safety word. Say it with us, "Klaatu barada nikto." That phrase appears three times in this book. Let's see J.K. Rowling do that!

Presented in the original 1.33:1 aspect ratio, *The Day the Earth Stood Still* looks incredible. Fox has really outdone itself with this transfer. The print is restored and looks almost flawless — the beautiful black and white texture is stunning. If only every classic film could look like this on DVD.

The audio is in both the original mono as well as a brand-new Dolby Digital stereo mix. Both sound wonderful, with no flaws, and Bernard Herrmann's score really grabs you. You really couldn't ask for a better presentation of this film.

Part of Fox's Studio Classics line, *The Day the Earth Stood Still* is filled with wonderful extras. First off, on Side One, we're treated to an informative and easy-going commentary by director Robert Wise, joined by fellow *Star Trek* movie alum Nicholas Meyer (who directed the best *Trek* flick, *Star Trek II: The Wrath of Khan*). You'll also find a 1951 MovieTone newsreel, the film's trailer, and a THX Optimizer.

Turn the disc over to Side Two, and you'll find a whopping 90-minute documentary about the making of the film. You'll learn about different versions of the script, alternate casting choices, and the reactions of early test audiences. There's also interview after interview with the cast and crew. Filmed in 1995, this is a superb historical overview. Also on Side Two is a *Restoration Comparison* showing the work put into preserving this film, six still galleries (including production photos, blueprints for the alien ship, poster art, model shots, images of Gort, and more), the full shooting script for the film, and even a couple of trailers.

The Day the Earth Stood Still is an awesome sci-fi flick, and this is an equally awesome DVD. Plus, you get Gort—only one of the greatest robots in cinema history. It'd be a shame not to have him in your DVD library, wouldn't it?

If You Like These Films . . .

. . . also worth adding to your collection are *The Iron Giant*, Ridley Scott's *Blade Runner*, John Carpenter's *The Thing*, and Howard Hawks's original *The Thing From Another World*. And don't forget Twentieth Century Fox's nine-disc *Alien Quadrilogy* (coming soon to a video store near you).

Swords and Six-Shooters on DVD

Here at *The Digital Bits*, there's only one thing we like more than DVD, and that's cowboys, samurai, and tales of epic revenge on DVD. Give us swords or six-shooters on disc and plop us in front of a TV and we're happy campers. That said, we have to admit, this was a very hard list to make. If it were up to us, the whole book would be reviews of these types of films. But since we had to limit ourselves, we tried to pick standouts from all the different subgenres: western, medieval, sword and sandal, and samurai. Remember though, if there's a cowboy on the cover of the DVD, or the words *Lone Wolf and Cub*, it's probably worth spending your money on.

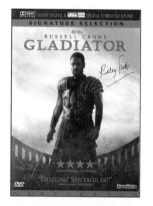

Gladiator: Signature Collection

DreamWorks

There is a certain kind of film that we are, almost invariably, unable to resist—the epic story. It usually starts with a simple, noble character of humble origins that suddenly loses everything that is most dear to him or her. Then the character finds him or herself thrust into the center of an epic conflict or struggle—some kind of dire situation whose outcome will have consequences far greater than they can possibly imagine. For such a character, the choice is simple—rise to meet the challenge or fall in defeat. Heroes are born in this way. So too are the best movies. Many of our favorite films share this common thread—*Star Wars*, *Lawrence of Arabia*, *Braveheart* . . . the list goes on. There's just no better drama than a good epic struggle.

So here's an epic tale for you—a noble Roman army general named Maximus has just won his greatest victory against the barbarian hordes in the North, ensuring the continued security of Rome for ages to come. The battle has taken years to fight, and all Maximus wants to do is return home to farm his land and be with his wife and son. But his dying friend (and Emperor) has one last task for Maximus. Corruption is running rampant in the Empire and the Emperor's son, Commodus, is not an honorable man. Fearing the worst for Rome upon his death, the Emperor asks Maximus to succeed him, and to eventually return Rome to its people—to make it a democracy again. Meanwhile, Commodus learns of his father's intentions and, feeling betrayed, he kills his father and takes control before the Emperor's plans for Maximus were widely known. Commodus then demands Maximus's pledge of loyalty and, when the general refuses, has Maximus stripped of his command and taken away to be executed. Maximus escapes, but Commodus' wrath falls heavy upon his family before he can save them. With nothing left to lose, Maximus soon finds himself sold into slavery as a gladiator. But in this seemingly desperate situation, Maximus finds an opportunity for vengeance. For it seems that the new Emperor is holding gladiatorial games in the Coliseum in Rome to engender the love of his subjects. And the very best gladiators are given an audience with the Emperor himself.

How could a story like that not grab you? Did we mention the film stars the always terrific Russell Crowe as Maximus? And this particular gladiator movie is directed by none other than Ridley Scott, the cinematic genius behind such stylish and influential films as *Blade Runner* and *Alien*. How frickin' cool is that? Very frickin' cool—that's how cool.

Director Ridley Scott's amazing attention to detail and his command of cinematic style and process has never served him better. Given time, *Gladiator* will come to rank among his best works in the eyes of fans. The production design here is superb—the glory of Rome at the height of its power simply comes to life onscreen. The cast supporting Crowe is terrific and includes the likes of Joaquin Phoenix as Commodus and Richard Harris as the Emperor, as well as Oliver Reed (who died during this production), Derek Jacobi, Djimon Hounsou, and Connie Nielsen. The script is well written and is easy to lose yourself in. And the gladiator action is intense and unrelenting, while managing to leave a lot for your own imagination to fill in. As violent as this film is, you don't see a lot of actual blood. This is on par with *Braveheart*, if even a little less bloody.

That's not to say that this is a perfect movie. First of all, the opening battle scene suffers from a little too much Spielbergitis (that heavily processed in post-production look), much like the opening D-Day scenes of *Saving Private Ryan*. Given that this is a DreamWorks film, that's no coincidence. Whereas in *Braveheart*, you get a real sense of geography and flow to the battle scenes, *Gladiator* gets a little too abstract for its own good at times. The film also doesn't humanize a couple of its characters enough. The Emperor's children, Commodus and Lucilla, suffer the most. Commodus isn't given enough development to really make him an effective villain (just *wait* until you see the deleted scenes on this disc — hold that thought) and Lucilla isn't given enough humanity to make you empathize with her as much as the film would like you to. But those are small criticisms. Taken on the whole, *Gladiator* is a passionate and gripping film. And surprise — the filmmakers did something right, 'cause this prototypical "guy" flick is a big favorite among many women too.

Starting with Disc One, the anamorphic widescreen video is terrific. The contrast is excellent, with good, deep blacks and sufficient shadow detail. The color scheme of the film is muted, but the transfer renders it accurately at all times. Rome has definitely never looked so glorious on-screen. The print is nicely clean and you'll see little in the way of dust and dirt. The sound is even better. You get your choice of Dolby Digital 5.1 (EX compatible) or DTS 6.1 ES, as well as a Dolby 2.0 Surround mix. The Dolby Digital 5.1 audio is surprisingly atmospheric, with very active rear channels, a nicely wide forward soundstage, and plenty of low frequency. Dialogue is clear and well centered, and Hans Zimmer's aggressive and haunting score is beautifully presented in the mix. As expected, the DTS track is even more expansive and natural sounding, squeezing out every last ounce of subtlety for you to enjoy. Just listen to the wisp of arrows in the opening sequence, or the metallic ringing of swords being sharpened in Proximo's dungeon. Whichever track you choose, this is great film audio on DVD. Also available on Disc One is a first-rate audio commentary track with director Ridley Scott, director of photography John Mathison, and editor Pietro Scalia. It's indexed by topic and you'll be fascinated listening as the trio discusses the psychology of character, historical accuracy, and the "logic" of the fictional world. Scott is one of the best sculptors of fictional film worlds around, and he proves that here.

So that's Disc One. Disc Two will make you happy as well. Let's start with the best part — included here are some 11 deleted scenes, all with optional Ridley Scott commentary, and a very cool, 5-minute *Treasure Chest* video, featuring even more unused footage cut to music specifically for this DVD by Scalia. Did we mention the music is a piece of Zimmer's score that doesn't appear on the soundtrack CD? These deleted scenes amount to some 25 minutes of footage in all, and a few are really terrific. There are two that involve the character of Commodus that we absolutely can't believe were cut. They really round out his character and would have only added three or our minutes to the film's running time. One in particular ("The Execution") is completely frickin' cool. It's an amazing scene.

Next, we get a trio of documentaries. These are no fluff pieces. The first is *HBO First Look: The Making of Gladiator* and it runs about 25 minutes. It's actually one of the best *First Looks* we've seen and includes great interview clips with the cast and crew. You also get a 20-minute piece called *Scoring Gladiator*, where we get a look into the creative mind of the film's composer, Hans Zimmer. He talks about searching for the themes that will convey the emotions in the film and his fear that he'll never be able to compose another great score. It's fascinating. So too is the third piece, a 50-minute Learning Channel documentary, called *Gladiator Games: Roman Blood Sport*. The producers interview historians and experts on the real-life gladiators of old, visit the ruins of real arenas, and were even allowed to shoot on the *Gladiator* sets, using props from the film, to create realistic reenactments of battle. Once you start watching, you won't be able to stop. Also on Disc Two are extensive storyboards for eight major sequences in the film and four deleted or unused sequences. You get a stills gallery with dozens of behind-the-scenes images from the production. There are several pages of production notes and detailed cast and crew bios. The film's teaser trailer and theatrical trailer are here, along with four rarely seen TV spots for the film. You also get a nifty little text essay, called *My Gladiator Journal*, written by young Spencer Treat Clark (who played Lucius). It's basically a day-by-day diary of his experiences on the production (illustrated by photos), starting from the day he got off the plane for filming until the final wrap party. It's surprisingly interesting and gives you a perspective on the making of an epic film that you usually don't have access to. And all of these materials are accessed via tasteful and classy animated menu screens, using imagery and music from the film. One last note—there are three Easter Eggs hidden in the menus on Disc Two. One is a credits page and the other two are brief (but cool) video clips.

Gladiator is a great film, and you can thoroughly enjoy it at home in a great two-disc DVD edition. The video and audio quality should satisfy hard-core home theater buffs and film lovers will appreciate the variety and comprehensiveness of the extras. Can you imagine the pitch session for this film? "We wanna make this kick-ass gladiator film with Russell Crowe and Richard Harris. We're gonna have lots of combat and we're gonna use CGI to bring the Roman Coliseum back to life. Oh . . . and Ridley Scott's gonna direct." Any studio executive that doesn't wet their pants upon hearing those words should be fired. But that's just our two cents. Enjoy!

Tombstone: The Director's Cut— Vista Series

Hollywood Pictures (Buena Vista)

When you think of the great "guy" flicks of recent years, several titles come quickly to mind. There's Michael Mann's *Heat* and John Frankenheimer's *Ronin*. There's Luc Besson's *Léon: The Professional* and Ridley Scott's *Gladiator*. But one of our absolute favorites is a campy, Hollywood star-powered, spaghetti western wanna-be . . . George P. Cosmatos' *Tombstone*.

The story, which you may be surprised to learn is a very accurate depiction of the real historical events, follows ex-lawman Wyatt Earp (Kurt Russell) and his brothers Virgil (Sam Elliott) and Morgan (Bill Paxton) as they attempt to start a new life and make their fortunes in the booming but lawless mining town of Tombstone, Arizona. Not long after the Brothers Earp and their wives arrive and settle in, their reputation catches up with them. Local law enforcement wants Wyatt to carry a badge and help bring order to Tombstone. And the local outlaws, the Cowboys (led by Curly Bill Brocious, gun-slinging ace Johnny Ringo, and the Brothers Clanton), want to make sure he doesn't. Wyatt and his brothers do their best not to get involved, but conscience and events get the better of them, and it's not long before they're on a collision course with the Cowboys for control of Tombstone. Fortunately, that infamous gentleman killer Doc Holiday (Val Kilmer) is on their side. As Wyatt himself says, "The bad guys are less apt to get nervy with Doc on the street howitzer." Throw in Dana Delany as a free-spirited actress who vies for Wyatt's affections and a whole host of great cameo appearances (from Billy Bob Thornton to Charlton Heston), and you've got a barn burner from beginning to end.

Tombstone absolutely rocks. Kurt Russell was simply born to play Wyatt Earp, as he infuses the role with the perfect mixture of testosterone gusto and subtle humanity. Val Kilmer's devilishly poetic Doc Holiday is the perfect friend and foil to Russell's straight man. And the various Cowboys are played with delicious menace by the likes of Powers Boothe, Michael Biehn, Stephen Lang, and others. In fact, there's not a badly cast part in this film. Everyone delivers the goods. And the dialogue! *Tombstone* boasts some of the best period dialogue you've ever heard in a western. Great scenes, big and small, abound here. We could spend this entire review listing the film's many "enchanted moments." But among these, Billy Bob Thornton's confrontation with Wyatt (as Wyatt "acquires" a quarter interest in the game at the Oriental) is a true gem. This is good stuff!

You should also know that this DVD features a new director's cut of the film, which reinstates approximately four minutes of previously deleted footage. There's an additional scene between Wyatt and Mattie (dealing with her opium addiction), an interchange between Doc and Kate (in which he leaves her to back Wyatt's effort to rout the Cowboys), a moment of Doc drunk in his hotel room, and a scene in which McMasters meets with the Cowboys and is betrayed. The inclusion of each adds additional weight to later moments in the film. Plus, it's just cool to have four more minutes of this fun film.

Tombstone was one of Buena Vista's very first live-action DVD releases, back when the studio decided to climb aboard the format bandwagon, and it didn't look very good. Thankfully, this new Vista Series Director's Cut is exactly what the doctor ordered. The new anamorphic widescreen video is a vast improvement over the old disc, mastered from a digital, high-definition transfer. It's not perfect, but if you've had to suffer with the old disc, you'll be thrilled with this. There's a hair too much edge enhancement and the print is occasionally a little soft. But it's far smoother and more natural looking than the previous disc, with a greater sense of depth and much deeper blacks. The colors are lush and accurate without bleed, and this print is definitely cleaner. You'll barely notice film grain and there's hardly a spec of dust or dirt to be seen.

While the audio on the original DVD was about as good as you could expect of a Dolby Digital 2.0 Surround track, this new disc blows it away with not just newly remixed Dolby Digital 5.1 sound, but DTS 5.1 sound as well. Both tracks provide a nicely wide soundstage, with good dialogue imaging, solid low-frequency support, and adequate ambiance from the surround speakers. This film, surprisingly, isn't highly active surround-wise, even given this new mix. You'd expect the bullets to zip around your head during the O.K. Corral gunfight, but they're mostly up front. But listen to the panning during the title shot, as the Cowboys race right by you from the distance into the left-rear corner of your home theater—very nice. And Bruce Broughton's thunderous score is well blended into the mix. The DTS track, as one would expect, provides a slightly greater measure of clarity, along with smoother, more natural panning. It's the preferable option, but the Dolby Digital track still delivers the goods plenty well.

The new DVD, being a two-disc set, also delivers some nice extras. To start with, you get a full-length audio commentary track by director George P. Cosmatos. The guy isn't overly charismatic, talking as he does in his appropriately gruff, heavily Italian-accented, baritone voice. But he delivers some good anecdotes about the production, his cast, and the tremendous effort that was made to stay true to historical detail during the making of this film. Disc Two provides a trio of short behind-the-scenes featurettes, which total about 27 minutes in length (when you select the "play all" option). These are a bit fluffy but manage to be surprisingly substantive. They're filled with interview excerpts with almost all the major cast members (even Charlton Heston, who has only a small role here) and provide a look at the real historical events the film is based upon. There's an interactive timeline of the real events, in which (when you select each date) an appropriately cheesy "old timer" voice reads the onscreen text to you. This would have been cool, except that it's isn't nearly detailed enough. And we couldn't find anything in here indicating when the actual O.K. Corral gunfight took place. That's okay, however, because one of the best extras on the second disc is a high-resolution scan of all four pages of the actual issue of *The Tombstone Epitaph* in which the report of the gunfight appeared. You navigate around the paper (using your remote) looking at a section of each page at a time. The scan is good enough that you can actually read the real eyewitness accounts of the gunfight. There's even testimony by the real Sheriff Behan, along with a map of the starting positions of the gunfighters when the shooting began. It's very, very cool and a nice touch.

Also on Disc Two is a video of the director's own storyboards for the gunfight scene (set to music from the film), the film's teaser and theatrical trailers complete with music by Peter Gabriel (sadly, neither is anamorphic), TV spots for the film, and a DVD-ROM interactive card game called *Faro at the Oriental*. Finally, there's an easy-to-find Easter Egg on the second disc that allows you to access a still gallery of poster art and production design sketches. All in all, it's a decent little package of extras.

The *Tombstone: Director's Cut—Vista Series* DVD is a must own. *Tombstone* is a great western and a modern classic, and with this DVD you can finally watch it in the quality it deserves. If you like the film even half as much as we do, our advice is to get this disc the minute you're done reading this review. Give it a chance, and you'll be glad you did.

Braveheart

Paramount

"I shall tell you of William Wallace. Historians from England will say I am a liar, but history is written by those who have hanged heroes . . ."

And so begins Mel Gibson's epic *Braveheart*, a story of love, tragedy, revenge, and bravery, as the sons of thirteenth-century Scotland rally against English tyranny. William Wallace (Gibson) is born a common highlander, the son of a lowly farmer and patriot, but he soon finds himself at the very eye of the storm. Scotland suffers under the rule of the ruthless English king, Edward the Longshanks, and any attempt to resist his cruelty is met with the harshest punishment. When young Wallace's father and older brother are killed one day in a failed bid for freedom, Wallace is taken away by his Uncle Argyle, who raises and educates him. Years later, Wallace returns home seeking a peaceful life as a farmer. He finds his childhood sweetheart, Murron (Catherine McCormack), and secretly marries her, hoping to start a family. But in so doing, he's already broken the law. Longshanks, in a bid to strengthen his control in Scotland, has given his lords there the right of "Prima Nocta"—the right to sleep with any new bride on their wedding night. Wallace's resistance soon becomes apparent, and as punishment, the local English lord takes Murron's life. Overcome with grief and rage, Wallace leads the locals in a revolt, wiping out the English presence in his village entirely. When the English retaliate, the situation quickly escalates, and Wallace soon finds himself the unlikely leader of a massive rebel uprising determined to free Scotland from the English forever or die trying.

Braveheart became the Best Picture of 1995, sweeping the Academy Awards that year with a total of 10 nominations and 5 Oscar wins. And it certainly deserved them. *Braveheart*'s epic story has drawn comparisons to David Lean's *Lawrence of Arabia*, but it lacks the latter film's visual mastery and polish. *Braveheart* is by no means the best film ever made—not even close. But what it does right, it does very, very well. The storytelling is emotionally honest, exciting, and even funny at times. It draws you in—every loss of these characters becomes your own, every victory a personal one. The film also holds some of the best large-scale battle scenes you'll ever see on the big screen, making other such attempts pale in comparison. But Gibson's talent as a director really shows itself here—the film is unquestionably violent, but you never actually see as much violence as you think you do. Most of the carnage is suggested by quick cuts and skillful editing. The film's story is based on semi-real historical events, the screenplay having been written by the real Wallace's descendent, Randall Wallace. The supporting cast simply shines, including Patrick McGoohan as the deliciously evil Longshanks, Brendan Gleeson as Wallace's childhood friend, Hamish, and Sophie Marceau as the Princess of Wales. Tie it all together with a stirring score by composer James Horner—arguably his best work to date—and you've got a great film experience.

The anamorphic widescreen video on this disc looks very good, if not quite great. This is a new transfer of the film, but it would have been nice if Paramount had selected a better print for the task. There's a little too much dust and dirt visible. In addition, the contrast seems a little too high. The black levels and detail are wonderful, but the brightest picture areas are a little too hot. The color timing on the DVD transfer appears rather cold and harsh looking, with muted colors—no doubt a stylistic choice. You'll notice this right from the opening shots of the Scottish highlands. It certainly works for the film, and most DVD fans will be quite happy with this video.

The Dolby Digital 5.1 audio features a nicely wide front soundstage. Dialogue presentation is clean and full, and James Horner's score sounds rich and is very well presented in the mix. There is occasional activity in the rear channels, particularly during the battle scenes, although not as much as you might expect. The bass, however, is thunderous. When the English heavy cavalry charges the fields of Stirling, you'll feel every hoof strike the ground.

The extras are where you'll be most impressed with this DVD, not for their quantity but their quality. To start with, there's an excellent 27-minute featurette on the making of the film entitled, *Mel Gibson's Braveheart: A Filmmaker's Passion*. It may seem that 27 minutes is too short for a film like this, but the featurette tells you everything you need to know and includes great interviews with Gibson, Wallace, the producers, and other cast members. There are some very funny moments—look for a shot of Gibson, in costume on the set, consulting a battered, leather-bound tome called *A Beginner's Guide to Directing the Epic*, which looks like it was printed in the thirteenth century. Very funny. Gibson's passion for this film is obvious, even more so in the filmmaker's commentary on the disc. His wit is engaging and while there are a few gaps in the track, you'll have a blast listening to this. It's easy to imagine that you're sitting in a dark theater, having a beer with the guy, as he tells you about his experience making the film in quiet little asides. And really, if you could have a beer and watch a great flick with any guy, Gibson's the perfect choice. Finally, there are two interesting theatrical trailers on the disc, both non-anamorphic but worth looking at.

Braveheart is a great flick, and one which easily falls into the ranks of the very best Guy Films of all time. It's definitely not for the kiddies, but for what it is, this film works perfectly.

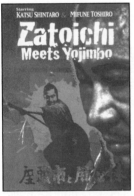

The Blind Swordsman Series on DVD

Home Vision and AnimEigo

Ichi may be blind, but he can see better than any of us. He also happens to be one of the best samurai out there. He'll cut you six ways to Sunday quicker than you can unsheathe your sword. Cast from every village he's ever stepped foot in, because of his killing or his gambling, Ichi is a loner who wanders from place to place, helping those in need and gaining the respect of other code-bound ronin and yakuza that he meets along the way. In short, the *Zatoichi* series of samurai films utterly rocks.

The role was created by Shintaro Katsu, the great Japanese actor and producer, who played the character in some 26 feature films and over a hundred episodes of a popular TV series as well (that *has* to be some sort of record). Home Vision, sister of Criterion, is releasing 17 of the early *Zatoichi* films, with AnimEigo releasing 7 more of the later titles. All of the films from both companies look great and are definitely worth hunting down and adding to your collection.

Presented in anamorphic widescreen, the only quibble you can have is that there are some minor print issues here and there related to age. These DVDs are transferred from old Japanese master prints. But they really do look as good as you could reasonably expect. The audio is also solid. Sometimes the discs have a canned, mono sound, but that's mostly because they were never really dynamic soundtracks in the first

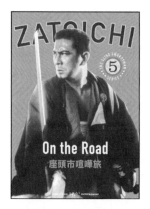

place. In terms of extras, all of these discs are pretty light. But you do get the occasional still gallery, trailer, and text essay, as well as sheets of collector's cards and poster reprints in the early Home Vision releases.

We're huge fans of these films at *The Digital Bits*, and we consider just having them on DVD a gift. We couldn't be more excited that they're available and we highly recommend them to you. *Zatoichi* films are a fun and seriously addictive guilty pleasure. We dare you to stop at just one.

The Samurai Trilogy

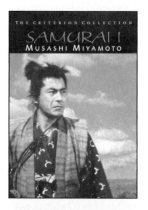

Samurai I: Musashi Miyamoto

Samurai II: Duel at Ichijoji Temple

Samurai III: Duel at Ganryu Island

Toho (The Criterion Collection)

There is nothing better than a samurai film. When we were kids, we'd talk all our friends out of being cops, robbers, cowboys, Indians, or superheroes, and into acting like samurai warriors. We'd get sticks and have sword fights, dress in full black, and run around the woods hunting each other down like dogs. All in all, our childhood was pretty good.

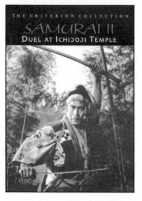

Many of us had our first experience with samurai films in *Star Wars*. You might not quite grasp just how much *Star Wars* is like a samurai film when you're a kid, but looking back on it, you have to admit that it's a fair summation — even more so with the new *Star Wars* movies. You realize this even more when you start to experience other real samurai films, such as the Akira Kurosawa classics *Seven Samurai*, *Yojimbo*, *The Hidden Fortress* (the film upon which *Star Wars* is loosely based), and *Sanjuro*. Once you watch those, you'll no doubt be hooked on the genre, and you just have to seek out the likes of

The Sleepy Eyes of Death, *Lone Wolf and Cub*, and the *Zatoichi* series. All of that, of course, will lead you to this set of films: *The Samurai Trilogy*.

Except for the fact that you have to "read" them, *The Samurai Trilogy* is very much like a series of American costume films. They are very watchable, very understandable, and very entertaining. The films are based on the book *Musashi Miyamoto*, which has been called the Japanese *Gone With the Wind* — a comparison that, although true, is based mostly on popularity and sheer scale. Musashi Miyamoto was a real historical person. He was a master swordsman, philosopher, and painter. He was well regarded both through his life and after his death. Even today, many business students follow his writings.

Do these films deliver? You betcha.

The Samurai Trilogy isn't from Kurosawa, but that's okay. The director was Hiroshi Inagaki, who went on direct two other samurai masterpieces, *Chushingura* (1962) and *Samurai Banners* (1969). The films follow Miyamoto's life, from a ballsy young kid to his early retirement, all the way battling those who stand against him. The first of the films starts with Miyamoto as a young man (then known as Takezo), who goes off to fight in the Japanese civil war, around the year 1600. He survives his first battle, but he and his friend are both hurt and tired, so they take refuge with a woman and her daughter. Things really start there for Takezo. You have to pay attention, because virtually all of the characters introduced in the first film will follow Takezo through the later films in various ways. To tell you too much more about the story would be a disservice to the films themselves — they really should be watched by all fans of big epic movie making. Suffice it to say, Takezo goes from headstrong young man to brute thug to enlightened Buddhist to an undefeated warrior . . . and eventually to content farmer. It takes three films and a little over five hours to tell his story. And what a story it is.

The cast of characters are all intriguing and well drawn. Even though we have never read the book these films were based on, the entire film plays out like a beautiful novel. It also helps that Takezo is played by

Toshiro Mifune, one of the greatest film actors *ever*. His performance is so good here—he goes through such a range of expression, it's a marvel to behold. Mifune was just so damn good at what he did. God bless him.

Of course, samurai films wouldn't be anything without sword fights, and there are some great ones in this series. The first film has plenty of battles, often with a great number of people involved in each. The second film opens with a bang, as Takezo battles a warrior armed with a chain and sickle, and it just gets better from there. The series ends with a battle on the beach featuring Takezo against his longtime fan and biggest rival, Sasaki Kojiro—a rivalry that grows out of the second film. The final battle, even if it's not as elaborate as the others, is breathtaking, mainly because it's the last one, and the angel of death hovers so close to our main character.

As they stand on DVD, the films look and sound wonderful. You should keep in mind that these are old movies from another land. The quality of older films differs from country to country. Too many pre-1950s films have been lost, thanks to . . . well, us Americans. As a rule, though, Japan was pretty good when it came to storage, so while the film stock here tends to be grainy and washed out in parts, overall it looks really good. The films are in color, which sometimes the color looks slightly off. That said, the films have been transferred and restored to the best standards possible today. *The Samurai Trilogy* looks as good as you can expect on DVD.

The audio is in the original Japanese mono and sounds fine on disc—no problems there. The extras are light, but you can't really get caught up with extras on a set of films like this. Just getting a trailer is a plus. It's nice to see how film trailers looked back then, and compared to early Hollywood trailers, these Japanese trailers seem much more like our modern-day variety. All in all, Criterion's given us a neat package with this trio of discs.

If you get the chance, take a road trip down the film history highway, and check out this set of films on DVD. If you like the samurai genre, or have never experienced it, this is a great place to start. Hopefully, the samurai genre will be rejuvenated after Tom Cruise appears in *The Last Samurai*. In any case, *The Samurai Trilogy* is a very important set of films, packed with enough thrills to make a modern-day director wish he'd stayed in film school—or behind the video store counter.

Unforgiven: Special Edition

Warner Bros.

"It's a helluva thing, killing a man. You take away all he has . . . and all he's gonna have."

This is a simple statement, and it means nothing really, but . . . Clint Eastwood was born to play William Munny. His grizzled face, slow delivery, and blank stare, it all feels so right as portrayed by Eastwood. You see, Munny was a bad man. He killed men, women, and children, and simply didn't care because he was an outlaw and a drunk. But then, Munny found love in a nice Mormon girl who saved him from himself and gave him a reason to change. She gave him two healthy children; he gave her a home and a pig farm. And then she died.

Staying true to his wife, Munny forges on with life, raising his kids and his pigs. But things get tight for ol' Munny, and when "the Kid" rides into his life with an offer of partnership, Munny has a hard time saying no.

The deal the Kid has is simple. It seems that a young prostitute was cut up by a john because she laughed at his . . . ahem, small member. When the sheriff, Little Bill (Gene Hackman, also in a role he was born to play) is light on the offender and his buddy, the prostitutes' friends take up a collection to put a bounty on their heads—a bounty big enough for a few takers, like the Kid. But Little Bill's a big man, and he's not about to let a bunch of bounty-hunting dogs come into his town, roughing up his people. So he puts a stop to it right quick. And that's where Munny and Little Bill cross paths.

Unforgiven has been on the list of DVDs badly in need of revisiting since it first came to us as a movie-only edition a few years back. Thankfully, Warner went back to it with a vengeance and has released a fantastic two-disc special edition. The film itself is presented in an utterly beautiful anamorphic widescreen (2.35:1) transfer. Eastwood's vista shots, his close-ups, his color scheme—all of it is wonderfully captured on this DVD. The sound has also been revitalized in a fantastic Digital 5.1 mix.

Running along with the film is a brand-new audio commentary track by renowned film critic and Clint Eastwood biographer Richard Schickel. Schickel goes into everything you'd want here, from the characters' motivations, to the subtext of what's really going on between them. It's a fascinating listen, worthy of your time. Also on Disc One are the theatrical trailer, production notes, cast and crew bios, and lists of awards the film won.

Disc Two goes a little more in-depth on the video side. Here you will find the newly produced documentary, *All on Accounta Pullin' a Trigger*, which features new interviews with Clint Eastwood, Morgan Freeman, Gene Hackman, and screenwriter David Webb Peoples. This is a nice look back on the film from the "10 years later" perspective, themed around the disc's anniversary angle, and everyone seems to acknowledge its importance with a bit of humility. Next is another documentary, *Eastwood & Co.: Making Unforgiven*, which was made in 1992 and focuses on the behind-the-scenes workings of the production. *Eastwood . . . A Star* covers some of this ground as well. The biggest and best extra is a TNT documentary, *Eastwood on Eastwood*, which looks at the whole of the actor's career. Produced, written, and directed by our commentary host, Richard Schickel, in 1997, this is a great look at both the filmmaker and the man. It's one of the better TNT filmmaker documentaries (right up there with their look at John Huston). Finally, rounding out Disc Two is an old episode of *Maverick* entitled *Duel at Sundown,* which is a fun addition to the set.

Unforgiven isn't the best western ever, but it *is* one of the best films starring cowboys in a western environment ever made. It says much about the genre and, honestly, it's a great finale to the cowboy role career of Clint Eastwood. If he never puts a cowboy hat on again, that would be all right by us, because he's done it better than anyone. *Unforgiven* is also one of the best meditations on violence ever made. It talks about the deadening of the soul, and the unnecessary purpose in the violent life of gunslingers. It says that these men had to take responsibility for their actions—that killing isn't easy and, ultimately, you can't walk away from it.

All in all, *Unforgiven* is just a great frickin' movie and is absolutely deserving of your attention on DVD.

If You Like These Films . . .

. . . you may also want check out AnimEigo's *Lone Wolf and Cub* series, Warner's *Excalibur* (a great flick but another bare-bones disc), Columbia TriStar's *Crouching Tiger, Hidden Dragon* and *Desperado*, Paramount's *Once Upon a Time in the West,* and MGM's *The Man With No Name* Trilogy.

The Classics on DVD

What makes a film great? We're not asking that question looking for the casual, critic's-blurb type of answer, tossed off with alarming regularity alongside such hyperbole as "Oscar-worthy," "superb," and "master-piece" these days. We're actually wondering aloud about what makes a film so great that, for example, for 50 years *Citizen Kane* has been widely considered one of (if not) *the* single greatest motion picture ever made. How does one even begin to approach watching a movie with such a reputation of brilliance?

For most people, it would seem the answer to that second question would be: with great reverence. But classic films should also be enjoyed as much as they're revered. We should be able to watch a classic film and have as much fun as we would watching a new release. The fact is, for many of these films, people are discovering them for the first time on DVD. And if they don't hold up here, how are they expected to last for all time?

We've picked a nice selection of classic films for you to check out, both old and new, that we feel *will* hold up for decades to come. More often than not, the quality of their treatment on DVD helps them to stand on higher ground.

Editor's Note: *Bits* writer Adam Jahnke contributed to this section.

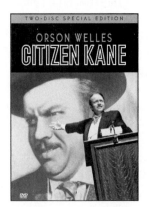

Citizen Kane: Special Edition

Warner Bros.

Every single person of the current generation (or younger) who's seen Orson Welles's *Citizen Kane*, didn't originally watch this film because they wanted to. They watched it in school, college usually, where it was discussed, analyzed, or held in the same esteem as Shakespeare or any other canonical literature. In other words, the joy of the movie is stripped away long before most people have seen a single frame. This may be one of the greatest losses in cinema history, because *Kane* is, above all else, a movie to be enjoyed. Yes, it's a dense and complex film that challenges the intellect and demands repeated viewings in order to be fully appreciated. But at least once, you owe it to yourself to forget all of that and just sit down to watch *Citizen Kane* like you would any other movie. Then and only then can you begin to discover what makes this film great.

Is *Citizen Kane* anyone's "favorite" movie? Probably not. Other movies have touched us deeper, have thrilled us more and have resonated with more personal significance than this one. *Citizen Kane* is like one of the jig-saw puzzles that Susan Alexander is forever assembling in Xanadu. You know what the picture is going to be when it's put together, but it's fasci-nating to see the infinite number of ways that picture can be constructed. There are some movies you watch until you've memorized every shot, every cut, and every line of dialogue. *Citizen Kane* seems to defy memo-rization. That is part of what makes it a great film. Every time you see it, your attention is drawn to a different aspect of the film, but you can't pre-dict or control where that will be. Welles somehow manages to say every-thing there is to say, and then some, in just under two hours. Many of today's filmmakers would do well to learn from this element of *Kane*, rather than trying to ape Gregg Toland's amazing cinematography or mimic the complex structure of the screenplay by Welles and Herman J. Mankiewicz.

Film buffs can take heart in the fact that Warner Bros. has released *Kane* in a stellar two-disc set, giving the film the treatment it deserves. First off, the transfer of the film itself is nothing short of revelatory. This cleaned-up print is without question the finest we have ever seen this film look. The blacks are gorgeously detailed, with shafts of light and plumes of smoke standing out in tactile relief. If anything, the movie looks too good. This can mainly be seen in portions of the *News on the March* seg-ment, which seem to be spruced up a little beyond what Welles may have intended. He wanted that part of the film to look older and in worse shape than the rest of the movie. With the cleaned-up picture, it's a little bit too obvious what is the real stock footage and what was shot for *Kane*. On the other hand, we don't envy the people whose job it was to decide what scratches were intentional and what were just the results of old age. On the audio side, Warner has wisely played it very safe, cleaning and remas-tering the sound in its original glorious mono. No Dolby Digital 5.1 remixes here, since the only reward Warner would get from that endeavor would be the loud and furious complaining of a legion of cinephiles. For what it is, this track is excellent, free of hiss or the tinny sound that often affects dialogue in films of this vintage.

Warner Bros. pulled out all the stops on extras for this release. The centerpiece *is The Battle Over Citizen Kane*, a feature-length documen-tary made for *The American Experience* on PBS that takes up the entirety of Disc Two. This is no ordinary making-of piece. *Battle* traces the lives of both Welles and the man who would become the inspiration for Kane, as well as Welles' primary real-life antagonist, William Randolph Hearst. The program alternates between Hearst and Welles on their way up, leading inevitably to their clash over *Kane*. This is a fascinating documentary, full of rare archival footage and interviews with those who knew these two brilliant, powerful men. Rightly nominated for a Best Documentary Academy Award, *The Battle Over Citizen Kane* alone turns this DVD into a must-have.

In addition, Disc One comes fully loaded with bonuses of its own. Leading the pack is a pair of audio commentary tracks. One is by filmmaker/Welles biographer/raconteur Peter Bogdanovich, and the other is by film critic Roger Ebert. Taken together, the commentaries are a study in contrasts. Those who've never had much use for Bogdanovich (endlessly annoyed by his tendency to work stories about "Orson" and "Hitch" into interviews that seem to have nothing to do with Messrs. Welles or Hitchcock) will find more of the same. Bogdanovich is like the film school professor everybody hates. He sounds faintly bored throughout, as if he's watched this movie a zillion times and can barely be bothered to pass on what he knows. For the most part, he simply describes what we're seeing or repeats dialogue that he finds particularly choice. Every once in a while, he'll say something interesting . . . but by then, you'll have probably tuned out. Ebert, on the other hand, is the film school professor everybody loves. It sounds like he's seen this just as much as Bogdanovich, but he never gets tired of it. His enthusiasm and love of the film is infectious and his comments are well organized and keenly observed. Quite simply, if you don't already love this movie, you'll want to love it after hearing Ebert.

The rest of the disc is jam-packed with info and film-related ephemera. Beautiful, richly detailed storyboards, advertising campaign art, rare photographs from on the set and of a deleted scene, studio call sheets, correspondence, footage from the New York premiere and, as they say on TV, much, much more. Of course, the disc also includes the original trailer, considered one of the best and most famous trailers of all time. The *Kane* trailer doesn't show a single frame of footage from the movie. You don't even see Orson Welles at all. It's a striking piece of ballyhoo that's just as innovative as the film it advertises.

With *Citizen Kane*, Warner Bros. has raised the bar for the presentation of classic films on disc. This is one of the most significant DVD releases to date and it belongs in the library of every serious movie fan. As the advertising copy for *Citizen Kane* itself reads, "It's terrific!"

The Third Man

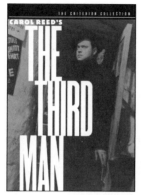

The Criterion Collection

"In Italy, for 30 years under the Borgias they had warfare, terror, murder and bloodshed. But they produced Michelangelo, Leonardo da Vinci, and the Renaissance. In Switzerland, they had brotherly love—they had five hundred years of democracy and peace. And what did that produce? The cuckoo clock."

The BFI, England's answer to the Academy of Motion Picture Arts and Sciences and the American Film Institute, has crowned *The Third Man* the greatest British film ever made. But in our book, it's one of the greatest films . . . ever.

The Third Man follows Joseph Cotten as Holly Martins, a down-on-his luck pulp fiction writer, who has come to Vienna at the request of his best college buddy Harry Lime (Welles). It's right after World War II, and Vienna is a military mess, occupied by Russia, France, America, and England. Post-war business is good for ex-patriot Harry Lime, and Harry has offered Holly a job doing publicity for him. There's only a slight problem -- Harry was just killed in a pedestrian car accident. The day Holly arrives is the day Harry is being laid to rest. Mourning the loss of a good friend, Holly sits down with Calloway (Trevor Howard), who turns out to be a military major investigating Harry. Seems Harry was a racketeer dealing bad penicillin, which has resulted in the deaths of many children. Holly doesn't believe this and starts his own investigation, which leads him into a seedy underworld of human nature. The film is just a spiral, which coils tighter and tighter on its way in. It's a ride that is as fun today as it must have been back in 1949. *The Third Man* also produced Harry Lime as one of the silver screen's greatest anti-heroes.

Most of the success for the film falls on director Carol Reed's shoulders. It was Reed who insisted upon shooting in Vienna, against producer David O. Selznick's wishes. It was Reed who fought hard to cast Orson Welles, considered to be box office poison, in the role of Harry Lime (again over Selznick's wishes—he wanted Noel Coward). Hell, Joseph Cotton, even though he was Reed's only choice, was almost replaced by Jimmy Stewart. Now, imagine if Noel Coward and Jimmy Stewart were in this movie—both are fine actors, no doubt, but neither has the menace that Cotton and Welles contain. Each character is a flawed human, and as the film stands, it's almost a documentary (not surprising, considering Reed's past as a documentary filmmaker during the war). If Selznick had had his way, *The Third Man* would have been a very Hollywood-looking film that would have fallen between the cracks of film history, lost and forgotten. As it eventually turned out, this is a brilliant film. Reed may have won an Oscar for *Oliver*, but he really deserved it for this film (for which he was also nominated).

The Third Man is obviously an old film, but you'd never know it looking at the video on this DVD. A lot of love went into transferring this film to disc, and when you look at the restoration reel, you'll see just how bad it originally looked. The picture shows a bit of grain (which just adds to the noir look anyway) but overall it looks great. The black and white image is full of personality and free of artifacts. The audio is in its original mono and does the job just fine. The Ferris wheel scene is really full sounding, with a great many mechanical audio effects that will fool you into thinking that the track is more than just mono.

On the extras side, Criterion went crazy with this film. And we mean crazy. Just about everything you could want is here. It all starts with a video introduction by Peter Bogdanovich, writer/director/film historian, who spent a lot of time with Orson Welles in his later years. The best moment in the intro is when he definitively lets go of the rumor that Welles directed himself in this film. He did write the "Cuckoo Clock" dialogue, but that was about it. Reed was firmly (and capably) in charge. There's a commentary track, which is kind of different when it comes to commentaries. It features actor Richard Clarke (*A Night to Remember* and *Midnight Cowboy*) reading from Graham Greene's original story treatment. Greene's treatment was a story overview, never really meant to be seen by the public, and served primarily as a skeleton to hang the meat of the screenplay on. The way it was recorded for this commentary is very cool, because it fits over the action and sometimes dialogue of the scenes it's talking about. Another nice extra (actually, it's more a set of extras) is a pair of radio broadcasts from the 1950s. One is Orson Welles as Harry Lime, in an adventure set years before the events occurring in *The Third Man*. Here, Lime is an adventurer with a conscience, who loves and foils with nary a thought. Apparently, there were a series of radio broadcasts that explored the past of everyone's favorite anti-hero—there's only one archived on this disc, but it's of note because Welles himself wrote it. Another broadcast available here is a radio performance of *The Third Man*, with Joseph Cotten reprising his role from the film. It's condensed, but it gives you an idea as to what it must have been like to experience the golden days of radio drama. A still of an old radio serves as the visual that accompanies both broadcasts onscreen. Also included here is a restoration demonstration (showing just how much work was put into this film—massive is a word that only halfway explains it) and footage of Anton Karas (the man behind the music of the film), along with news footage of the sewers that play a major part in the film.

How much would you pay for a special edition of this magnitude? Don't answer yet—there's more. You also get the original British opening of the film (which is attached to this print on the DVD), as compared with the American version. Carol Reed reads the opening with a dark air about him—Cotten's version is less ambivalent. There're also trailers for both the U.S. release and the recent re-release, and a briefly explored production history by Charles Drazin (writer of the book *In Search of The Third Man*) with photos and notes.

Do yourself a favor and check this disc out. If you've never seen the film, you're in for a treat. And if you're as big a fan as we are, then this is the DVD special edition you've been waiting for.

Superman: The Movie

Warner Bros.

Every good American should know the mythology of *Superman* by now—he's been a cultural icon in this country since his debut in *Action Comics* #1 in June of 1938. *Superman: The Movie* takes great care to tell his story right from the start. As the film opens, all is not well on the planet Krypton. It seems that the planet's orbit is shifting, and only one man knows that this spells disaster—the great scientist Jor-El (played here by Marlon Brando). He knows Krypton is doomed, but none of his fellow council members agree. They make Jor-El swear that he won't leave the planet himself, thus starting a panic. But he's decided to send his only son, Kal-El, away to the planet Earth to spare him. Sure enough, Krypton is destroyed. And after traveling through space for many years, little Kal-El crashes to Earth in Kansas and is found by a farm couple. Jonathan Kent (Glenn Ford) and his wife Martha adopt the little boy, name him Clark and raise him right, with good old-fashioned Midwestern values, in tiny Smallville. But it's not easy—Clark's got abilities far beyond the average human. He's faster than a speeding bullet and can leap tall buildings in a single . . . well, you know the story. As Clark gets older, he feels the inexorable pull to learn his true heritage and his purpose for being on Earth. And when Clark finally discovers his destiny, and heads off for sprawling Metropolis, the human race gets its very own superhero. That hero, of course, is none other than Superman (played by Christopher Reeve), who fights for Truth, Justice, and the American Way. In the big city, Superman will find both love (in the form of Lois Lane, played by Margot Kidder) and a villain to match wits with (the infamous Lex Luthor, played by Gene Hackman). And the world will never be the same.

Superman: The Movie is really a delight, with a grand story, wonderful production design, and an amazing ensemble cast. In addition to Brando, Hackman, Ford, Reeve, and Kidder, you've got Jackie Cooper, Ned Beatty (he's *never* been funnier than he is here), Terence Stamp, Valerie Perrine . . . the list goes on. The film never takes itself too seriously, managing plenty of laughs, but still treats its "fantastic" subject matter with great respect. The special effects in this film were groundbreaking in their day — no one had any idea how to make a man fly in a convincing way back in 1978 (there was no such thing as a CGI effect). And then newcomer Christopher Reeve played the title character with so much straight-laced humanity and innocence (not to mention his funny Cary Grant-riffs as the bumbling Clark Kent) that he simply *became* Superman for an entire generation of moviegoers. Warner may be currently developing a new *Superman* franchise, but this film will *always* define the Man of Steel.

Superman is one of those films you just can't help but remember fondly. Sure, the special effects are dated and the script is a little campy. Two of its three sequels are just plain awful (the only good one, *Superman II*, was written and filmed mostly at the same time as the original, with the same cast and crew). And for a while there, the film itself was deteriorating so much that it was in danger of being lost forever to the ravages of time. But against all odds, *Superman* continues to endure. And the film has thankfully gotten a major boost, in the form of Warner's new special edition DVD. This disc is really a treat. There are certainly DVD special editions out there that are more comprehensive and loaded. But for sheer fun factor alone, *Superman: The Movie* is hard to beat.

Let's start with the video . . . the new anamorphic widescreen film transfer is tremendous. This isn't reference quality, but the film has absolutely never looked better. In fact, it's doubtful that this film *could* look better. It's been lovingly restored and, while it does look a little soft on occasion, the contrast and overall clarity are excellent. And the color! Given that this is a comic book adaptation — one of the first of our generation — color is absolutely critical. And what you get here is extraordinarily vibrant color. Right from the opening credits, lush color just blasts off the screen. Flesh tones are always accurate and Superman's big red-and-yellow "S" has never looked so stunning. You will notice some grain given the film's age (particularly on some of the restored footage), and there is a measure of dust and dirt visible on the print. But the overall video quality is quite impressive. As fans, we're more than pleased.

Better still, the newly remixed Dolby Digital 5.1 audio actually manages to one-up the video. We can't tell you how cool it is to hear the opening percussion in John Williams' amazing score building as the titles fly in from behind you in full 5.1 sound. And when that big red "S" crashes onto the screen, your subwoofer will kick in with force. This audio may not be quite as good as the best 5.1 tracks on DVD, and there *has* been some tinkering with some of the film's sound effects, but it's going to blow most fans away. The sound stage is big and wide, with very active surrounds, plenty of nifty directional effects, and lots of deep bass. The result is a completely encompassing audio experience. Listen, for example, to the subtle ambience in Chapter 14, as Jor-El is tutoring young Clark in the Fortress of Solitude. Very nice. Occasionally, the dialogue will sound a little bit flat (particularly later in the film, in Luthor's underground lair). But on the whole, this is fun DVD sound.

And the extras are every bit as good as you'd expect. First of all, director Richard Donner has restored some 8 minutes of footage to the film that was unseen in its theatrical release—10 scenes in all. The best of these involves Superman telling his father, Jor-El, how much he enjoys helping humanity. There's also a nifty bit with a young Lois Lane seeing teenage Clark running alongside the train she's riding in (the girl was in the original cut, but we never knew it was Lois before). Better still, you can access each restored scene directly from the menus.

Next up is a very good audio commentary with Donner and creative consultant Tom Mankiewicz. Mankiewicz actually rewrote much of the script for Donner, adding humor and creating a number of unique supporting characters, but was prevented from getting a screenwriting credit by WGA rules. You can tell here just how much he was actually responsible for, and how much affection he and Donner have for each other. Clearly, these are two longtime friends and collaborators, and it's a treat to listen to them go back 22 years in their memories to recall the stories behind the filming of *Superman*. And they do have some great, and even funny, stories to tell. There's a wonderful bit (about 49 minutes in) where they argue about why Superman, if he's so powerful, doesn't simply save everyone (the classic superhero conundrum). The track really draws you in and keeps you interested for the entire length of the film.

And here's an extra that's worth buying the disc for all by itself – the disc features an isolated, music-only track, presenting John Williams's entire score in full Dolby Digital 5.0 sound! This is really a coup. It's getting harder and harder for studios to convince composers to license their movie soundtracks for isolated use on DVD. How they managed to convince Williams to allow it, we'll never know. But what a treat! *Superman* has long been one of fandom's most favorite film soundtracks, and to have it like this is absolutely wonderful.

Also included here are the film's theatrical trailer (in anamorphic widescreen) and a good text essay on the history of the *Superman* character. And everything listed so far is only on Side One of the DVD! Side Two gives you two more deleted scenes that weren't restored to the film (involving the feeding of Lex Luthor's "babies" – we'll say no more). You get almost 20 minutes' worth of screen test footage, featuring Christopher Reeve and Margot Kidder, along with Anne Archer, Stockard Channing, and other Lois Lane wannabes. The film's teaser trailer and a TV spot are also included (again, all in anamorphic widescreen). And get this – you even get some 8 unused music cues from Williams' score (almost 30 minutes worth in all), all in Dolby Digital 5.0! These include variations on the "Main Title," along with the infamous pop version of "Can You Read My Mind." And via PC DVD-ROM drives, you can access storyboards from the film and other online sites. About the only thing missing are the various versions of the screenplay – Mario Puzo's original 500-page monster and the various rewrites.

But we're *still* not done. Side Two also contains a trio of very good (and original) behind-the-scenes featurettes hosted by actor Mark McClure (who played young Jimmy Olsen in the film). The first two are roughly 30 minutes in length (the third is about 24 minutes). *Taking Flight: the Development of Superman* covers the conception of the project and describes the effort to convince Warner to back it, the effort to get the screenplay completed, and the search for the perfect actor to play Superman. *Making Superman: Filming the Legend* looks more in-depth at the actual production and covers the many difficulties (including the fact that many of the special effects had never been tried before) of the 19-month effort. Finally, *The Magic Behind the Cape* looks at the actual effects process and includes some great screen test footage shot in the effort to figure out how to make a man fly convincingly on film. Some of this stuff is *very* funny.

Best of all, all three featurettes include brand-new interviews with Donner, Mankiewicz, John Williams, and various members of the production team, as well as actors Christopher Reeve, Margot Kidder, Gene Hackman, and more. These aren't just the quick, EPK style interviews—what's included here is much more substantial and valuable. The result is that these documentaries really give you an excellent and well-rounded look behind the scenes at the making of *Superman*. You'll come away having learned a lot you didn't know, and you'll enjoy every minute of it.

This really is a wonderful DVD special edition. It's not as comprehensive as some, but we haven't had this much fun with a disc in quite some time. It absolutely deserves a place in the collection of every DVD fan. So don your best cape and fly . . . don't walk . . . to your local retailer. Whip up a bowl of hot buttered popcorn, sit back in your favorite comfy chair, and enjoy an evening of great DVD entertainment.

The Classic Monster Collection

What's more classic than these spooky monster tales? Universal's classic monster films are American mythology at it's finest. *Dracula, Frankenstein, Creature, The Wolf Man*, and *The Mummy*—merely mentioning these titles in a hushed voice can cause a chill to run down anyone's spine. And these DVDs are pretty stellar presentations. Overseen by horror film historian David J. Skal, all the big films in this series are given deluxe, showcase treatment on disc. All of the films are available separately on DVD, but for the true enthusiast, a definitive collection is also available featuring all the major titles in the series.

And if the films we're about to take a closer look at aren't enough for you, also available from Universal are *Invisible Man*, *Bride of Frankenstein*, *Phantom of the Opera*, *Abbott and Costello Meet Frankenstein*, *Abbott and Costello Meet the Mummy*, *Dracula's Daughter/Son of Dracula*, *The Mummy's Hand/The Mummy's Tomb*, *Son of Frankenstein/Ghost of Frankenstein*, *The Mummy's Ghost/The Mummy's Curse*, *Werewolf of London/She-Wolf of London*, and *Frankenstein Meets the Wolf Man/House of Frankenstein*. In other words, there are enough classic monsters on DVD to make any fan deliriously giddy with home theater thrills and chills.

Frankenstein: Classic Monster Collection

Universal

There probably aren't many people over the age of 10 out there who don't know what *Frankenstein* is. If that happens to be an incorrect statement in your case, then you'd best just go out and either read the book or add this disc to your collection right now. We can't—nay, we *won't*—go into the story in detail here. The briefest of summaries is this—*Frankenstein* is about a scientist who messes with Nature, only to have Nature bite back and take a huge chunk out of his ass. Boris Karloff (who played the monster only after Bela Lugosi backed out), with his Jack Pierce makeup in this role, has become one of the most identifiable images in all of twentieth century cinema. This film made him one of the most bankable horror actors of all time. It began a legacy of silver screen horror that still endures today. And the movie itself became an instant classic.

In terms of extras, the supplemental information on this disc is presented in basically three ways. First, there is a very enjoyable commentary track from film historian Rudy Behlmer, who really knows his stuff. He obviously loves this film, and you can hear his passion for the subject in his voice as he throws nuggets of information about. He talks about the creation of the monster, the history of the book, and all of the theatrical stage plays and short film versions that preceded this production. It's an invaluable track, and one of the best to be featured on a classic film DVD. On top of the commentary, there is also a 45-minute documentary, written by historian David J. Skal. Finally, the disc boasts a nice section of stills, posters, cast and crew bios, and production notes.

Okay, so the movie is super, and the special edition is super. What about the quality of the disc video and sound-wise? That's a tougher question. It's an old film, first of all. Based just on that knowledge alone, this DVD presentation looks better than any format we've seen this film in before. But there is that one monster that nothing can escape, especially film—time. Time has not been kind to *Frankenstein*. The film "flashes" on occasion from faded areas of the print, excessive hairs and dust specks are visible, and a few tears on the emulsion are very clearly apparent. Did the restoration crew do the best job they could? We betcha they did. Could a better job be done? Maybe, but it would have cost a pretty penny. Don't get us wrong, this is an outstanding print, but the faults are there to be seen. They're faults that may not be too distracting visually, but they're still faults that could have been fixed with time and money (based on what we've seen with other old and even more damaged prints out there, suchas *The 400 Blows* and *The Most Dangerous Game*). The audio on this disc is straight and classic mono, although the documentary is in stereo. Both sound just fine. The audio quality is perfect for this film, very natural, and surprisingly vibrant for such an old classic.

Literally everything you will ever need to know about *Frankenstein* the movie, the book, and their effect on Hollywood, is contained on this disc. All in all, you're gonna walk away an expert on the monster after you watch this disc all the way through. This DVD is a resource for film lovers and a must-own disc for DVD fans everywhere.

Dracula: Classic Monster Collection

Universal

There can be little doubt that the original 1931 version of *Dracula* is a classic and important film in the history of cinema. Surely most of you know it well, so it's not worth retelling the plot or talking up the film too much here . . . except, of course, to say that Bela Lugosi rocks. Period.

Dracula is presented on DVD in two different versions. The first version is the original, restored English version, which is quite good. Like *Frankenstein*, *Dracula* is obviously an old film with flawed source material, giving us a picture that can be dark at times and less than stellar in terms of picture quality. There's a slight bit of a flickering going on here and there, but it's nothing too alarming. It's worth noting again that these issues have to do with the original source print, which is old, and has nothing to do with the DVD production or the restoration. The audio is presented two ways—the original mono, which sounds very good, as well as a newly recorded Philip Glass/Kronos Quartet score, which is quite cool. Being film purists, we're not usually too excited when people fool around with a film like this and try to update the soundtrack. But Philip Glass's music fits the film perfectly. This new score is weird and freaky and everything you'd want and expect to go along with this film. A separate title card introduces this version, so you'll know right at the beginning which version you're watching.

The other version of the film on this disc has got to be one of the coolest extras ever put on a special edition DVD—the Spanish version of *Dracula*. It's on the second layer, with a nice, long video interview/intro with Lupita Tovar Kohner (she plays Eva, this version's Mina). If you go to the menu, you can also access a separate production notes section that explains more about this version of the film (it's also discussed in great detail in the documentary, but "more on that later"). The Spanish version was filmed simultaneously with the English language version in 1931, using a separate cast. It feels like pretty much the same exact movie, with perhaps a bit more flavor. It runs about 30 minutes longer, and some of the acting is actually a bit better (save for Carlos Villarias as the Count) than what was seen in the version most of us are familiar with. What really sets this version apart is that some of the camera tricks used in the Spanish version leave Browning's version in the dust, while others strangely fall flat. It's hard to say if one version is better than the other, but it sure is fun to watch and compare the two. When it comes to film quality, it's funny to see that this version seems better preserved (and a bit better looking) than the original English version. Maybe sitting in a vault (not being seen) all these years did a service to this version. In any case, the blacks are richer and the source print is mostly free of flaws.

David J. Skal, the producer, writer, and director of the special edition material in this series, participates pretty hardcore with this release, giving his own take on the film in a really nice commentary track. Skal has a very enjoyable, NPR-type radio voice, and he fully covers everything (and we mean everything) that you'll ever need to know about this film. He walks us through *Dracula*'s origins as a novel, its true-life historical roots, the stage versions, and the overall influence the film has had on the genre and cinema in general. You'll find a few long gaps in the commentary and a bit too many "more on that later comments" early on, but the information eventually does pour out. If you're patient, it's definitely worth a listen.

As for the other extras, they're pretty great as well. The documentary *The Road to Dracula* is very well done. It's hosted by Carla Laemmle, niece of producer Carl Laemmle, and speaker of the first words in the film. It covers a broad range of topics, serving as a nice little overview to the film and the information you'll hear in the commentary. All the standard interviews with historians and living relatives that we've come to expect from the Classic Monster Collection are featured here, along with examples of the silent 1931 Browning version, and a bit of "lost" footage originally tacked onto the end of the film. Also included on the disc are a poster and photo montage set to music, some production notes, cast and crew information, and a re-release trailer. About the only thing we didn't like about this DVD was that the only way to switch between the two versions of the film is from the initial startup menu on the disc. And you can't go back to that once you've chosen, so you have to keep restarting the disc in order to select the other version. Still, it's a very minor complaint about an otherwise stellar DVD special edition.

The Wolf Man: Classic Monster Collection

Universal

"Even a man who's pure in heart and says his prayers by night may become a wolf when the wolfbane blooms and the autumn moon is bright."

The monster at the heart of *The Wolf Man* is different from most of the other "nine" Universal Monster standards (those being *The Hunchback of Notre Dame* [whose greatest incarnation, the 1923 Lon Chaney version, is available on DVD, but not as part of this Universal Collection], *The Phantom of the Opera, Dracula, Frankenstein, The Mummy, The Invisible Man, The Bride of Frankenstein, The Wolf Man*, and *The Creature from the Black Lagoon*). He's different because he's a monster, but not by choice. Sure, Frankenstein didn't ask to be reanimated and *The Bride of Frankenstein* is hardly a monster, considering she never really monstered out. But good ol' Larry Talbot is a monster and he doesn't even know it. He just wakes up with dirt on his feet, and cuts and bruises all over his body, with an "Oh my God, what have I done?" look on his face. That must be a bitch. For Larry, life is a nightmare, and all he wants to do is die. It makes him a sad monster—one that deserves our pity.

The Wolf Man begins with Larry Talbot coming home to see his father somewhere in Europe (it's unspecified, with English lords, Gypsies, modern-day cars, and German castles). Larry is the new heir apparent to his father's lordship, after his older brother was killed in a hunting accident. Larry has spent so much time away in America that he's become an outsider in a world he was once a part of. Lon Chaney, Jr. plays Larry as a lumbering child-man, who doesn't understand anything he can't put his hands on. He chases girls, he seems shy around authority figures, and the first thing we see him do is build a nice telescope for his father's enjoyment. Legendary actor Claude Rains plays Larry's father and, try as they might, the two look nothing like father and son. It's kind of funny, but it also adds a sort of sadness to the relationship they share. Larry seems out of place in every aspect of this world, and what might first be perceived as a casting mistake takes on an importance. Lon Chaney/Larry Talbot just doesn't belong in this world, and when things begin to happen to him, it's that much more sad.

While testing his father's new telescope, Larry spies the beautiful Gwen (Evelyn Ankers) in her apartment above an antique shop and immediately high-tails it right over to her. In a bit of uncomfortable "stalker" dialogue, Larry asks her out and ends up buying a silver cane with a wolf's head on top. They make plans to get their fortunes read by a traveling Gypsy clan. Unfortunately, by the light of the moon that night, Larry is bitten by a werewolf and soon becomes one himself. Forever cursed to hunt the night for victims, Larry does everything he possibly can to prevent the inevitable, while everyone he loves thinks that he's slowly going crazy.

The transfer for this DVD is remarkably well done. The video quality is clean, with only a few moments of the damaged source print apparent. The sound is nice as well. The mono track is full, with nice dialogue and music. Both represent the best quality we've ever experienced for this film.

Unfortunately, the extras here are a bit lighter than those of the *Dracula*, *Frankenstein*, and *The Mummy* DVDs. The documentary is a little more than 30 minutes, but it focuses on werewolves in movies in general, along with this film's makeup and script. These are good points, and they're well covered here, but there's more to this movie that should have been discussed. With *Frankenstein, The Mummy*, and *The Bride of Frankenstein*, the special edition material was layered with history about the people who made the film, the alternate drafts of the screenplays, and how the films influenced cinema. Those points are barely mentioned here and are never explored. It feels like there was a lot more that wasn't being told. Thankfully, the commentary track by historian Tom Weaver does go into more detail. Weaver spends most of the time talking about the makeup and the legend of Jack Pierce, making the track a great listen.

The other extras include production notes, and a cast and crew section that sadly is nothing but regurgitated information from the commentary. There's also the "archive" of stills and poster art (set to *The Wolf Man* theme music) and a badly preserved trailer.

The Wolf Man works on many different levels, despite some of the flaws in the story, logic, and production. It has a nice symbolic underpinning, thanks to writer Curt Siodmak, that includes Freudian study, political commentary, and issues of human nature. If you look at the film with a critical eye, you'll see a handful of production gaffs and reused shots. All that said, *The Wolf Man* is a great monster film that looks great on DVD. Fans might be slightly disappointed by the special edition material here, but the disc is still well worth having in your collection.

The Mummy: Classic Monster Collection

Universal

"Oh, Amon-Ra! Oh, god of gods! Death is but the doorway to new life! We live today—we shall live again! In many forms shall we return! Oh, mighty one!"

Brought to life in 1932 by Boris Karloff (a.k.a. "The Uncanny"), *The Mummy* became yet another flagship in the Universal hall of horrors. Karloff endured eight hours in Jack Pierce's makeup chair to be made up as the newly unearthed mummy Imhotep, only to have maybe three minutes of combined footage in the costume in the final film. That's dedication. That's the stuff screen legends are made of. And despite all the information you'll ever read about Karloff, not one thing says he ever complained about it. That's pretty impressive by today's standards.

The film begins in Egypt in 1921 (one year before the discovery of King Tut's tomb). An English archeology team from the British Museum has just unearthed the tomb of Imhotep, a temple priest who dared to fall in love with a sacred vestal virgin. For this act of treason, he was buried alive along with the mystical Scroll of Thoth, which carries the curse of death to whoever unlocks it from its box. Of course, as curiosity killed the cat, it also brings the dead back to life sometimes. Out comes the Scroll, up comes the mummy, and a young archeologist named Ralph Norton (Bramwell Fletcher) goes laughing all the way to his death bed.

Flash forward to 1932. A new team of explorers, lead by Frank Whemple (David Manners), is about to leave the dig site empty handed, when a strange man named Ardath Bey suddenly introduces himself and offers the team the location of the lost tomb of Anck-es-en-Amon. These guys are not stupid, and they indeed find the tomb, making Ardath Bey a very happy man. Why? Well, Ardath is really Imhotep (looking pretty good for a guy that's some 3700 years dead). Anck-es-en-Amon was that vestal virgin Imhotep was buried alive for, and the years haven't dulled his desire for her. All he needs now is the body of Anck-es-en-Amon's latest incarnation, Helen Grosvenor (Zita Johann), to raise her spirit back from the dead. It would seem love doesn't die—it just gets dusty and walks around with one leg dragging behind. As you can imagine, Imhotep'll use all of his incredible power to get what he wants, and murder isn't something he's gonna think twice about.

The Mummy is a pure horror tale, although dated. It's also a love story. You sympathize with Imhotep's love for his beloved princess—in a way you want him to be reunited with her. But you can also see the menace in his eyes. The scene when the mummy first awakens is a classic. And even though the film has lost a bit of its punch over the years, it works. *The Mummy* is simply a great classic horror film, expertly directed by Karl Freund (the German expressionistic cinematographer who helped bring *Metropolis*, *Dracula*, and *Murders in the Rue Morgue* to life). Freund knows exactly where to put the camera to help stir up his audience, and his visual mastery is quite apparent in this film.

For more information about Freund (and his problems with actors), look no further than this DVD. This special edition disc looks at both the making of this film and the filmmakers themselves. Karloff and Johann are addressed quite well here in both the commentary and the accompanying documentary. The commentary track features Paul Jensen, a man that definitely knows the ins and outs of *The Mummy*. You have to struggle somewhat to get through his play-by-play commentating, but the nuggets of real information he provides make it all worthwhile. The documentary, on the other hand, is very informative, funny, and light, which is what you'd expect from a film titled *Mummy Dearest*. It's too short, at just 30 minutes, but it's well worth watching. Also included on this disc are a collection of stills put to music, production notes, cast and crew bios, and the trailer.

Something interesting that might catch your ear, in both the documentary and the commentary, is the talk of how the theatrical cut of the film was edited down. Unfortunately, there are no deleted scenes on this disc, and no one explains why that is. We understand that sometimes stuff gets lost or destroyed, but the experts talk about this stuff like they've seen the deleted material recently, so it gives the impression we, as DVD-fans, are missing out. Oh well.

As we're sure you know by now, because this is an old film, it has a tendency to look pretty old. Still, the film benefits from being on DVD. The transfer shows little to no artifacting, and there are only a few spots of noticeable edge enhancement. The source print is scratched, worn, and beat up in spots, but overall, it looks pretty darn good for its age. The sound is a mono track that does great justice to the film. There is only minor hiss audible, and we didn't hear any pops or cracks at all.

All things considered, this is a pretty decent presentation for such an old film on DVD. The extras might be a little wanting, but you definitely shouldn't let that stop you from enjoying *The Mummy* on disc.

Creature from the Black Lagoon: Classic Monster Collection

Universal

"I can tell you something about this place. The boys around here call it the Black Lagoon. Only they say nobody has ever come back alive to prove it."

It's not easy being green. Imagine that some glory-hungry scientists come ripping through your home looking for your dead relatives' remains. Once they find you, they decide, "Hey, look—a rare living thingy that we can kill and study!" What are you going to do? Like *The Wolf Man*, *Creature from the Black Lagoon* is a bit different from the other Universal monsters in that the creature really isn't that bad of a guy. Sure, he attacks humans with no remorse, but he's just doing what anyone would do if they found people in their backyard with shovels—he's going to disembowel them. The *Creature* is one of the greatest monster films of all time and is definitely one of the best designed (well . . . at least next to *Alien*). Considering it's close to 50 years old, that's saying a lot.

David Reed (Richard Carlson) and his girlfriend Kay (Julia Adams) are the aforementioned scientists. They're hunting down a rare lungfish in the Amazon when they learn of the nearby discovery of fossilized remains of a strange fishman. Uniting with David's boss, Mark Williams (Richard Denning), the group heads down the river to the unexplored Black Lagoon, with hopes of finding more remains at the bottom of the riverbed. But menacing them is a creature that represents a living, breathing version of what they're looking for. This Creature is not so keen on the newcomers tramping around his neck of the jungle . . . until he gets an eyeful of the beautiful Kay in a nice one-piece swimming suit. With a newfound taste of love (or lust, either way it's sick), the Creature wants Kay all to himself. But he'll have to go through her boyfriend to get her, and David's not about to just give her up.

Presented full frame (in its original 1.33:1 aspect ratio), the film looks pretty damn good on disc, even if it's not (thankfully) presented in its original 3-D. It may upset a few people that a 3-D version isn't also included on the disc, but keep in mind that 3-D on home video rarely looks good . . . even on DVD. 3-D aside, this picture is generally crisp and heavily detailed. Unfortunately, it isn't the best source print ever used for a DVD transfer. It's a theatrical print (complete with "cigarette burns") and shows heavy dirt, white density, and a noticeable residual artifact caused by combining the separate 3-D filmed elements to make a standard, 2-D film image (this is particularly noticeable during the "Creature on fire" scene). In many ways, however, there isn't a whole lot that could be done with this, so while this video isn't the best, it's still pretty decent considering.

The disc's sound is a cleaned-up mono, and every annoying crescendo heralding the Creature's screen time comes through loud and clear. It's interesting to note, since we're talking about sound, how much this track influenced the music of *Jaws*. Sure, everyone knows Spielberg lifted the worm's eye shot of the girl swimming, but did you know he had Williams lift elements of the score as well? Take a listen to the familiar bass line, isolated in the special features section of the disc. It's unmistakable all throughout the film. Gotta love it.

Extras-wise, included here are trailers, production notes, a photo gallery, an audio commentary track by B-movie expert Tom Weaver, and a documentary on the making of the film. Weaver's commentary is fast and furious. He has so much to say here that the 79-minute running time of this film just wasn't enough. He crams information in here as fast as he can, so you might want to listen to the commentary a couple of times just to make sure you get everything. The documentary showcases the talents of historian David J. Skal. He opens up a world of information here on the making of the film, the people involved, and the politics behind it all. It's very fun and, surprisingly, adds even more information to the already dense commentary track.

A lot of people are fond of this film, and that makes sense when you consider that it was arguably the most accessible monster film of the baby boom generation. *Creature from the Black Lagoon* was a huge hit for kids during the 1950s. *Creature* was a throwback to the halcyon days of *Dracula* and *Frankenstein*, and it stands tall among them despite having been made close to 20 years after they lit up the silver screen. *Creature* is really the perfect monster film, and it works largely because it's fun. The film is simple, it's got great characters, and the monster is appropriately creepy and fun to look at — rubber suit and all. Thankfully, it's been given wonderful treatment on DVD. It's not to be missed.

The Films of Akira Kurosawa on DVD

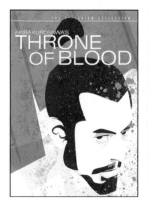 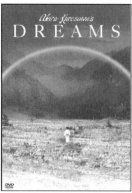 

If a DVD says "directed by Akira Kurosawa" on the cover, it's absolutely worth a spin. Criterion has in its collection a number of his greatest samurai epics, including *Seven Samurai, Sanjuro, Yojimbo,* and *The Hidden Fortress*. In addition, you can also find Kurosawa's life-affirming *Red Beard* on Criterion DVD, along with his retelling of Shakespeare's *Macbeth* in the classic *Throne of Blood*.

Once you've given these a spin, you're ready to watch some of Kurosawa's later films, including his retelling of *King Lear* in *Ran*, the Russian film *Dersu Uzala,* and his three last films, *Rhapsody in August*, *Dreams*, and *Madadayo*. All are available on DVD from a number of different studios, including Warner Bros., MGM, and Wellspring. The bottom line is that you simply can't go wrong in watching any of these films directed by one of the greatest filmmakers of all time.

Let's take a closer look at one of his best works, the classic *Seven Samurai*.

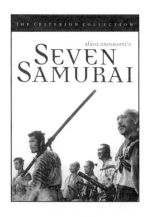

Seven Samurai

Toho (The Criterion Collection)

When the American Film Institute put out its list of the 100 greatest American films, it was met with some controversy. People argued over which films deserved to be there and which ones didn't. Everyone had their own ideas. Well, imagine if that list were the 100 greatest films *ever* made. Half of the top 10 American titles in the AFI list wouldn't even be there. Sure, you'd still see *Citizen Kane* up near the top, but would it be number one? Not in our opinion. If we had our choice, Akira Kurosawa's magnificent epic *Seven Samurai* would hold that lofty spot.

Seven Samurai is the most watchable 203 minutes of film ever made. Where else are you going to get such amazing character development, action, adventure, and this level of human comedy, without having to live it yourself? *Seven Samurai* has all of those things, plus it stars some of the best actors who have ever graced the silver screen, Toshiro Mifune and Takashi Shimura among them. And it was directed by a man who was simply born to make such great films, the legendary Akira Kurosawa.

Samurai follows the trials of a village plagued by bandits. Knowing that the bandits plan to strike when their next crop is harvested, the villagers "hire" down-on-their-luck samurai to protect them from the upcoming attack. At first, the villagers seem to have no chance of finding anyone willing (or good enough) to take the job. That is, until they bump into Takashi Shimura's character Kambei. In a brilliant scene that brings up so many different cultural questions and issues, we see Kambei save a young child who's been taken hostage. His sense of honor leads him to help the villagers, and it's through Kambei that the rest of the samurai are pulled together.

After Kambei recruits his fellow samurai, we begin to learn about the other men. Primarily, we learn about Kikuchiyo, played by Mifune. Kikuchiyo's past allows the group of samurai to eventually identify with the farmers. They're able to gain their trust and fortify the village better through him. He eventually becomes the heart and soul of both the samurai and the film. It also helps that he is the easiest of the seven samurai to identify with. Mifune is wonderful in this role, and it sums up his own, real-life personality well. Wild, angry, funny, caring — Mifune was all those things and more. The character is also a surrogate for Kurosawa himself, having compassion for the farmers, but also disdain. He wants to be part of his peers yet strives to be an individual. Kurosawa's career shows these aspects of his own personality.

The quality of the video on this DVD is good, but it's not perfect. The film has been digitally restored and retains a bit of a digital look as a result. It's also a film that was released in 1954. As a result, restoration or not, age is going to play a part in the look of the transfer. Still, this is as good as we've ever seen the film looking on home video. The audio has also been remastered. It's a loud and bright mono track that preserves the original sound presentation nicely. Most of the analog tape hiss has been removed, leaving a wonderfully clean-sounding track.

The extras on the disc are pretty sparse but they include a commentary track by Michael Jeck. When he's good, he explains many interesting aspects to the film. When he's bad, he's boring. Most of the time he simply gives play-by-play commentary and it gets on the nerves sometimes. But he's also informative enough that serious fans will want to give the track a listen. Also included on the disc is a trailer for the film.

This film is as close to perfect as they come. The way that Kurosawa moves his camera and sets up his scenes, his heavy use of deep focus — it's just beautiful. *Seven Samurai* stands as Kurosawa's greatest effort . . . and his most accessible. While the DVD extras here are somewhat lacking, if you watch only a single film on disc that we recommend in this book, this should probably be the one.

Ben-Hur

Warner Bros.

During the late 50s, MGM and the other Hollywood studios were faced with a problem — people simply weren't going to the movies like they had in the past. The average Hollywood movie was making less from box office receipts, and only big, epic pictures were making any kind of a sizable profit. The reason was television — more and more Americans were bringing their first TV sets home and were being entertained in their living rooms. So MGM decided to roll the dice on a big, DeMille-style epic film using one of their most successful properties — the story of *Ben-Hur* (which was based on a successful novel and had been made onto a highly successful MGM film once before in 1925). Director William Wyler (*The Best Years of Our Lives*, *Wuthering Heights*) was hired to helm the massive effort, and actor Charlton Heston was cast in the lead role. And the pressure was mounting — as costs grew on the production, it quickly became clear that the success or failure of MGM as a studio was riding on it.

Ben-Hur may not be the best epic film ever made, but it certainly was the largest. At a cost of some $15 million, its 212 minutes feature more than 8,000 extras, 300 sets, and one of the most dangerous and thrilling action sequences ever captured on film — the great chariot race. MGM's gamble paid off handsomely when *Ben-Hur* grossed some $80 million at the box office and went on to sweep the 1959 Academy Awards, winning 11 Oscars (including Best Picture, Best Actor, and Best Director).

While many consider *Ben-Hur* to be a biblical tale (indeed the film's full title *is Ben-Hur: A Tale of the Christ*), its story only occasionally intersects with the story of Christ. The film is set in the Middle East, beginning in the time of Christ's birth. Judah Ben-Hur (Heston) is a Jewish prince of Palestine. As a boy, he was best friends with Messala (Stephen Boyd), the son of the local Roman governor. Messala left Palestine when his father was reassigned, but years have passed, and now Messala's returned as the commander of the Roman legions in Palestine (second in command only to the new governor). He and Judah hope to rekindle their friendship and restore peace to the troubled region. But Messala wants that peace to remain firmly under Roman control, while Judah wants freedom for the Jewish people. When it becomes clear that Messala only hoped to use his friendship with Judah to climb the ladder politically, Judah renounces his friendship. Shortly thereafter, an accident that injures the new Roman governor is blamed on Judah's family. So his mother and sister are thrown in prison by Messala's order, and Judah Ben-Hur is sold into slavery as an oarsman aboard one of the galleys of the Roman navy. Fate and, perhaps, divine province afford Judah an opportunity to return to Palestine years later to find and free his family. But will Judah choose revenge or forgiveness for his former friend?

This single-disc special edition release features an amazingly good film transfer. *Ben-Hur* was shot in the widest aspect ratio ever used theatrically, 2.76:1. That ultra-wide image has been perfectly preserved here through a high-definition, fully digital transfer. Right from the start of the film's Overture, you'll be stunned. The transfer features startlingly vibrant colors, rich and accurate flesh tones, tremendous contrast with deep blacks, and very good fine detail. There's just a few instances where the color seems a little washed out (look at the start of Chapter 4, about 18 minutes in), and there's maybe a hair too much edge-enhancement on rare occasions. But those are very minor complaints. Other than the occasional tiny nick on the emulsion, this is a rock-steady image. You'll also be pleased to know that the film is presented in full anamorphic widescreen on DVD, meaning that you've simply never seen this film looking so good at home before.

One note here—we defy *anyone* to look at the chariot race scene in this film on DVD, compared to the pan-and-scan versions we've seen before, and tell us that letterboxing isn't the best way to view widescreen films. *This* is the way widescreen films were meant to be seen! And after watching the chariot race, you'll be stunned at just how shamelessly George Lucas copied it for the "pod race" sequence in his recently released *Star Wars* prequel. It's almost shot for shot in some places.

The audio on this disc is also very good, in fully remixed and remastered Dolby Digital 5.1 surround sound. The mix is surprisingly dynamic, with good ambience and solid bass. Dialogue is always clear (although you will occasionally hear a little remaining mono hiss under some of the longer spoken lines). This isn't 5.1 audio to die for, but it's very good given this film. Better still, the Academy Award-winning soundtrack by Miklos Rozsa is wonderfully presented. While it's almost a guarantee that most viewers will skip past the Overture and Entr'Acte, they shouldn't. This is a sweeping score, and listening to it isolated like it is here is a real treat.

This film is presented on a DVD-18 disc, so the first half of the film, up to the Intermission, is found on Side One (with an RSDL layer-switch in Chapter 15). Once you're done viewing up to that point, you'll have to flip the disc over to Side Two, where you get the rest of the film (on one layer) and the disc's supplemental materials (on the other layer—there's no layer switch on Side Two). Those extras include theatrical and teaser trailers for the film, both in very good-looking anamorphic widescreen. There's a small gallery of production photos, a couple of pages of cast and crew information, and a listing of the awards won by the film. You also get rare, newly discovered screen test footage, which features Leslie Nielsen (yes . . . as in the bumbling star of the *Naked Gun* films), along with Italian actor Cesare Danova, auditioning for the roles of Messala and Judah respectively. Israeli actress Haya Harareet (who landed the role of Esther) is also shown in a costume test. The best of the supplements, however, are a good, hour-long documentary on the history and making of the film, *Ben-Hur: The Making of an Epic*, and a newly recorded commentary track featuring Charlton Heston himself. It's a so-called "interactive" commentary, because he doesn't speak continuously. Rather, he'll talk for three or four minutes, and then you'll be prompted to press the **>>|** button on your player's remote, which will skip over a few more minutes of the film until you get to Heston's next bit of commentary. It seems awkward at first, but it actually works surprisingly well. And the commentary is well worth listening to, even given the film's length. Heston repeats himself a few times, but he's also got some great stories to tell and plenty of interesting information to relay. *Ben-Hur* was arguably the most important film of his career, and as you listen to him speak, you can really tell he appreciates that fact.

Thankfully, Warner's given this massive, sprawling epic their best kid-glove treatment and it really shows on this disc. If you're a fan of the cinema, *Ben-Hur* absolutely demands to be added to your DVD collection. And be sure to watch it on a big, widescreen display to get the full effect.

Dr. Strangelove or: How I Learned to Stop Worrying and Love the Bomb—Special Edition

Columbia TriStar

Just in case you haven't guessed by its title, *Dr. Strangelove or: How I Learned to Stop Worrying and Love the Bomb* is black comedy at its finest. The film wraps itself smartly in the trappings of the Cold War and just as smartly reveals how truly absurd it all was. But when this Stanley Kubrick film first premiered in 1963, it took an unsuspecting public by surprise. The arms race was in high gear. People took those silly "duck, cover, and don't look at the flash" civil defense films seriously. And home bomb shelter construction was considered a growth industry. So it goes without saying that folks didn't quite know what to make of *Dr. Strangelove*. It was nominated for four Academy Awards, but it was many years before this classic came to be widely appreciated. The more we've come to understand the sheer magnitude of the military and political folly of the time, however, the more brilliant this film seems to become.

Here's the plot in a nutshell. During the height of the Cold War, the U.S. Air Force maintained an around-the-clock airborne strike force of B-52 bombers, poised to deliver nuclear annihilation upon the Soviet Union at a moment's notice. Aboard one of these bombers, Major "King" Kong (Slim Pickens) and his crew receive the unthinkable: the "go" code ordering them to initiate Wing Attack Plan R. It seems that, back at good old Burpelson AFB, their wing commander, General Jack D. Ripper (Sterling Hayden), has gone stark raving mad. Believing that the government is ignoring a Communist plot to poison the water supply with fluoridation, Ripper sets out to "protect our precious bodily fluids" by launching a nuclear first strike. This, he hopes, will force ineffectual President Merkin Muffley (Peter Sellers) into action—when Muffley learns that he can't recall the bombers, he'll have to order an even bigger strike to overcome the inevitable Soviet counter-attack. Of course, when confronted with this news, Muffley does nothing of the sort, preferring instead to inform a drunken Soviet Premier that one of his commanders "went and did a silly thing." He even invites the Soviet ambassador into the top secret War Room, much to the chagrin of his gung-ho military advisor, General "Buck" Turgidson (George C. Scott). The ambassador quickly reveals that the Soviets have developed a Doomsday Machine, which will automatically destroy all life on earth if it detects an American attack. As things unravel in Washington, back at Burpelson, a British exchange officer (Captain Mandrake — also Sellers) attempts to reason with Ripper, knowing that he's the only person with a prayer of stopping the attack. Meanwhile, aboard his B-52, no-nonsense Kong is determined to complete his mission, come hell or high water. And in the end, it's up to the film's infamous Dr. Strangelove (Sellers yet again) to devise a last clever plan for "preserving a nucleus of human specimens."

Sellers is in great form here as Muffley and Mandrake, but it's as the off-kilter Dr. Strangelove, a wheelchair bound ex-Nazi scientist, that he really shines. With his lop-sided hair, Strangelove is an obsessive, maniacal figure, for which Heil-Hitlering is an involuntary response. George C. Scott's blustering performance is equally entertaining as the gumchewing Turgidson. Fans of classic Disney live-action films will quickly recognize Keenan Wynn as Colonel "Bat" Guano. And yes . . . that is James Earl Jones among the bomber crew (in his first feature film role). But it's for Slim Pickens's goofy turn here that we really love this film. He's absolutely hilarious as cowboy-turned-pilot Kong: "If this thing turns out to be half as important is I figure it just might be, I'd say that you're all in line for some important promotions and personal citations when this thing's over with. And that goes for every last one of ya, regardless of yer race, color, or yer creed!"

Dr. Strangelove is simply loaded with sly, tongue-in-cheek jokes. Almost every character name in the film is some kind of clever sexual innuendo or pun. Look for the "Peace Is Our Profession" sign at Burpelson AFB . . . as American soldiers engage in a firefight all around. The pin-up girl in Kong's *Playboy* is Turgidson's secretary. Among the books in front of Turgidson in the War Room is one labeled *World Targets in Megadeaths*. Even the Soviet Premier's name is a joke . . . Dimitri Kissoff.

Dr. Strangelove was among Columbia TriStar's first DVD releases, but has since been replaced by the new *Dr. Strangelove: Special Edition*. The black and white film print used for the new DVD exhibits some scratches and dust, but it's still fairly sharp, relatively clear, with a decent gray scale. Kubrick used a variety of aspect ratios in the general range of 1.33:1 when filming (during the original release in theaters, this was matted to 1.85:1). Fortunately, the DVD presents these changing aspect ratios unmatted as intended by the director. As a result, blurry edges can sometimes be seen on the top and bottom of the frame. The English Dolby Digital soundtrack is presented in its original mono and is of average quality. Dialogue is generally clear and crisp and the film's soundtrack is well integrated in the mix. The audio does exactly what it needs to do but won't blow you away.

The special edition release delivers a nice batch of extras, accessed via simple (but wonderful) animated menus, based on the film's poster artwork. A good, 14-minute featurette has been added, *The Art of Stanley Kubrick: From Short Films to Strangelove*, which looks at the director and his career up to the time of the making of this film. Better still is *Inside the Making of Dr. Strangelove*, a 45-minute documentary that examines the film's production in much more detail. There are interviews with many of Kubrick's collaborators, fascinating stories (among them how Slim Pickens was cast at the last moment when Sellers, who was originally going to play Major Kong as a fourth role, broke his leg), and some great behind-the-scenes photos (including shots of the film's original ending, which featured a cream pie fight in the War Room—sadly, the Kubrick estate wouldn't allow this to be included on the DVD). It's well worth watching if you love the film. There's a pair of "split-screen" interview clips featuring Sellers and George C. Scott answering scripted questions (news organizations could film their reporters asking the questions later to get a one-on-one interview effect for TV broadcasts). Also included are a gallery of poster and advertising artwork, the film's bizarre and brilliant theatrical trailer (along with trailers for *Fail Safe* and *Anatomy of a Murder*), talent files, and an Easter Egg (featuring another teaser trailer).

Dr. Strangelove is a terribly funny film and is definitely an all-time favorite. It's not for everyone, but if you like dark comedies and biting satire, absolutely don't miss it. Given that it's arguably one of Kubrick's best works, it's worth seeing for that reason alone. Just be sure to pack your survival kit and watch out for "deviated pre-verts." And remember . . . there's no fighting in the War Room!

Lawrence of Arabia

Columbia TriStar

Modern movie audiences have lost their grip on what a movie epic is supposed to be, misapplying the term to movies that are merely long or expensive looking. Real epics are vast, sweeping affairs. And no one mastered the epic form as well as David Lean. Lean's early films tended to be smaller, character-based works like *Brief Encounter* and *Summertime*. But in the second half of his career, he worked almost exclusively on a much larger canvas, creating epics both great (*Lawrence* and *The Bridge*

on the River Kwai) and . . . well, a lot less great (anybody for *Ryan's Daughter*? Didn't think so). Lean's epics fill the screen with landscapes never before seen on film, teeming with hordes of people and constant motion. There is never a completely still shot in *Lawrence of Arabia*. That isn't to say that the camera is moving all the time, but rather that there's dynamic activity within the frame. Even if the movement is so slight as to be nearly imperceptible, as in the famous shot of the sunrise in the desert, Lean and his key collaborators (the most indispensable of whom are art director John Box, cinematographer Freddie Young, and composer Maurice Jarre) manage to draw your attention directly toward that motion in fascinating ways.

What sets *Lawrence of Arabia* apart from other gargantuan spectacles is its central character. T.E. Lawrence is one of the most complex, elusive, and ultimately unknowable figures in history and, as played by Peter O'Toole in a legendary, star-making performance, one of the most multi-dimensional characters to ever anchor a film. As the film begins, Lawrence is a mapmaker for the British army in Cairo, looked upon as a dilettante and a half-wit by his superiors. Lawrence is sent off to act as an observer in the camp of Prince Feisal (Alec Guinness). Instead of merely observing, however, Lawrence gets involved, soon becoming committed to Feisal's fight to liberate Arabia. He leads a successful, if arduous, campaign to seize the key city of Akaba, securing his image as a savior in the eyes of the Arabs. But because Lawrence is a man who really doesn't know who he is, he soon starts believing his own legend ("Didn't you know? They can only kill me with a golden bullet."). At the same time, Lawrence is both appalled and thrilled by the bloodlust this war has awoken in him. As the Arabs look to Lawrence for guidance and the British expect him to serve their interests, Lawrence struggles to figure out who he is. Is he a hero? A savage? A British soldier? An Arab revolutionary?

Columbia TriStar has done an outstanding job with this release, starting with the extremely handsome book-like packaging. This is (by and large) the 1989 restored director's cut of *Lawrence of Arabia* and it looks absolutely breathtaking. Since so much of the cinematography consists of tiny figures dwarfed by and emerging from unending landscapes, it's absolutely essential to watch this on as big a screen as possible. Just try to not be blown away by the clarity of such famous shots as Omar Sharif's unforgettable entrance, emerging from a mirage, or Lawrence, his face caked white with sand, seeing a ship seem to cruise through the desert as he reaches the Suez Canal. Even better than the video, however, is how amazingly good this disc sounds. Remixing older films to take advantage of 5.1 surround sound can be a dicey proposition. But done properly, as it is here or on Universal's *Vertigo*, it can be a revelation. From the second we hear those familiar rolling drums in the *Overture*, Maurice Jarre's score has never sounded as good as it does here. Battle scenes envelop you with galloping horses and camels, earth-shaking explosions, and whizzing bullets. And the scene where Lawrence, riding alone, discovers an echo in the hills brought a big grin to my face as O'Toole's voice reverberated from speaker to speaker. Most importantly, the new 5.1 mix retains the essential character and feel of the original track. A strong Dolby Surround 2.0 option is also included, as are French, Spanish, and Portuguese language tracks. But if you have access to 5.1 equipment, you'll be more than satisfied.

It's important to note that, as good as *Lawrence* looks on this special edition, a new Superbit edition will be out about the same time this book hits shelves. And it should look even better. Robert A. Harris, who did the original restoration work on the film under David Lean's supervision a few years back (and also a regular columnist for *The Digital Bits*), has overseen a new transfer for DVD with a keener eye towards what Lean was trying to accomplish. If you're a film purist, and a fan of *Lawrence*, it'll be a must have. Of course, you'll still need this special edition to reap the benefits of all the extra features.

And those features, relegated mostly to Disc Two, are extremely good, befitting a film of this stature. Laurent Bouzereau's hour-long documentary is simply one of his best, high praise considering the consistently high quality of his behind-the-scenes features. This was not an easy film to make and Bouzereau's documentary shows us how it was done step by step. Also included in the documentary is valuable on-set footage of director David Lean at work, which is simply fascinating. More of this footage is seen in the four featurettes. These newsreel pieces promoting *Lawrence* are kind of the early '60s version of *Entertainment Tonight* segments. They're interesting and reasonably well preserved, and give some idea of the hype that must have accompanied this film. Speaking of hype, the *Advertising Campaigns* feature is presented in a novel way. Instead of simply consisting of a gallery of posters, lobby cards, and promotional items, we see a short mini-documentary with narration explaining the different rationales behind each campaign, from the initial 1962 release to the much-shortened re-releases, to the 1989 restoration. Even the booklet is interesting, reprinting the text of the 1962 souvenir book. This essay goes into a surprising amount of detail about the historical T.E. Lawrence.

The only feature that you might not have much use for is the *Conversation with Steven Spielberg* featurette. It's not that it isn't interesting; it's just that it could have been edited into the documentary itself. There was no real need to highlight it as a separate feature, except of course to trumpet Spielberg's participation in the DVD. Also included in the set are the expected talent files for the cast and crew and theatrical trailers (though we were surprised that there was not a trailer for the '89 restoration and re-release).

DVD-ROM features are included on both discs and are of definite interest. *Archives of Arabia* divides the screen into three quadrants: the film in the lower left, text in the lower right, and behind-the-scenes photographs across the top half. There's a wealth of information in this feature. Because the photos correspond to their place in the film itself, the *Archives* are spread over both discs. Disc One also contains *Journey with Lawrence: Interactive Map of the Middle East.* This allows you to see the changing borders of the Middle East up to the present day. With so much international attention focused on the region today, this is a fascinating feature, lending historical perspective and making *Lawrence of Arabia* one of the few DVDs that can actually teach you about something other than movies.

Lawrence of Arabia is an astonishing achievement and, watching it today, you can easily believe that it took over three years to make. It's thrilling that Columbia TriStar gave this movie the treatment it deserves. Movies of this size and scope are virtually impossible to make today for a variety of reasons, mainly economic. Also, we no longer have many directors capable of bringing this level of complexity and sophistication to bear on a subject. Since we are unlikely to ever see another movie quite like *Lawrence of Arabia*, it's extremely gratifying to finally have the real thing immortalized on DVD. This release comes as close as anything will to capturing the thrill and grandeur of truly epic cinema.

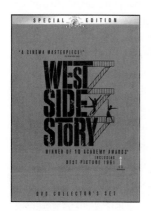

West Side Story: Special Edition Collector's Set

MGM

Musicals have come back in a big way these last few years, but in our minds, no musical says *musical* more than this classic film. *West Side Story* is so much more than just a great pop cultural reference, and it's even more than a redo of Shakespeare's *Romeo and Juliet*. It's an absolutely perfect melding of the Broadway style and Hollywood cinema. By simply taking a beautiful Broadway production (by Jerome Robbins and Arthur Laurents with music by Leonard Bernstein) and throwing it into a massive real-life city, the filmmakers essentially turned the world into a stage. Slightly dated, but full of passion, *West Side Story* is hard to beat when held up to film musicals young and old, great and small.

In this vibrant world, two gangs of kids (the Jets and the Sharks) don't get along much. The Jets are of full themselves and just a little bit (okay, a lot) racist towards the Puerto Rican members of the Sharks. When we first see them, they're chasing down a lone member of the opposite gang. That is, until he hooks up with his buddies and a rumble looks imminent. The cops break up the fight, and the teens go their separate ways. We're then slowly introduced into the worlds of these two rival gangs, and start to see them for who they really are—crazy, mixed-up kids, who are chasing dreams and wanting more out of life. Then we meet Tony. Not so much a Jet, but still a part of their group, Tony's a good kid. He has a job and, left alone, will go on to big things in life. At a party one day, he spies Maria, a beautiful woman who just happens to be the sister of the Sharks' leader. They make eyes, fall in love, and the stars cross—leaving the couple in an inevitably difficult situation.

West Side Story is remarkably beautiful. Co-directors (see the documentary for more on this) Robert Wise and Jerome Robbins join forces incredibly well, blowing life into Ernest Lehman's script. It's a shame the production went south behind the cameras (again, see the documentary). Happily, those problems didn't show on film, because what we see is pure magic. The cast is utterly perfect, especially the spunky charm that is Rita Moreno and the achingly beautiful Natalie Wood. We can gush, we can postulate, but ultimately time has spoken best—this *is* arguably the greatest musical of alltime.

The film is preserved in its original 2.35:1 anamorphic aspect ratio on this DVD, and it looks gorgeous. Colors are bright and deeply rendered. You won't hear any complaints from us on this transfer. The soundtrack is a newly created Dolby Digital 5.1 mix that, for the most part, sounds wonderful. It's not as dynamic as you would expect for a 5.1 track, but the film is over 40 years old, so the sources weren't too dynamic to start with.

This special edition contains a nice selection of features to please fans of the film. First, on Disc One, we get the original intermission music. Back when films were a bit more epic, audiences would often get a little more bang for their buck in limited or "road show" engagements. These engagements were considered quite special and had a very fancy, operatic feel. The film would have opening music or an Entr'acte, which culled together major themes from the film and presented them along with images. Depending on the length of the film, there might also be an intermission. On this DVD, special music was created to play with the intermission. In the wide theatrical release, the film didn't have an intermission — it just played in its entirety uninterrupted. Via seamless branching, this DVD lets you choose which version you wish to see — a very nice option.

Disc Two is where the bulk of the extras live. First up is a new documentary, which clocks in at about an hour. *West Side Memories* looks at the film's past, present, and future, courtesy of interviews with Robert Wise, Stephen Sondheim, Hal Prince, the original Broadway version's producer, as well as Rita Moreno, Richard Beymer, and Russ Tamblyn among others. Much ground is covered, and it turns out to be a fascinating look at the production. Also included are a whole slew of trailers, the original intermission music isolated without the film, and many photo galleries. One of the coolest extras here isn't even on the disc — it's a small book, included in the packaging, that contains a new introduction by screenwriter Ernest Lehman, the film's original script, reproductions of lobby art, and even some memos from the production. It's a hefty tome, and it's almost worth buying the set for this alone.

West Side Story is a beautiful reminder of how a musical can be powerful without being self-absorbed and unwieldy. Thankfully, this disc shows just how well the musical format can live on DVD. No movie library should be without it.

If You Like These Films . . .

. . . other classic titles worth checking out on DVD include *Apocalypse Now*, *Cleopatra*, *North by Northwest*, *Bridge on the River Kwai*, *Dr. Zhivago*, and the *Walt Disney Treasures* series just to name a few. One of the great things about DVD is that so many great older films are being released on the format in better quality than most people have ever seen them before. So don't pass up an opportunity to check them out when you can.

DVDs to Make You Wet Your Pants

Humor is in the belly of the beholder, they say. And since this is our book, you're going to see what makes us laugh. We're picking what we think are the funniest films with the bestest DVD treatment out there. If you like to laugh, be it ha-ha laugh or tee-hee laugh, we got your back. This is humor everyone can fall in line with. And if you don't laugh at these flicks, then check your pulse, 'cause you're dead, fella. That or you're Steven Segal. We don't think that guy ever laughs.

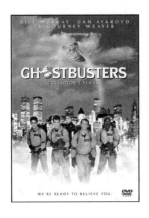

Ghostbusters: Collector's Series

Columbia TriStar

"Who you gonna call?"

Bill Murray, Dan Aykroyd, Harold Ramis, and Sigourney Weaver star in this hit 1984 comedy about a team of rough-around-the-edges paranormal investigators, who figure out a way to make big bucks by ridding people of their ghosts, spooks, and specters. Murray plays Dr. Peter Venkman, a university psychology researcher, who spends much of his time "testing" his prettier female students and dabbling in para-psychology. When he and his partners, Dr. Stantz (Aykroyd) and Dr. Spengler (Ramis), get kicked off the university gravy train, they're in desperate need of a way to make a living and continue their work. Using their newly developed ghost-hunting equipment, the trio decides to go into business as professional Ghostbusters ("We're ready to believe you," their TV ads say). Enter Dana Barrett (Weaver), an upscale New York musician whose apartment is becoming the epicenter of spook central, and the Ghostbusters suddenly have a client that leads to something *big* (cosmically speaking).

Murray, Aykroyd, and Ramis are absolutely in their element here, with glib one-liners and funny, off-hand comments dropping like flies. Aykroyd and Ramis, who also wrote the script, set Murray up time and again for perfect punch lines. And sight gags and physical humor abound as well — just try not to laugh as Murray tries to wipe ectoplasm off his hand and ends up flicking it in his eye. This is absolutely great stuff. Rick Moranis even manages to pull off a very funny subplot, as Ms. Barret's geeky, health-conscious neighbor, who gets possessed by the spirit of the "key-master" and runs around looking for the mysterious "gatekeeper." We won't blow any gags here, despite the fact that we're pretty sure most of you have seen this film. But on the off chance that you haven't, there's never been a better time to check it out . . . or a better way.

To be fair, the video and audio quality on this Columbia TriStar DVD isn't outstanding. Which is not to say that they're bad either. The video is presented in 2.35:1 anamorphic widescreen, and you can tell that the folks at Columbia really worked hard to get it looking as good as possible. The color is accurate, if a bit muted, with only light film grain visible. The contrast is generally good, but there's somewhat of a lack of detail in the blacks. The reason for this is there's a lack of detail in general. The print used for this transfer was of fairly good quality (and we're sure it was the best available), but it exhibits that softness that a lot of films from the early '80s show. The result is that there isn't a lot of crisp detail here – this is not reference quality. That said, the picture is still generally good and remains entirely watchable.

The audio is on about the same level of quality as the video. It's been remixed for Dolby Digital 5.1 surround, but it still comes off a bit lifeless. The mix is very front-centric, with only occasional rear channel use (and none of it is especially aggressive) during special-effects sequences. The bass seems adequate, but the dialogue comes across flat sounding and is at times overwhelmed by sound effects and music. Still, while this is not an especially immersive mix, it's not really bad either. It's adequate.

But these issues aside, it's the extras that make this a truly great DVD. And you get tons of extras, some of which are really unique, and hadn't been tried on DVD before they were tried here. To start with, you get all of the usual stuff, such as theatrical trailers, a photo gallery, deleted scenes, and various production featurettes. Just the fact that *so* much of that sort of material is included here would be enough to make any DVD fan excited. But Columbia TriStar has gone a step further, with a host of nice touches. For example, the gallery of conceptual art is set up so that as you look at each drawing, it appears to be laid out on a studio artist's drafting table. You also have the option to compare several scenes in the movie with and without special effects, and it's been set up so that, by using your remote's angle button, you can jump back and forth from the work print to the final film. It's even been frame-matched, so the jump is seamless – you can go back and forth several times during the scene while watching on the fly. You can look at the film's original storyboards with script excerpts, or as compared to the final filmed scene, in split screen. And that's not even the best of it. One of the subtitle tracks contains production notes on the film, excerpted from Don Shay's book *Making Ghostbusters* – almost the whole text of the book is here.

If you're like us, and you love commentary tracks, you're in for a real treat here. Not only can you listen to commentary with Ivan Reitman, Harold Ramis, and Joe Medjuck (the director, writer/actor, and producer respectively), you can actually *see* them as well. Columbia has shot video of them watching the film against a screen in silhouette and has encoded this visual information on another subtitle track. In this way, by using your remote's subtitle select and on/off buttons, you can turn on the "video" of them at any time during the movie, as well as the audio. The effect is rather like watching an episode of *Mystery Science Theater 3000* — you see them pointing at the screen, while you hear them laughing and talking. It's an extremely impressive piece of work.

These big features aren't the only things that stand out here. It's also the little touches that make you happy with this DVD. Many people don't think of menu screens as terribly important to the DVD experience, but the very best DVDs use animation and sound in their menus to really immerse you in the experience of the film. And they also make exploring the disc's contents easy and intuitive by streamlining navigation. The menus here are absolutely outstanding. You start out hovering above Dana Barrett's building in New York City, complete with traffic on the streets below, and the StayPuft Marshmallow Man stomping between the buildings. From here, you can opt to play the movie, activate the "live" commentary track, or jump to a specific scene or extras. Choosing the "scene selection" option, for example, then leads to a 3-D fly-through animation, where you zoom closer to the building, and several of its windows open up to reveal moving images of the various scenes, which you can then select. Want to go to a later scene? No problem — the "Next" button is already highlighted, so there's no need to press three or four buttons just to get to the next page (a *very* nice, and rare, ease-of-navigation feature). And let's say you're deep into the extras and want to go to another area of the disc. These menus are extremely well planned and implemented.

Ghostbusters is one of the funniest movies you'll ever see. Whether you're a fan of the movie, or just the kinds of innovative features DVD can offer, this classy disc is absolutely not to be missed.

This is Spinal Tap on DVD

The only way to rate this film properly is by giving it an "11." This is the all-time mother of all rockumentaries. Forget *Biography, Behind the Music,* or Lifetime's *Intimate Portrait* — this is what it's all about. *This is Spinal Tap* made a minor dent in the Heavy Metal armor of youth culture when it hit theaters in 1984. Over the years, that dent has grown bigger, as *Spinal*

Tap has become a phenomenon on video. *This is Spinal Tap* was, and is, a major cult favorite. Everywhere you turn, someone's either ripping it off or paying homage to it in some way.

This is Spinal Tap amusingly tracks the fall (forget rise — there *was* no rise) of England's least important metal band, Spinal Tap, from their days as a teeny-bop hair band, through their reinvention as a hippie love quartet and on to their present incarnation. Rob Reiner makes his directorial debut here (and also stars in the film as documentary filmmaker Marty DiBergi). The overall plot goes like this. DiBergi hears that his favorite band is doing a "farewell tour" and decides to document the experience as the band claws their way through the Americas in pursuit of money and success. All of their triumphs and catastrophes are caught on tape . . . and you won't stop laughing as you watch each and every one of them. If you haven't seen *Spinal Tap*, do yourself a favor and check it out soon. If you have, then you know what we're talking about.

Now . . . there are two versions of this film on DVD, and both are worth tracking down . . . even if one of them is out of print. You can still find the Criterion Collection version on eBay, and it *does* contain more and different extras than the recent MGM release. To give you a better sense of this, we'll break both down separately for you.

This is Spinal Tap

MGM (The Criterion Collection)

The Criterion Collection didn't disappoint with this one. Even the animated menu screens are well thought out — we get hellfire lying at the bottom of the screen, licking and spitting at the onscreen options. The disc looks wonderful for a non-anamorphic transfer. The film fits nicely at 82 minutes on one side of the disc. Flip it over, and the other side holds the extras. The audio is available for the film in the original 2.0 stereo and it sounds fine.

As for the features, let's just list them off for you, shall we? You get two separate audio tracks (one with the boys of *Spinal Tap* showcasing their comedic genius, and another track, almost equally funny, featuring Rob Reiner and some key talent behind the camera). These both run the length of the film and can be found on the movie side. The highlight of this disc, though, is the supplemental section. There are the hilarious promotional trailers showing an Octoberfest celebration that has absolutely nothing to do with the movie. There's also the original, 20-minute demo reel (*Spinal Tap: The Final Tour*) that features some of the better bits of the film, shot with different supporting actors, and some scenes that didn't even make the theatrical release. This was put together to raise money for the production, and it gives a completely different look to some of the characters. They also throw in a fake commercial for a best-of album entitled *Heavy Metal Memories* and the video for "Hell Hole." Both look real and are really funny. Finally, you get about an hour of deleted scenes. A whole hour of material that didn't make the final cut! It's all really funny, but some of the best bits are the more in-depth scenes with Billy Crystal's ranting mime, along with some great character work by Bruno Kirby as a chauffeur that can't get enough of Frank Sinatra.

This is Spinal Tap: Special Edition

MGM

Held against the Criterion edition, it's clear that this edition features a much cleaner-looking video image over the previous release. While I like the more grainy "film" look the Criterion edition has, the colors, blacks, and overall feel of the MGM edition is stronger. A lot of that comes from the fact that the MGM release features a new anamorphic widescreen transfer, which is a big improvement, plain and simple. There is very little artifacting anywhere to be seen. The image is framed at 1.85:1 as compared to the 1.70:1 of the Criterion edition. In an A/B comparison, you don't see much difference. You're not missing any information, so there shouldn't be any complaints there. Sound-wise, the MGM disc is also much improved, with a remixed Dolby Digital 5.1 track (compared to the 2.0 stereo track on the previous release). You won't care much about that in the straight dialogue sections of the disc, but it makes a big difference in the concert footage. The remixed sound is crisp, clear, and rock solid. You'll be very happy if this edition is the only way you have a chance to experience *This is Spinal Tap* on DVD.

The reason we still chose to mention the outdated Criterion edition in this book is because of the extras. The better extras, in terms of video and audio, are those on this MGM edition. But true *Spinal Tap* fans are still going to want to get the Criterion edition. That's not to say that these extras are bad, because they're not. They're just different. First up here, there's only a single commentary track on this edition. But the shtick here is a pretty good one—the commentary was recorded with the *actual* members of Spinal Tap (yes, the same actors as on the Criterion track . . . but they're actually in character here). And although it's frickin' funny, it's definitely different than the Criterion edition. So right there, you have one reason to get both DVDs if you're a fan. The second reason you'll want both is that both versions feature some different deleted scenes. The deleted scenes on this MGM version are much better, in terms of video and audio quality, than the ones on the Criterion disc. On the other hand, the Criterion edition has *more* cut footage. In addition, this new MGM release lacks the demo film *Spinal Tap: The Final Tour*, which appeared on the Criterion disc.

All that said, the MGM disc includes some new things we haven't seen before. On top of the character-based commentary, you get a nice teaser of it when you first pop in the disc—the band members talk over the menu screens. We won't spoil it for you here, because some of what they say is hilarious. Next, there's a short interview with Marty DiBergi (Rob Reiner in character) where he talks about what he's been up to since the film was released (he designs mouse-pads for promotional releases). Included in the interview is a snippet of a press conference where Tap talks about the film negatively. It's pretty funny and looks great on DVD. Also exclusive to this disc is a very short clip of Spinal Tap on *The Joe Franklin Show*. It's not very funny, but it's not on the Criterion disc. There's also the film's original theatrical trailer; one of the "cheese-rolling" promos (the Criterion DVD had two); some TV spots; music videos for "Big Bottom" (exclusive to this MGM release), "Hell Hole" (which is also on the Criterion disc), and two other songs; the fake commercial for *Heavy Metal Memories*; and three commercials for an Australian food pocket snack called Rock n' Rolls.

If you had to make a choice—and it would be a tough choice to make—you pretty much can't turn away from the new MGM version over the Criterion. The movie itself looks better and this is just a better special edition than the previous release. It's generally more fun with some nicer touches fans will appreciate. But for those of you who *are* fans, adding the Criterion edition of *Spinal Tap* to your collection, along with this MGM disc, is like having the ultimate two-disc edition. So our advice is to grab both if you can.

The Blues Brothers: Collector's Edition

Universal

"Orange whip? Orange whip? Three orange whips!"

Funny men Dan Aykroyd and John Belushi star in this raucous comedy about a pair of musical con men "on a mission from God." Based on a routine the two comedians developed just for laughs (and later portrayed on NBC's *Saturday Night Live*), *The Blues Brothers* begins with motorhead Elwood Blues (Aykroyd) picking up his brother Jake (Belushi), who's just been released after serving three years in Joliet. Their first stop: their old home, the Saint Helen of the Blessed Shroud Orphanage. Elwood reminds Jake that he promised to visit the head nun (Sister Stigmata, a.k.a. "The Penguin") when he got out of the slammer. There they learn that the Church has decided not to pay $5,000 in back taxes owed to the county. Unless the money is found, the orphanage will be closed. Determined to help but unable to decide how to obtain the money legally, Jake and Elwood take in a high-energy sermon by the Reverend Cleophus James (none other than James Brown), and Jake suddenly "sees the light." He's got the answer to their problem: put their old R&B band back together and play a few gigs to earn the cash. So Jake and Elwood set upon their "holy mission" to track down their former band members, who have all gone their own ways. What follows next has got to be one of the most bizarre and funny series of musical numbers, car chases, and general misadventures ever captured on film.

Where else can you find rip-roaring comedy, fast-paced action, 50-car pileups, and performances by some of the finest rhythm and blues musicians ever assembled, including Cab Calloway, Ray Charles, Aretha Franklin and John Lee Hooker? *The Blues Brothers* has definitely got a little of everything, and there's certainly something for everyone. Heating up the action is the fact that everybody in this film is out to get Jake and Elwood for one reason or another. There's a rocket launcher-packing ex-girlfriend hell bent on revenge (Carrie Fisher), Jake's portly parole officer (John Candy), a band of Illinois Nazis, and even an R.V. full of good old boys (a country band called, appropriately, the Good Old Boys). Throw in Jake and Elwood's ragtag band of musicians, hundreds of Chicago police officers, Illinois state troopers, tank-driving National Guardsmen, one ninja nun, and a dash of the *Peter Gunn* theme, and there's just no stopping the take-no-prisoners mayhem.

The real beauty of this collector's edition DVD is that director John Landis has edited some 12 minutes of footage back into the film — footage which hasn't been seen since the first preview screening back in 1980. In the production notes, provided in the booklet that accompanies the DVD, Landis explains that the original edit of the film contained even more footage than what was restored here, but unfortunately that original print has been lost. Too bad — we would have loved seeing Jake and Elwood singing "Sink the Bismark" at Bob's Country Bunker! Nonetheless, the 12 minutes that *were* found are a real treasure. This isn't just bits and pieces — several scenes have been expanded or restored. Finally, we learn what gives the Bluesmobile its power, we see Elwood quit his job (and learn where he got that can of glue) and almost every musical number is longer, including John Lee Hooker's street performance. This is *The Blues Brothers* better than you've ever seen it.

The film is presented in 1.85:1 letterboxed widescreen and is enhanced for anamorphic displays (even the restored footage). The video quality itself is generally good, although it's occasionally a bit grainy, and this print is high in contrast. The added footage can usually be identified, because it exhibits slightly less contrast. There is also some artifacting to be seen (particularly in the murky clouds at the beginning), although it's minor. It really isn't a problem anyway — the film has always had a bit of a grungy, south-side of Chicago look to it anyway. And the color of this print is very good.

The soundtrack has been remixed in full Dolby Digital 5.1 surround sound. With all the action sequences, there's some cool use of the surround channels, but it's the music that benefits most here. This film has some terrific musical numbers and they've never sounded better. Listen to the scene at the Triple Rock Baptist Church if you doubt it (Chapters 5, 6, and 7). The congregation can be heard calling out all around, and the organ thrums majestically from the rear channels. In Chapter 10, during the car chase inside the shopping mall, there's a shot inside an upside-down, spinning police car, and you can here a brief spinning effect moving around the room. And just listen to the bullets fly when Ray (Ray Charles) scares off "another kid going bad." This is some very nice remixing work.

The extras on this disc are equally good. There's a gallery with dozens of production photographs (be sure to look for a section of Jake's "engagement" photos). There's a theatrical trailer, and pages of production notes and cast and crew bios. Best of all, there's a great behind-the-scenes documentary on the making of the film. It's almost an hour long, and it's full of great stories and anecdotes. There's a very funny (and touching) moment, where Aykroyd recalls a night during filming when Belushi disappeared. A quick search of a nearby neighborhood revealed that John had simply invited himself into some stranger's house, helped himself to a sandwich and a glass of milk, and then crashed out on the guy's couch! It's a great story to listen to, and you can really see how much Aykroyd still appreciates and misses his friend.

The Blues Brothers: Collector's Edition DVD is a blast. There's no other way to put it. With all of the expanded footage, and with the extras provided, there's just no better way to watch this movie. As big fans of *The Blues Brothers*, we're completely satisfied by this DVD. Just one recommendation . . . avoid the sequel, *Blues Brothers 2000*, like the plague. That should just *never* have happened.

Monty Python's Flying Circus on DVD

Okay, we should confess it right here and right now . . . we're huge fans of those chaps from Monty Python at *The Bits*. Whether it's the Ministry of Silly Walks, the years-long game of Olympic Hide and Seek, the Fish Slapping Contest, the Dead Parrot Sketch, the Spanish Inquisition, or the Lumberjack Song, there's nothing that gets us laughing faster than an old episode of *Monty Python's Flying Circus* on DVD. Well . . . except maybe one of the Python films on DVD. Hint, hint . . .

Remember . . . a nod's as good as a wink to a blind bat!

Monty Python and the Holy Grail: Special Edition

Columbia TriStar

Bedevere: "And that, my liege, is how we know the Earth to be banana shaped."

Arthur: "This new learning amazes me, Sir Bedevere. Explain again how sheeps' bladders may be employed to prevent earthquakes."

Bedevere: "Oh, certainly, sir."

It's the first, it's the best, it's . . . *Monty Python and the Holy Grail*. It's basically a retelling of King Arthur's tale, except with coconut-carrying sparrows, killer bunnies, and a group of more than slightly askew knights. First there's King Arthur, played by Graham Chapman, who is the straight man of this topsy-turvy world (and even that aspect is funny when you think about it, an openly gay man playing the film's only "straight" man). His right-hand man is Sir Lancelot (John Cleese), who generally saves the day by killing everyone and everything in sight. Next up is Sir Robin, the Not-Quite-So-Brave-as-Sir Lancelot (Eric Idle), a cowardly knight with his own minstrel group giving him guff for his shortcomings. There's Sir Galahad the Pure (Michael Palin), who finds himself an inch away from a life-changing event most men in their right mind would beg for. There's also the scientific Sir Bedevere (Terry Jones), whose mastering of duck physics knows no equal (except maybe Arthur). And, of course, there's Arthur's faithful "steed" Patsy (Terry Gilliam).

Each of the Pythons also throw in for additional roles, most notably Cleese as both the Black Knight and a French solider who "taunts" our heroes, Palin as a politically savvy peasant named Dennis ("Listen . . . strange women lying in ponds distributing swords is no basis for a system of government. Supreme executive power derives from a mandate from the masses, not from some farcical aquatic ceremony."), Gilliam as the gatekeeper of the Bridge of Death (be sure to know your favorite color when riddled by this guy), and just about everyone as the fan favorite Knights Who Say Nee!

The film looks damn fine on DVD. It a great transfer. It's an anamorphic transfer with great color representation and good detail. For the record, this version of the movie incorporates a few seconds of film added back in (a funny sequence), but you'd never know it—it's that quick. The audio is in both the original mono and a new Dolby Digital 5.1 remix, which both sound great for their age. Can't fault this disc's quality for anything.

And that doesn't change with the extras. This is a two-disc set and it shows. Disc One includes the movie and several commentaries. There's a directors' commentary with Terry Gilliam and Terry Jones, who go into the history of the troupe, the production, and their careers in general. Next is the cast commentary with the surviving Pythons: Eric Idle, John Cleese, and Michael Palin (we lost Graham Chapman to cancer sadly). Attention is focused on the actors and the writing, and it's clear these men love the film.

Disc One continues with the screenplay running as a subtitle feature. It's a very neat thing that we haven't seen again, and it's a shame, especially for those of us who dislike DVD-ROM. We also get some jokey extras, like Subtitles for People Who Don't Like the Film, which are nothing but mangled bits of Shakespearean prose. There's also menus for the hard of hearing with a voice screaming the disc's contents at you in 5.1. Also on Disc One is the Follow the Killer Rabbit feature, poking fun at *The Matrix*'s Follow the White Rabbit option. This gives you access to notes, sketches, and storyboards while you're watching the film.

Disc Two has some of the more interactive things. There are sing-a-longs (featuring excepts from the film with subtitles and a bouncing ball), an interactive cast directory (with photos from the film of each character), and a whole heap of video-based extras. There's a new documentary, *The Quest for the Holy Grail Locations*, where Terry Jones and Michael Palin head out on a tour of the Scottish shooting locations. *How to Use Your Coconuts* is a new short starring Palin as a bureaucrat instructing the audience on how to use coconuts to make horse hooves' sounds — just like in the movie. There's also a Japanese version of a scene from the film that proves to be funnier than you'd expect because of syntax and cultural differences. Rounding out the disc are trailers, stills galleries, a BBC news report, poster and conceptual art, and a reenactment of the Camelot sequence done with LEGOs. Seriously.

You should also know that by the time this book is in stores Columbia TriStar will have released a second version of this film on DVD — the *Monty Python and the Holy Grail: Collector's Edition*. Basically, it's the same two-disc set we've reviewed here, simply packaged in a new, black collector's slipcase with a paperback copy of the script and a collectible film "seni-type" cell.

So that's *Holy Grail* in a coconut shell. You want brilliant comedy? Look no further than this. But hold tight, 'cause we're just getting started with the Pythons . . .

Monty Python's Life of Brian

The Criterion Collection

"No one . . . is to stone ANYONE . . . until I blow this whistle!"

Here's how the story goes. Way back in ancient Judea, at the same time as the birth of Jesus, another boy is born in the next stable over. His name is Brian, the bastard son of a Roman foot soldier (but don't call him a Roman). As Brian (played by Graham Chapman) grows up, he finds that his life is inexplicably intertwined with the Messiah's. And as for plot . . . well that's about all there is really. This is just the insane story of his life. Along the way, Brian will get mixed up with the People's Front of Judea, an ex-leper, countless Roman centurions, Pontius Pilate, several wise men, a wannabe middle-manager named Reg, a pair of one-eyed aliens from outer space (no, we're not kidding), and a whole host of other loonies.

There are some truly great moments in this flick. In one of the most memorable, Brian is caught painting "Romans go home!" on the side of a wall by a Roman soldier. The soldier gets mad . . . not because what he is doing is a crime, but because Brian's Latin is wrong! So he proceeds to correct Brian's grammar, and then makes him paint it hundreds of times correctly. We could go on and on, but we don't have enough space. There is just one absurdly funny moment after another in this film.

Brian is the sole Python film that's been released by the Criterion Collection on DVD, and it looks and sounds as good as can be expected (considering its shoestring budget). Criterion retooled their previous laserdisc release for this new disc. We get not one, but two commentaries comprising the whole of Python, minus the late great Graham Chapman, of course. Although he is sorely missed from these tracks, what we get is still funny. The first track is director Terry Jones, with Terry Gilliam and Eric Idle. The second is John Cleese and Michael Palin. In addition, there's a very well produced documentary on the making of *Life of Brian* and the history of the legendary comedy group. If you're a Python fan, you need the disc just for this. Filling out the DVD are five deleted scenes, a trailer, and some British radio spots for the film. All in all, it's some very funny stuff.

Monty Python's And Now for Something Completely Different

Columbia TriStar

So you say you like Monty Python as much as we do, huh? Well, until you have this disc in your library, we won't believe you. You see, there's no better way to enjoy the boys' onscreen antics than with *And Now For Something Completely Different*.

Watching the film, you'll probably recognize much of what you see here as familiar. That's because *And Now For Something Completely Different* is made up entirely of the best sketches the group did on British TV, simply restaged for the big screen. The film was shot in the early '70s, for American college audiences, who hadn't quite discovered the Python boys yet in syndication. The result is absolutely hilarious and just downright silly (for which one of the Pythons apologizes profusely several times during the film). You get The Lumberjack Song, The Dead Parrot sketch, and A Man with a Tape Recorder Up His Nose. You get Upper Class Twit of the Year, Nudge, Nudge, and Silly Army Drills. Hell, you even get The Funniest Joke in the World. There are some 28 bits in all, and lots of Terry Gilliam's goofy little animated interludes are there along the way. How's that for a comedy-loaded film? If you're a Python fan, you probably remember every bit we just mentioned. But if any of this sounds unfamiliar to you, then you absolutely don't know what you're missing.

On DVD, this film is presented in fairly good quality. You get anamorphic widescreen on one side of the disc and full frame on the other. The video is very good looking, given the age of the film, and its low-production value. It's a little on the soft side, with some light film grain and print dust visible. But the colors are accurate (if muted), and the contrast is generally good. The audio is in Dolby Digital 2.0 only, but that's just fine with us—this isn't a surround sound-fest anyway. Extras on the disc include . . . well, there really aren't any except bios of director Ian MacNaughton and the rest of the Pythons. And the scene selection menu is organized by sketches, so that you can jump from one funny bit to the next as you please—not really an extra, but convenient.

It's obvious that *And Now For Something Completely Different* isn't going to win any DVD awards. But who cares? It's in our book anyway, 'cause it's Python and therefore a must own.

We should note that by the time this book is out, you'll also have a new snazzy edition of *Monty Python's The Meaning of Life* from Universal. It'll be in anamorphic widescreen video, with both Dolby Digital and DTS 5.1 audio. Extras are slated to include a new introduction by Eric Idle, audio commentary, deleted scenes, *The Meaning of Making the Meaning of Life* documentary, and a slew of featurettes (including *The Songs, Snipped Bits, Un Film de John Cleese, Songs Unsung, Education Tips, Fish, Re-mastering a Masterpiece, Song and Dance, Virtual Reunion*, and *Selling the Meaning of Life*). Sounds like it'll be worth picking up, so watch our web site for a review.

Well? What are you bloody well waiting for? Pick yourself up some Python on DVD and give it a spin! You'll be glad you did. Wink, wink, nudge, nudge . . . know what we mean?

If You Like These Films . . .

. . . be sure to check out *O Brother, Where Art Thou?*, *Blazing Saddles*, *Young Frankenstein*, *1941*, *Caddyshack*, the *Naked Gun* series, *Airplane*, the *Austin Powers* films, *Galaxy Quest*, *Office Space*, *A Christmas Story*, and *National Lampoon's Vacation* series as well.

DVDs to Make You $#%! Your Pants

Turn down the lights, cuddle close to your significant other, and pop into your DVD player some of the scariest films on disc out there. And after you've pooped your pants, you can learn how the filmmakers gave you such great scares . . . and poop your pants again. In case you haven't guessed, it's all about pooping your pants in this chapter.

Editor's Note: *Bits* writer Adam Jahnke contributed to this section.

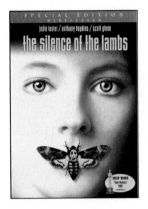

The Silence of the Lambs: Special Edition

MGM

Based on Thomas Harris's brilliant novel, *Lambs* focuses on the twisted journey taken by a young FBI trainee named Clarice Starling, played with an unflinching honesty by Jodie Foster. The journey Starling makes is, in fact, a quest. She is asked by her superior, Jack Crawford, to question the infamous Dr. Hannibal "The Cannibal" Lecter. Lecter is Starling's mentor in a way. She's trying to glean from him information about a current rash of serial killings occurring around the state of Illinois. Nicknamed Buffalo Bill, the serial killer is so evil, criminal profilers claim there couldn't be a person evil enough to commit these crimes. What we know (and the FBI doesn't) is that Bill is making himself a suit. But not just any Brooks Brothers job, no siree, Bob. Our nasty friend is making a suit out of human skin — dead, overweight, and female human skin. It's a nasty idea, but it works in the context of this film.

Clarice and Lecter form a bond of sorts, and he agrees to help her, but only if she tells him a story from her childhood. She agrees, and that's the movie in a nutshell. We usually don't like giving things away, so if you haven't seen the film for yourself, you should — and don't let anyone's opinions stand in your way. This is one well-made film, and if the subject matter doesn't appeal to you, just keep in mind that it's done very tastefully. Hell, we doubt Foster would be involved if it was exploitative.

Cinema was given one of its most endearing anti-heroes in *The Silence of the Lambs*, and somehow we're all the better for it. Anthony Hopkins injected so much life into the Thomas Harris creation of Dr. Hannibal "The Cannibal" Lecter that it earned him a Best Actor Award at the 1991 Academy Awards. Hopkins took what was a stomach-turning, revolting, little goblin and made him into one of screens sexiest characters. If it weren't for the tongue eating and the explosive temper, Lecter would be someone we'd all love to have over at our house when entertaining guests. He's calm, cool, and witty . . . which makes sense. There's a reason why serial killers get as close as they do to their victims. You'd have to be a charming, witty, and intelligent guy to be able to pull off some of the weird crap that goes on in this world.

Despite her pedigree and many awards, Jodie Foster plays her best character in this film. Clarice Starling is the classic screen hero. She experiences emotional growth, follows her own mind, and saves the day — eventually. Foster took something that could have been a standard female role and made it into a three-dimensional character that people all believe in, all with a look, an accent, and a quiver. That, my friends, is acting, acting that won Foster her own award at the Oscars.

Surprisingly, Hollywood and the mass public alike embraced this tale of serial killers, red herrings, and a frumpy FBI agent with a good bag and cheap shoes. This MGM special edition presents the film in anamorphic widescreen pulled from a newly created high-def master. The color representation is beautiful, showcasing the wide and bold palette used by Demme and his cinematographer, Tak Fujimoto. The grain has been kept low and the blacks deep and solid. This transfer really shows what the film should look like in your home. *Silence* comes with a solid Dolby Digital 5.1 soundtrack on DVD as well, and it's quite good.

Considering this is a special edition, the focus is on the extras. Sadly, the audio commentary with director Jonathan Demme and actors Anthony Hopkins and Jodie Foster (that were on the previous Criterion DVD release) is missing from this set. That's too bad, because it was a great track. Most of the other extras from Criterion's edition are also missing. But that doesn't mean this edition sucks. Far from it. There's a new hour-long, making-of documentary entitled *Inside the Labyrinth: The Making of The Silence of the Lambs*. It features interviews with just about every person who made the film (sadly missing are director Demme, actress Jodie Foster, and actor Scott Glenn, although archival footage bring them into the mix) and even a few people who only saw the film and liked it a lot. You'll find appearances from Hopkins, actresses Brooke Smith and Diane Baker, producer Ron Bozman, screenwriter Ted Tally, composer Howard Shore, costume designer Colleen Atwood, bug wrangler Ray Mendez, and even cult filmmaker Roger Corman. These interviews are illustrated by stills and production footage, giving light to the film from concept to release. It's a great documentary.

Next up is the original 1991 featurette, *The Making of The Silence of the Lambs*, which is a little better than your standard *electronic press kit* (EPK) fare. Also included are 22 deleted scenes running around 20 minutes in total, some outtake footage, the theatrical trailer, about nine TV spots, the *Hannibal* DVD trailer, a phone message from Lecter (done by Anthony Hopkins), and an extensive still gallery.

If you're interested in following Clarice Starling into the pits of Hell, let this DVD be your guide. Its wonderful print, good sound, and bountiful extras will keep you occupied for hours. Just FYI, this DVD is best served with fava beans and a nice Chianti. And no, we're not going to close this review with that creepy slurping noise, so don't ask.

A Nightmare on Elm Street

New Line

"One, two, Freddy's coming for you . . . "

Wes Craven, a filmmaker who seems to live for reinventing the horror genre at least once every decade, made magic in the mid-80s when he gave life to maniac Freddy Krueger. Freddy is known the world over. He's a veritable pop culture icon at this point, with dolls, T-shirts, and tattoos popping up everywhere, even years after he first entered our movie-going collective consciousness.

But Craven didn't just reinvent horror. *Nightmare* became one of the most successful film franchises in the history of the medium, making fledgling distributor New Line Cinema a major player in the film world. With seven feature films, a television series, comics, and videos to his name, Freddy was long one of the most requested characters for release on DVD (save for a couple of guys named Indiana and Luke).

The king *Elm Street* flick is the first one. *A Nightmare on Elm Street* is a creepy "haunted house of the mind" film. Four teenagers (that we see at least) are tormented by a mysterious man in a red and green sweater, with "long fingernails" (which we know from the opening credits to actually be knives). The man only comes to them in their dreams, and he's not trying to be their friend. He's out to get them, and the reason why is only revealed at the end. Nancy (Heather Langenkamp) is the lead teen, a girl who has her share of typical teen problems, such as divorced parents and a mother who drinks too much. The film is seen through her wide-open eyes as she comes to grips with the fact that she could die in her dreams at any moment—and that means dying for real. Who is this mysterious man in the red and green sweater? Why does he want to kill Nancy? Where can we find that sweet, talking Freddy doll Matchbox put out a few years back?

When it comes to the overall horror factor, this flick blows most other such films out of the water. Here, Freddy is creepy, his body is thick with maggots and bile, and his skin is Kentucky-fried crispy rather than the melted cheese sandwich look he was given in most of the rest of the series. The film also has a layered, mythical quality to it. It's not an easily dismissed film—it has an intelligence behind it that sadly has been undermined by the rest of the films in the series (at least up until *New Nightmare*). Robert Englund gives so much character to Freddy that we find it hard to believe he's such a nice guy in person (he is). The guy *has* to have a few bodies stacked up in a closet somewhere. Seriously, though, Englund is a really talented actor, and to be able to emote as well as he does under all that latex is testament to that. The rest of the cast is also good, in that B-movie sort of way. Who didn't have a crush on Heather Langenkamp as a kid? *A Nightmare on Elm Street* is one of the best horror films ever made. If no sequels had ever been made, you'd find this film listed among the scariest of all time. Sadly, the sequels hurt this film.

There are two ways to get this great film on DVD: as part of the complete *New Line Platinum Series: The Nightmare on Elm Street Collection* or individually as a stand-alone edition. Since most of the films in the set aren't terribly good (unless you're a Freddy fan) and the box set has since fallen out of print, the stand-alone version is your best bet.

The stand-alone DVD contains an anamorphic, 1.85:1 widescreen version, as well as a full-frame 1.33.1 edition. Both look really good. This DVD transfer is quite comparable to the fantastic laserdisc special edition put out by Elite a few years back, except for one thing: The transfer for the LD was supervised by cinematographer Jacques Haitkin. If you look carefully at the sequence after Freddy is brought into the real world, he takes a tumble down the stairs. On the widescreen version you can clearly see the stunt mat covering the stairs, which was a gaff. On the LD, Elite and Haitkin worked around the gaff and it was covered up quite nicely. On the DVD, it's there in all its glory. Oh well. The good thing about that is at least we know that this is a brand-new transfer to DVD. Other than those *minor* complaints, the picture is pretty sweet. The blacks are solid, there are no artifacts to be found, and the grain isn't exacerbated by the MPEG-2 compression. Just as a note, the box set edition contains only the anamorphic widescreen version, and it looks exactly like the movie-only edition.

On the audio side of things, *A Nightmare on Elm Street* is super sweet. There's a new 5.1 track done for this DVD, and when you're in the boiler room . . . *you are in the boiler room*. The surround effects are very well done here. The original mono is also available, and it's fun to compare the two — the mono track is nicely presented as well. It's pretty full-sounding for a mono track, but doesn't hold a candle to the 5.1. Both sound options are available on both the box set and movie-only discs.

The special edition materials are only slightly different for the two discs. The movie-only edition and box set DVDs both contain the commentary track with Wes Craven, Heather Langenkamp, John Saxon, and Jacques Haitkin. It's the same track produced by Elite Entertainment for their laserdisc, and it's incredibly informative and fun to listen to. Craven and the gang all seem to really enjoy themselves, and they bounce loads of stuff off of each other. You'll have fun listening in. Another fun extra that appears on both discs is a set of cast and crew bios. Taken from the film's original 1984 press package, they read like a blast from the past. Although not incredibly useful, this is an interesting approach to press information, historically speaking. The original press lists omitted actors Johnny Depp (who made his screen debut in this film) and Robert Englund (who plays Freddy — New Line wanted a slight mystery to surround the main character and didn't want to give a human face to the onscreen evil). Both discs also feature DVD-ROM capabilities. The movie-only edition has a longer trivia game, the theatrical trailer (which is available on the Encyclopedia disc in the box set), and a link to the series' web site. Both discs also include the ability to read the script while watching the film at the same time (through the DVD-ROM).

A Nightmare on Elm Street made true horror fans out of a boatload of people back in the 1980s. It helped bring intelligence to the genre and was good enough to make New Line Cinema a film industry powerhouse. Check it out for yourself on DVD, but be warned . . .

Halloween: Limited Edition

Anchor Bay

No one who has seen *Halloween* will ever forget it. Who would have guessed that an awkward girl with a long, sad face, a scraggly, pasty-white filmmaker, and a guy in a Captain Kirk mask (no kidding — that's what it is), would change horror history? *Halloween*, film scholars remind us, is one of the highest-grossing independent films ever made. Film scholars like to throw that around, because it validates the film's power somehow. It makes it legitimate to be a fan of the flick. It's like enjoying *Sweet Sweetback's Baadasssss Song*, which was arguably the first blaxpoitation film (and therefore is important, even if it *is* cult trash). Frankly, we don't see the point in validating it. If a film is worth watching, it's worth talking about. *Halloween* is a true classic, and even if it made nothing and influenced no one, most of us would still be here singing its praises. The simple fact that it *did* make loads of cash and influenced a generation of filmmakers just nails that point home all the more.

As the film starts off, we find ourselves looking through a six-year-old child's point of view. It's Halloween 1963, and a young boy pulls on a clown mask. At first we think, "Oh, boy! We're going to see some trick or treating from this kid's point of view." No. What we do get to see is some serious knife stabbing. There's no blood, but that icky knife-in-a-melon sound fills in so well for a knife in a human chest. Cut to 1978. It's once again Halloween, and the little boy is all grown up and freshly escaped from a mental hospital. His keeper, Dr. Sam Loomis (played by the all-time best Carpenter character actor, Donald Pleasence) is hot on his trail. He knows that the boy/man (named Michael Myers) will return to the scene of the crime — Haddonfield, Illinois. What he wants in Haddonfield, only Loomis and Myers really know. The best guess is that he wants to kill everyone. And the one person he wants to ventilate with a kitchen knife more than any other is Laurie Strode (played by pop culture diva, Jamie Lee Curtis). This Myers guy, now sporting the aforementioned Captain Kirk mask, really has a jones for Laurie. And Loomis is slow to realize what is going on . . . or to get the police to help him.

The movie unfolds slowly, and when the chills and killings start, they're relentless. What really makes this film work is its pioneering use of Steadicam and its music. Carpenter used the Steadicam expertly, weaving in and out of houses and sneaking up on people with an unearthly flow. It's unnerving and makes you feel like you're right there in the room. And following Bernard Herrman's lead (based on his soundtrack work for the Hitchcock films), Carpenter fashioned a truly relentless score that still makes you want to check your closets for bogeymen each and every time you hear it. Scary.

Quite frankly, *Halloween* looks beautiful on DVD. It's anamorphic widescreen, which preserves the original Panavision widescreen aspect ratio of 2.35:1. Blacks are very solid, shadows are detailed, and there's only minimal edge enhancement. On the audio side, we're given three incredible English tracks — a remastered mono track that sounds nice, a Dolby Digital 2.0 track that sounds even better, and a wondrous Dolby Digital 5.1 mix. We don't think you could hear Carpenter's music sounding any better if you tried. Just pop in the DVD, crank up the stereo, and let go.

Because of video problems we've detected on the new 25th Anniversary Edition, we're sticking with this Limited Edition of *Halloween* (the one with the lenticular cover). If you're a completist, then you'll probably want to pick up the new edition for the stellar extras it offers, such as a newly produced documentary and the reissued Criterion audio commentary track. But in terms of what the film should look like, this Limited Edition is the way to go. Disc One includes all the standards (picture galleries, cast and crew info, trailers, and TV and radio spots), as well as an original documentary, called *Halloween Unmasked*. This edition includes a second disc featuring the television cut of the film. Containing about 11 minutes of additional scenes, this is an interesting keepsake. This version includes a few moments that only those of us who watched the original NBC broadcast have seen. The additional scenes don't add much and were actually filmed during the production of *Halloween II*, but they're cool to have anyway. The picture quality on the TV version is just as good as the theatrical version (including anamorphic widescreen), and the sound is given to us in a nice Dolby Digital 2.0 track.

So what are you waiting for? Get these discs a spinnin'! After all, it's getting dark. And when the babysitter's away . . . Michael comes out to play.

[Insert creepy theme music here.]

Night of the Living Dead: Millennium Edition

Elite Entertainment

Hey, hey, we're the zombies! People say we stumble around. But we're too busy moaning, and trying to keep the humans down . . .

That's not a quote from the film, we just think it's clever.

Ah . . . zombie films. You have to love them. Nothing is more gory or stomach churning than lumbering, putrid human corpses stalking the living, chewing stringy flesh, squirting blood, tearing muscle and sinew, and gnawing bone all over the place. Filmmakers who make zombie films know one thing – there's gotta be some screams. And all the zombie films of our era owe a debt to one man, George A. Romero. Romero is the one who gave zombies back their soul. He made the zombie film an allegory of our twisted times. He made zombie movies cool. *Night of the Living Dead* isn't the first living dead film ever made, but by most accounts, it's one of the best. The documentary style of *Night of the Living Dead* proves time and again to be quite influential. In fact, having been made in 1968, you'd think it was older than that – a fact that has fooled many a pirate who thought the film was public domain. Sadly, now it is – the copyright notice was inadvertently left off the film's title optical, seemingly leaving it free and clear for anyone to put on DVD, leading to many bad versions out there. In any case, *Night of the Living Dead* is a true gem of our cinema history, horror film or otherwise.

The film plays almost like a true account of the events concerning a plague of the living dead taking over a small town (and as we'll see in the film's many news reports – the world). At first we focus on Barbara, a young woman visiting a dead relative with her brother. After some serious taunting, Barbara and her brother are attacked by a zombie. Barbara barely escapes and makes it to a small farmhouse, where she slips into catatonia . . . and is joined by Ben, a rational guy who knows exactly what he needs to do to live through the night (and damn anyone who gets in his way). Together they board up the windows and doors and settle in. After a short time, two families that have been holed up in the basement come up to see if everything is okay. Ben and Barbara meet Harry and Helen Cooper, their young daughter Karen, and newlyweds Tom and Judy.

Together they try to survive until help can arrive. But when several different courses of action are set by the group, with no real plan in mind, things are bound to go bad. Do Ben and the rest of the living stand a chance . . . or are they destined to be zombie chow?

Night of the Living Dead is a great little film. It's claustrophobic, set up to be realistic and quite disturbing. For what it is and what it does, it definitely puts you in a very creepy mood. Some of the supporting acting is amateurish, but then again, this film is a drive-in flick made by a bunch of friends who worked in commercials for a living. It's a shock that *Night of the Living Dead* has stood the test of time at all, but it sure has and deservedly so.

As stated, *Night of the Living Dead* is on DVD in many forms, but the one you want is the Millennium Edition. It's in a red Amaray case with a gravestone on the cover. This is a rejuvenated version of a previous release by Elite — one that went off the market for a while but makes its triumphant return here.

The transfer is the same stunningly clear version used by Elite on the previous edition. Technically, the only difference is a new Dolby Digital 5.1 remix, which you would initially approach with some trepidation. This is a low-budget, down and dirty movie and the fear is that a 5.1 remix would distract from that quality, making the movie sound artificial. Well, the good news is that the remix isn't bad per se. It just doesn't really need to exist at all. There is virtually no difference between it and the original remastered mono track, apart from some added oomph in the bass during the rolling thunder effects. Otherwise, we detected no newly created sound effects creeping up in the surrounds or differences in the way the music is presented. The 5.1 mix adds a little more depth but is very respectful of the original audio.

This Millennium Edition ups the ante on the previous release by incorporating many of the extras from the original award-winning laserdisc. From the previous DVD extras, you'll find two commentary tracks. The first one features the filmmakers, including director George Romero himself, and you should note that this is the *only* DVD version to feature Romero's participation. The other track is with the cast. You'll also find scores of TV commercials and trailers, a spoof about living slices of bread attacking people, and a collection of retro commercials shot by the Image Ten group (the guys who made this film). This new edition brings to the table even more extra features. Elite has pulled these bonuses from their definitive laserdisc version because they didn't fit on the original, very early DVD. The best of these new features is the complete original screenplay by John

Russo and Romero, along with Romero's original treatment and Russo's unfinished treatment for a much different movie about ghouls from space. Also making this disc worth having is an audio recording of star Duane Jones's last interview before his death. Jones was never entirely comfortable with his status as a horror movie icon (he also starred in the underrated *Ganja and Hess*). He rarely gave interviews about his experiences on *Night of the Living Dead*, making this an invaluable extra. Additional features include a short video interview with Judith Riley (who plays Tom's girlfriend Judy); scrapbooks compiled by cast members Vince Survinski and Marilyn Eastman with a wealth of rare photos, ads, and correspondence; and a gallery of original props, posters, and collectibles.

When all is said and done, Elite's version of *Night of the Living Dead* remains the one to beat. You simply cannot find a better looking or sounding version of this movie anywhere else. And the newly expanded bonus features make this a no-brainer, so to speak. Other studios could learn a thing or two from Elite. The original DVD of *Night of the Living Dead*, released in the early days of the format, was outstanding. This new version is even better than before—not just a lame reissue, but a real improvement of the original with real added value. Now *that's* the kind of DVD reissue we can get behind.

If You Like These Titles . . .

. . . you may also want to give *Frailty*, *The Hitcher*, *The Blair Witch Project*, and *Scream* a spin on DVD. If you dare, that is.

Music on DVD

Sometimes the best music is *seen*. Concerts, videos, and musicals are all a beautiful combination of equal parts music and video, with a nice helping of energy thrown in for flavor. Part of the music experience is seeing a crowd light up with joy as their favorite band brings a song to life. Then there's the fun of catching that new mini-movie on MTV. Whatever the case may be, DVD is great for bringing the world of music into your home. There's a lot of great music on DVD, and we hope our choices will help get you started in experiencing more of it.

> **Editor's Note:** *Bits* writers Dan Kelly and Matt Rowe contributed to this section.

Talking Heads: Stop Making Sense

Palm Pictures (Rykodisc)

Movie going is probably one of the more subjective art experiences. The concert film is even more so than the average film. In a dramatic film, if there's a character or actor that you're not particularly fond of, there are other elements to hold your interest. If you don't like an actor, there are sure to be scenes in which he or she does not play a part. The concert film, on the other hand, is a different story. After all, if you don't like the band playing, it's likely that you're not going to even bother seeing the film. The focus of the film is, in fact, the band on stage.

Stop Making Sense may just be the exception to that rule. The Talking Heads are admittedly an acquired taste, but as a film, *Stop Making Sense* is so well put together, and the Talking Heads put on such a great live show, that there's enough here to keep even fans of passing interest entertained throughout most of the film. *Stop Making Sense* has deservedly earned its reputation as one of, if not *the*, best concert films of all time.

The stage is at first bare. David Byrne comes out wearing his trademark white canvas deck shoes and white suit, and he performs a stripped-down version of "Psycho Killer," with only a recorded drumbeat as accompaniment. He ends the song with the "spastic dance" (as it's called in the storyboards), tripping over parts of the stage being wheeled out behind him.

From there on, the performances, band, and stage pieces slowly grow from modest and understated into a full-blown, theatrical rock show.

Part of the excitement in this performance is watching it grow through each song. By the time Byrne, Weymouth, and company get down to doing "Slippery People," one of the more inspiring songs of the set, the players (including keyboardists and backup singers) are all on stage. Each performance is rousing and stimulating, and the rendition of "Burning Down the House" is one of the most high-energy performances ever seen on film.

Jonathan Demme (best known for directing films like *Silence of the Lambs* and *Beloved*) helped bring the group's vision of their stage show to the screen. The pace of the concert builds in complexity and the show becomes more involving as it progresses. In this aspect, it plays out a lot like a traditional film, and less like a concert. There are also no shots of the audience until the very end of the movie. The band members (all of them, not just Byrne) are always the center of attention. It's not often when you watch a concert film where you feel like you are actually at the concert and not merely watching one on film. *Stop Making Sense* is absorbing and entertaining from beginning to end.

Palm Pictures and Rykodisc have given us an across-the-board nice presentation of *Stop Making Sense* on DVD. This is a live show, so there is a lot of visual information to take in. The source print used in the transfer is mostly clean of defects, but the film does have a slight intentionally grainy look to it. There are many light-to-dark transitions between lighting segments, and these suffer from distracting edge enhancement at times. Backgrounds are filled with solid black levels that allow greater detail on the lighter foregrounds. The predominantly red backgrounds throughout the second half of the show are vibrant without looking overly grainy or edgy.

The audio is also good, though in certain areas, it's a disappointment. There are two Dolby Digital 5.1 mixes included here, as well as a Dolby 2.0 surround mix. Of the two 5.1 mixes, the second (the studio mix) is the strongest and makes the best use of the front speakers. The other 5.1 mix (a remix of the original feature film audio) makes good use of the surround speakers to incorporate the audience noise into the track, but this mix is otherwise somewhat distracting. Tina Weymouth's bass and Chris Frantz's drums are focused mainly on the center speaker, rather than spread across the front speakers where they'd be more effective. Both 5.1 tracks are sadly lacking in bass.

One other note about the audio mix — occasionally the audio is very briefly not in synch with what's going on onstage. This is not a problem with the sound mix. *Stop Making Sense* is a compilation of three different performances done at the Pantages Theatre in Los Angeles. If something went wrong with a part of the sound mix (hissing or static) from one night, portions of one of the other sound mixes were overdubbed to make up for it. This doesn't happen a lot, but it is definitely noticeable if you pay attention.

Feature-wise, this is an entertaining disc. The commentary (by director Jonathan Demme and band members Weymouth, Byrne, Frantz, and Jerry Harrison) is enlightening and gives detail not only about onstage goings-on, but also about specific songs and how they came to be. Perhaps the most amusing of the added features is David Byrne's self-interview. As far as video and audio quality goes, this feature would make cable access look like reference-quality material. Nonetheless, seeing David Byrne answer some of the more common Talking Heads questions asked by himself as several characters (of varying sexes, ages, and races) is amusing. The storyboards give insight into Byrne's ideas for the visual presentation of the show. Rounding out the meat of the features on the disc are two songs originally cut from the feature-length concert, "Cities" and "Big Business/I Zimbra." As stated on the back of the box, these aren't in the best condition but are the best they could prepare for the disc. The two additional songs are shown in full-frame 1.33:1 (they look like they were taken from an analog video source) with either a Dolby 5.1 or 2.0 sound mix. Those are the main features, but there are other more standard features as well, including the film's trailer and a band discography. It's a very nice package.

Stop Making Sense is a very entertaining film. It's got everything a great concert needs — groundbreaking songs, dedicated musicians, energetic (and admittedly comical) choreography and an enthusiastic crowd. Even if you're not a big Talking Heads fan, give the disc a go.

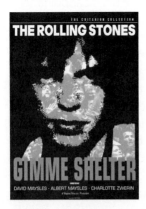

The Rolling Stones: Gimme Shelter

The Criterion Collection

The 1960s were a time of incredible turmoil and great social change. In 1969, two events unintentionally summed up the dual nature of the 1960s. The summer of 1969 saw Woodstock, an epic concert that served as the greatest love-in our country has ever seen. Peace, love, and understanding were summed up in a relatively violence-free environment. Then, in the winter of 1969, the Rolling Stones spontaneously decided they could better Woodstock — this time on the West Coast. They'd throw a free concert to end all concerts and . . . well, it was definitely the end all right.

The concert was at the Altamont Speedway, which was the third choice after the first two locations backed out. That information would serve the story little if it weren't for the fact that Altamont was settled upon at the last minute. A stage, scaffolds, and lighting had to be built overnight basically, and that left little room for security planning and space. The concert ended up being held on a smallish stage, maybe four feet off the ground with no barriers. Well, there were barriers . . . in the form of Hell Angels. There are two things you absolutely don't do to a Hell's Angel. First, you don't bother them while they drink. Second, you don't — and we mean *do not* — touch their bikes. So . . . here you have a situation where the Rolling Stones ask the California chapter of the Hell's Angels to act as security for the concert. They're told to hang out, drink what they want, and keep people off the stage. No problem, right? Wrong. There's not much room up near the stage for the Angels and, considering they aren't the type to park their bikes and walk, they decide to bring them along. As the crowd surges forward, people begin to climb on the bikes to get better views. This pisses the Angels off, and they start getting violent. This means zero tolerance for those who make it to the stage. If you *do* get to the stage, expect a fist, a boot, or a pool cue in your face.

Because of the violence going on in the crowd, both Jefferson Airplane and the Grateful Dead, who were both scheduled to perform with the Stones, were relegated to one or two songs each, before things got out of hand. A riot preempted the Dead and a fist to the face of Airplane's lead singer ended their show. What should have been a nice get-together to end the 1960s became a firsthand look into the future of American civilization as we know it. Did Mick Jagger, taking on the role of Satan in "Sympathy for the Devil" usher in the devils that plague us today? Maybe. But nonsensical, meaningless violence is what this event was all about — most notably involving a young man who whipped a gun out and was brutally beaten and stabbed by the Angels. Did he deserve it? Probably not, but then again, he did have a gun pulled out. The times, they were a-changin'.

Gimme Shelter is a beautiful and awe-inspiring look at a snapshot of a time many of us either never experienced or can't remember ('cause if you can remember the '60s, you weren't there, man). Over time, the facts surrounding the events contained in this film have become foggy. Thankfully, because of this film and the DVD supporting it, we have a better chance of saving the facts for the ages.

But with all the commentary on the time of the film's shooting, *Gimme Shelter* is more a concert film than it is a documentary. It's hard to separate the two visions of this film. It starts off like a concert film, then goes into documentary mode, then back into concert mode, and right back into documentary. This schizophrenic film style is banged home by the fact that at various times during the film, we're watching the Stones watch the same film we are. It's part of the documentary, but it gives us a chance to see the confusion of the situation even better. Mick Jagger and Charlie Watts watch the film in utter horror at what they caused. It's remarkably beautiful to see this. And the concert footage itself is incredible. Here we get to see the Stones before they went the way of corporate rock. Mick doesn't quite have his Tina Turner strut down, but he's trying and having a good time. They are at the point were they aren't quite immortal, but better than the average band. They hang out with their buds during the mixing stages of their new album, eat hamburgers, and sign autographs for eager fans. One word is all that can be uttered by the end of this film, and that is "Wow."

The DVD is even more "Wow." Fully restored from the original negative and best possible prints, this disc is beautiful. Colors are bright and the blacks are solid, with nice detail. Take a look at the restoration comparison on the disc to see exactly how much work was put into making this film look as good as it does now. But it doesn't end there. The sound too was restored, and restored so well that there's a beautiful DTS track on board. It sounds really, really good. The Dolby Digital track is good too, but the DTS is much fuller with a wider soundstage. For stereo nuts, there's a third Dolby Digital 2.0 track as well. Guaranteed, you won't see or hear this film in your home any better than on this DVD.

Now, if it were just that—a super presentation of video and sound— we'd already say this is a disc worth owning. But it just gets better. Criterion also slaps on five deleted scenes from the original work print, which include three songs excised from the Madison Square Garden concert footage, as well as some behind the scenes stuff, like the Stones mixing "Little Queenie" and Tina and Ike Turner hanging with Mick and Keith Richards. It's incredibly cool. There's also about an hour's worth of radio broadcast from a KSAN call-in show the next night, with new introductions by Stefan Ponek, who was the DJ on the show. The calls themselves are a brilliant insight on the times, as everyone tries to sift through the horrible aftermath. There's also a commentary track on this DVD with co-directors Albert Maysles and Charlotte Zwerin (who oversaw the editing) and their all-around Renaissance man on the production, Stanley Goldstein. Some of the information they provide is incredible, but most of it is cosmetic. You'll also find a selection of trailers including the original and re-release trailers of *Gimme Shelter*, as well as two other Maysles Film productions, *Grey Gardens* and *Salesman*. Two photo galleries from renowned photographers Bill Owens and Beth Sunflower, a filmography of the directors, and a 44-page booklet of essays round out this impressive single-disc DVD.

Gimme Shelter is a beautiful portrait of what it must be like to stand in the middle of a tornado. For a while, the sky is calm and gentle, but suddenly a house gets sucked into the vortex and the next thing you know, you're in there with it. You can't do anything but become part of the action. As Mick Jagger struts his stuff on stage, and the rest of the Stones bang out their primal dance music, you can't help but wonder how much they have changed over the years due to the events captured in this film. It must be hell to have something so beautiful represent such a horrible part of your career.

Beastie Boys Video Anthology

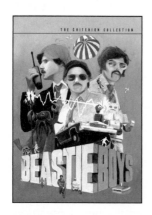

The Criterion Collection

This is, without a frickin' doubt, the absolute best DVD treatment a compilation of music videos has ever received. It's a benchmark that we don't think can be topped actually. Beastie Boys videos aren't mini-films, but they have always served both the music and the spirit of the performers themselves better than any musical group out there. It's probably because, if the videos weren't directed by Beastie Adam Yauch (a.k.a. Nathaniel Hornblower), they were directed by close friends and relatives of the band. It's easy to keep things close and on track when people who love you are involved. That same spirit has gone into this DVD.

The *Beastie Boys Video Anthology* is cutting-edge DVD and yet it's all so simple. Take 18 videos from the Beasties, break 'em in 2 equal halves, and lay them out with supplements. Each disc has nine videos split over two layers. You can watch the videos one at a time or all the way through. The videos are done up with full-on Dolby Digital 5.1 and/or 2.0 sound. You can also watch the videos with either audio commentary from the Boys or by the directors. Then, when you get done watching the videos, you can go to another area on the menu screen that sends you to the same videos with supplements. Now, keep in mind that these aren't the same exact videos you just watched and listened to. These are the videos presented with alternate angle features and remixed audio, so you can basically thumb around and create your own version of the video. It's incredibly cool and, for Beastie fans, this is going to be the ultimate item to have.

The video quality is great overall. There's all sorts of different media being used in these videos: 16mm, 8mm, animation, digital video, standard video — it's all over the place and could have fallen into a DVD dead zone. But don't worry — it's all been well taken care of by Criterion. Every one of these videos looks great. Colors are dead on and there are no artifacts anywhere (unless they were in the original source material). The audio quality is excellent. The videos themselves sound wonderful in both Dolby Digital 5.1 and 2.0, and the remixes are even better.

The supplements, on top of the video presentation, kick this set into the stellar zone. The commentaries are hilarious. The Beastie commentary is funny, because they have nothing to say and they know it. It's usually quite hard listening to tracks with uninterested participants, but these guys are so funny about the track itself. Ad-Rock keeps referencing how his comments are the best because he does play-by-play. The director's track is even funnier and actually has information in it. It's Adam Yauch, Evan Bernard, and Spike Jones in a room with a phone, talking about the videos and how they shot some of this stuff. They make fun of crewmembers and hangers-on. Then, if a video wasn't directed by one of the three, they call the director on the phone. Actually Spike Jones calls the director, playing Criterion production intern Ralph Spaulding, who does a pre-interview with them about how they got the job, how barn-raising plays an important part in Beastie Boy videos, and what the director is currently wearing. The directors they call have no idea what the hell is going on. Better still is Tamra Davis (wife of Mike D), who isn't around when they call, so it's just a track of phone messages back and forth. It's so funny to listen to this stuff. The video supplements themselves are deep and smooth. Along with the remixes and alternate angles, there's also stuff like storyboards for a couple of the more complicated videos, photo galleries, and other things like that. This two-disc set is *packed*.

This disc set speaks for itself, and it has every right to. Although this set doesn't include the entire lot of videos from the Beasties (the Def Jam material is not included), it's voluminous enough and worth whatever you have to spend to pick it up. Absolutely do not miss the boat. This DVD is the bomb.

THE CRITICALLY ACCLAIMED FULL-LENGTH FEATURE FILM

UH-OH.

SOUTH PARK

BIGGER, LONGER & UNCUT

WIDESCREEN DVD COLLECTION

South Park: Bigger, Longer and Uncut

Paramount

Well . . . what can you say about *South Park: Bigger, Longer and Uncut*? This film is rude, crude, and extremely funny. It's probably the most politically incorrect film you'll ever see. And it's refreshing as hell.

Despite all the profanity, there's still a story here (silly though it may be). One afternoon, Stan, Kyle, Cartman, and Kenny sneak into a showing of the Canadian-import movie *Asses of Fire*, starring Terrance and Phillip. Soon after, their speech can be measured in OPM (Obscenities Per Minute), and their parents aren't too happy about it. When they discover that a movie from up north started the problem, the mothers of tiny South Park, Colorado, declare war on Canada in retaliation (a standard theme on the TV show — parents never take responsibility for their kids if they can lay blame elsewhere). Meanwhile, Saddam Hussein and Satan are up to no good, with plans to bring on the end of the world. So naturally, it's up to Stan, Kyle, Cartman, and the ghost of Kenny ("You bastards!") to save the day. And along the way, we're treated to tons of laughs, at least one good Jar Jar Binks joke, and a surprisingly entertaining soundtrack, featuring hilarious songs like "What Would Brian Boitano Do?", "Blame Canada!", "It's Easy, M'Kay", and the infamous "Uncle Fucka" (in which the F-word is used many, many times).

The video on this DVD is presented in anamorphic widescreen, and it's of excellent quality. *South Park*'s cutout style animation isn't exactly high art, but it looks great here. There's good detail and contrast, vibrant color, and very little edge enhancement or digital artifacting visible. Heck — there are hardly even any flecks of dust visible on the print. On the other hand, this soundtrack isn't exactly the most active Dolby Digital 5.1 surround mix you'll ever hear on DVD. It's very biased to the front hemisphere, and only occasionally will you get much action from the rear channels. That's not to say it's bad — this is exactly how the film sounded in the theater. Just don't expect it to push your sound equipment much.

Sadly, the extras on this DVD leave a lot to be desired. You get two theatrical trailers of excellent quality, along with one of the film's teaser trailers. Unfortunately, here we get shortchanged — the studio could easily have included all three versions of the teaser trailer. Each was identical until the end, when Cartman said something different to the audience. This is a disc that Paramount could (and should) have had a lot of fun with in terms of extras. As Stan might say, "Dude, this is f**king weak!" One last note – the subtitles on this disc are rather odd. Whenever Kenny talks, you've got about a 50/50 chance of having his muffled words translated. It's almost like whoever did the transcription did the best they could to figure out what he was saying, and sometimes just gave up. And the disc includes a French subtitle track that only very rarely actually translates anything (as with the titles at the beginning of the film). Why it's there at all is tough to figure.

So is Paramount's DVD version of *South Park: Bigger, Longer and Uncut* worth buying? Lame extras aside, we'd have to say that the answer is yes. Get it on sale, but do get it. This flick is a blast. You'll laugh, you'll be shocked, and you'll learn once and for all that Cartman's mom is *definitely* a dirty slut. But do keep in mind that this film is not for the faint of heart, and it ain't for kids, m'kay? That much *should* be obvious.

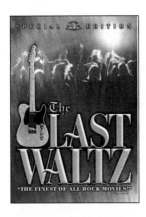

The Last Waltz: Special Edition

MGM

"THIS FILM SHOULD BE PLAYED LOUD!"

That line is emblazoned across the opening sequence of this film. Consider it the ribbon that wraps a DVD package that is, once unveiled, a graceful and wonderfully historic moment in the account of rock music. *The Last Waltz*, a farewell to one of rock's most prolific acts, is a concert film/documentary extraordinaire that ends up becoming a celebration for an era. Helmed by Martin Scorsese, this paean becomes something more than celluloid; it becomes a song unto itself. *Last Waltz* is a work so profound that few after it have replicated its intensity. It possesses a beating heart and a living spirit, creating a timeless entity.

Designed as a final show for a retiring, and tiring, group named (quite simply) the Band, *The Last Waltz* grew in size until it became much bigger than the sum of its parts. After spending a grueling 16 years on the road, 8 in bars and roadhouses and 8 in arenas, the Band decided to call it quits. They collectively determined to stage this final show at Bill Graham's Winterland on Thanksgiving Day, November 26, 1976. It's all history after that. The show, a distribution of talent with the Band at its core, showcases many of the popular names that embody this time and place. And it can still bring tears to our eyes.

The members of the Band, Robbie Robertson, Rick Danko, Levon Helm, Garth Hudson, and Richard Manuel, recount a varied history of their travels and experiences along the way. These anecdotal pieces, stitched between grand performances and memorable songs, become the reasoning behind their intent to close the curtain. Without them, the purpose of the film, as a statement, gets lost in the overwhelming beauty of its presentation. They allow a brief glimpse into the lives that made up the Band and, in so doing, allow us to become a part of them.

There are great songs performed by many here. There's Ronnie Hawkins (who gave The Band their start, more or less, with a rousing "Who Do You Love") and Mac "Dr. John" Rebennack (with his jazzy cool "Such a Night"). There's Neil Young (with "Helpless"), Joni Mitchell (with "Coyote"), Butterfield (with "Mystery Train"), Neil Diamond (with "Dry Your Eyes"), and Muddy Waters (singing "Mannish Boy"). There are other luminaries on hand as well, including Bob Dylan (who provided the forum from which the Band's popularity soared) and Eric Clapton.

Mixed into this incredible array are the songs for which the Band is best known, performed by them: "Don't Do It," "The Shape I'm In," "Up on Cripple Creek," and "The Night They Drove Old Dixie Down." And there are other diamonds found on this disc as well. There's an extremely satisfying performance of "The Weight," with the Band being joined by the Staples Singers for an unforgettable rendition. The selection of songs on this disc is nothing short of delightful.

But just listing the performances by these artists is a grave injustice. You must watch, and become a part of, their sets to fully realize the importance of *The Last Waltz*. There is a feeling of community that you identify with as you watch each successive performance. It's a complex state that you find yourself in: awe of style, reverence of talent, and a feeling of love as you meld with this show. You are not just watching this movie; you are spiritually intertwining with it. Their elation is yours and belongs to you as much as it belongs to them. You may never see a greater collection of performers together at one time again.

To say that this is simply a concert film depicting the musicians of that time is criminal. *The Last Waltz* is an absorption of the '60s and '70s—a culmination of an era. Although the '70s existed for three more years before giving way to the electronic pulse of the '80s, this film brings closure to a time of astonishing diversity in music.

The disc by MGM, presented in anamorphic widescreen, has been beautifully and lovingly remastered. The colors are vibrantly displayed and the detail is perfect. We're not surprised either, given the fact that Martin Scorsese was involved. This is concert footage shot in 1976, and yet you have to be impressed with the picture quality. The audio is equally spectacular. Mixed in Dolby Digital 5.1, as well as the requisite 2.0, the channels are all in good use. The 5.1 surround is particularly active and provides an immersive feeling that only serves to enhance the film.

The extras are sweet on this DVD. To start with, there is audio commentary—two sets actually. Band members and other musicians, along with filmmakers and historians, provide the first track. The other features Scorsese and Robbie Robertson. In addition, there is an incredible impromptu jam by many of the musicians. Imagine, after the show, to be treated to this spectacle. The visual portion of the tape gives out after 10-plus minutes, but the audio remains for an extra minute. There's also an incredible arrangement of gallery stills, four sets in all, with descriptions that inform you of every nuance of the event, from poster art to photo captures, from promo shots to lobby cards. Also included are the trailer and a TV spot for the film. And all this is topped off by a behind-the-scenes featurette—in effect, a devoted recollection. Finally, a collectible eight-page booklet written by Robbie Robertson himself is tucked into a case that is contained in a strikingly embossed outer sleeve.

We could ramble on and on about the necessity of this film's inclusion in every music fan's library. And if we stepped into a "twilight zone" of sorts for some of you younger music fans, that shouldn't preclude any of you from seeking out this film. For those not of this particular time or era, we suggest that you watch *The Last Waltz* with an open mind. You'll come away with admiration for the multiplicity of styles that defined a generation. And you'll see how it influenced, whether you know it or not, today's rock and roll.

U2: Elevation 2001—Live from Boston

Interscope (Universal Music)

For many of us, our musical educations began on August 1, 1981, when some guy in New York flipped a switch and MTV came blasting into our living rooms. There was Peter Gabriel, shocking his monkey. The Police were walking on the moon, complete with bleached blonde hair. A bar band named REM loudly proclaimed Europe radio free. And there were four punk-ass, Irish teenagers celebrating New Year's Day out in the snow. Our collective love of U2, and all of the above, began then and has lasted to this very day.

So, for any fan of U2, and for all those who experienced U2's Elevation tour up close and personal, we're thrilled to say that this DVD succeeds in bringing the experience home in a big way. This disc brings back all the fun and energy of the show very well indeed.

Can a concert performed in a sports arena reasonably be called inti-mate? Well, anyone who saw the show knows that that's exactly what this was — Larry, Adam, Bono, and the Edge mixing it up with their audience, right in the thick of the crowd. The concert on this DVD delivers U2 in a stripped-down version — gone are most of the glitzy trappings of the band's previous tours. Here the band is singing and playing their instru-ments with collective flair and naked honesty of emotion. Bono's mag-netic swagger is plugged in and lit up like the spotlight he uses to pick out faces in the crowd during "Bullet the Blue Sky." We get a fine mock bat-tle between Edge and Bono's dueling egos in "Until the End of the World." You get "In a Little While," performed in tribute to the late Joey Ramone, and an acoustic rendition of "Stay (Faraway, So Close)," with Edge and Bono right out in the middle of the arena. The band even rounds out the show with a ripping version of "The Fly", and closes with a new favorite, the stirring and hopeful "Walk On." But easily the highlight of the night is a terrific performance of "Bad," with its electrifying transition to "Where the Streets Have No Name," in which the crowd is chanting with almost religious frenzy. Experiencing this moment live, you could have powered Las Vegas for a month with the energy in the air. You'll reexperience the chills with this DVD.

The video quality on this DVD is generally very good. It's in full frame, which is our only real complaint. Still, color representation is excellent — lush and vibrant, as one would expect from a concert video. The contrast is equally good, with deep blacks and sufficient detail throughout the image. This doesn't appear to have been shot on high-definition, which is a shame. You should also note that the video's been processed to look like film. So, while it's not reference-quality video, it's not supposed to be — it looks exactly as was intended by the producers.

The sound on the DVD is also mostly good but will probably not impress more discerning audiophiles. This is not the best Dolby Digital 5.1 mix ever heard for a live music performance, but it's definitely ser-viceable. The center channel seems a little muddy, which means that it can be difficult to understand what Bono is saying on occasion (and it's not just because he's got an Irish accent either). The effect is mostly noticeable between songs, when he's talking to the audience. The front of the soundstage could also be a little wider and smoother, and there's not as much audience fill from the surrounds as you'd expect to see in most 5.1 concert mixes. As a result of these issues, the track doesn't sound as natural as it should. All that said, the music does sound very good throughout the entire performance. And the louder you listen to it, the more "realistic" it becomes in terms of recreating the live experience. Nobody ever said music recorded in a sports arena could sound like the Royal Albert Hall. Most people will be happy with the 5.1 mix.

This two-disc set isn't exactly loaded with extras, and some obvious ones have been left out (song lyrics and a band discography are obvious omissions). But if what you do get seems a little thin at first . . . take a deeper look. Most of what's included here is pretty cool. Disc One includes a 24-minute featurette on "the making of the making of" the concert. That sounds a little funny at first, but capturing this concert on video presented distinct technical and creative challenges. The featurette includes interviews, insights, and tons of behind-the-scenes footage with director Hamish Hamilton and his production team. The piece is a bit glossy in tone, but it's substantive and worth watching. Disc Two contains the real bulk of the extras. The most notable of these is called *Another Perspective*. It's basically 12 songs from the concert presented on the other disc (running about an hour), offered again here with the option to change to any of 3 different camera angles on the fly. There are two angles of the show itself (the final program and the show from the audience's perspective), and one of the director and his team doing their thing in the control booth. The quality here is not even close to what's on Disc One — all three angles are presented smaller and more compressed. You watch through the main window of a graphic interface that also shows you smaller picture-in-picture views of what's on the other angles. You switch by selecting the one you want with your remote. You can also skip ahead or back to any of the 12 songs. Remember that amazing transition mentioned above from "Bad" to "Where the Streets Have No Name"? Well, watch the director's angle during this moment. It's very clear that Hamish loves what he does and his enthusiasm is infectious. We should all be so happy at work.

The remainders of the extras are a mixed bag. There's a fun little time-lapse video (*Road Movie*) that runs about five minutes, showing the entire process of setting up, running, and tearing down the show. It's set to "Walk On" and it'll suck you in. There're also three additional tracks, including a time-coded video of "Elevation" shot live in Miami in March 2001 (a nice souvenir given that this was the first song performed on the first concert of the tour). You also get the band up on a rooftop again, this time in Toronto in May 2001, singing "Beautiful Day," as well as the band performing "Stuck in a Moment You Can't Get Out Of" in the studio (along with various behind-the-scenes antics) in Dublin in September 2000. Next up are brief video trailers for a pair of other U2 concert videos, namely *ZOO TV Live* from Sydney and *Popmart Live* from Mexico City. Finally, there are DVD-ROM web links to various online sites (including causes the band supports — a nice touch) and a DVD-ROM screensaver.

So there you have it. *Elevation 2001* on DVD does a very good job of capturing the emotional energy of the live performance. If you've seen the real deal in person, you'll be very grateful to be able to recapture the experience in your home theater, and share it with those who weren't there with you. We love U2 because these guys, in addition to being a great band, are undeniable optimists. They give us hope. And this DVD gives us a little bit of joy. Hallelujah.

If You Like These Music Titles . . .

. . . you may also want check out *Peter Gabriel: Secret World Live*, *Pearl Jam: Touring Band 2000*, *Sting: All This Time*, *Sarah McLachlan: Mirrorball*, *The Chieftains: Live Over Ireland*, *The Rocky Horror Picture Show*, and *Pink Floyd: The Wall*. And believe us when we say, that's just for starters. There's a lot of great music available on DVD.

Anime on DVD

Animation has experienced a distinct renaissance in the last decade in America. There's two reasons for this — the advent of *computer-generated imagery* (CGI) in animated films and mainstream audiences beginning to embrace Japanese animation, better known as anime. A few years ago, hardly anyone knew about the films we're listing here, and yet they're as good as any other film we've listed in this book. If you like animation, or simply good storytelling in general, take our advice and check these films out.

Akira: The Special Edition

Pioneer

Very rarely in cinema history does a film truly define its genre. It takes a lot for a film to singularly become the title people think of when trying to get their friends turned on to a particular style of movie. When fans in America are discussing the subject of anime, one film and one film only comes up more than any other. That film is, of course, *Akira*.

When it first roared its way onto American shores, animation fans knew instantly that it was special. Loyal followers of the film claimed that it blew anything by Disney out of the water (and at the time it did). It was cartoony, but ultra-violent. Drug-peddling biker gangs were the heroes of the film, government forces were the villains, and those caught in the middle seemed to be morally neutral, guided by both good and evil at the same time. Throw in the fact that each and every cel was hand-drawn, painted, and photographed (using minimal computer assistance) and when you look at the end product, you see a film that changed the way many of us looked at animation.

At the time it was released, *Akira* took place in the near future. It seems that during World War III (in 1988 to be exact), an explosion wiped out Tokyo. The film starts years later, after the city has been rebuilt. We meet Kaneda, the young leader of a gang of kids who wage war on the city while riding electro-powered motorcycles. During one of their nightly wars with a gang called the Clowns, they bump into a young boy in the middle of a closed-off portion of highway. That is to say, young gang member

Tetsuo literally bumps into the boy. But as it turns out, the boy is super-powerful as a result of a top-secret government experiment, so Tetsuo bounces off a force field projected by the kid and is thrown off his bike onto the road. When the rest of Kaneda's gang find Tetsuo, he's unconscious and surrounded by an elite government team led by the mysterious Colonel, who whisks Tetsuo and the young boy away and has the local authorities arrest the rest of the gang. As a side note (and an important plot point), while being released from custody, Kaneda meets a pretty young revolutionary girl and helps free her when he tells the cops that she's with his crew. And that's the introduction of all the players in the story. We have our government force, our secret project super kids, our biker gang, and the revolutionaries. As the story of *Akira* develops, all four parties face off against each other, while Tetsuo struggles to deal with the fact that the blow to his head has awakened a mysterious power in him, even greater than that of the other super kids in the experiment. And when Tetsuo goes mad from this strange power, his friends confront him, the government attacks him, and the secret of "Akira" is finally revealed in the film's mind-bending climax.

There's so much going on in *Akira* that you really have to experience the film multiple times before you'll truly understand it all. Even then, you'll have to probably go out and hunt down the phonebook-sized Japanese comics (known as manga) the film was based on. Created by Katsuhiro Otomo, who conceptualized and directed the film, the comic-based *Akira* defined the comic book epic. This film only boils down the key aspects of the original story, but it still manages to boggle the mind. It's part sci-fi, part-social commentary . . . and all of it kicks major ass.

And as a DVD, *Akira* is even more impressive. It's available in a deluxe, special edition, two-disc set. A bare-bones, movie-only edition is also available, but just forget it exists and pick up this one. Whichever version you choose, the picture quality is very fine indeed. Color representation is outstanding in this anamorphic widescreen transfer. There's a little bit of edge enhancement here and there, but overall this is a really nice video presentation. The blacks are solid and the print itself has been digitally cleaned up to get rid of blemishes and dirt specks. We seriously doubt anyone has seen *Akira* looking any better than this.

The sound is also excellent. You'll appreciate the audio right from the disc's menu screens. They'll blow you away. You'll hear motorcycle revs and then thumping, organic Japanese music will push you to make your selection. When you jump into the film, your ears (as well as your eyes) are in for a treat. Given to us in a brand-new English Dolby Digital 5.1 mix, all of the dialogue has been more accurately translated and re-recorded to give it a nice punch. There's good low frequency in the mix and there's plenty of activity in the rear channels. This track is very, very good. The Japanese Dolby Digital 2.0 is also good but is obviously not as explosive. However, companies should give more attention to the original Japanese audio on anime in the future. It would have been great to hear this film in the original Japanese, but with a new 5.1 mix. The fact that there isn't a Japanese 5.1 track here is a bit of a missed opportunity. We should note that there is a new movie-only version of this film on DVD that features the Japanese audio in DTS 5.1. It's a nice disc, but in no way replaces this set in terms of extra content.

This special edition doesn't just give us one of our favorite anime with beautiful sound and video. We're also treated to great special edition material — some old and some new. Let's start with Disc One, which is where you'll find the newer features. When you watch the film, you have the option of doing so in a special "capsule mode." This is a sort of "Follow the White Rabbit" feature that allows you to access special material whenever a small drug capsule icon appears in the lower corner of your screen. The information tends to be English translations of written Japanese on books, banners, and graffiti that appear in the film. It's a neat feature, but it's nothing earth-shattering. There's also a THX optimizer on Disc One that will allow you to quickly pseudo-calibrate your home theater system for proper enjoyment of the film.

Disc Two has the bigger extras, but they're also the older features for the most part. First up is the *Akira Production Report*. Running 48 minutes in Japanese (with English subtitles), this is the "making of" documentary that was originally released on the Japanese laserdisc. It follows the conception, making of, and release of *Akira* from start to finish. It's interesting, but these Japanese documentaries tend to be hard to digest because of the need for translation. Everything is pretty rapid fire, and it's hard to look at the images of pre-production material and read the translation at the same time. We also get the *Akira Sound Clip* by Geinoh Yamashiro Gumi, which has a running time of 20 minutes. This was also released on Japanese laserdisc a while back. *Sound Clip* is a fascinating documentary focusing on Gumi, the composer, and his ideas for the organic sound of the film. Next up is an interview with director Katsuhiro Otomo (with optional English subtitles). It seems to be an archived interview with Otomo, conducted during an earlier time than the release of this

DVD. As with most Japanese creators, Otomo is very humble. He discusses his inspirations, his art, and the production in detail. The last of our video-based extras are three featurettes newly created for this DVD. The first, entitled *Picture*, is about the new transfer and digital video restoration. *English Voice-Over* is all about – you guessed it – the English voice-over cast. And *English 5.1 Audio Mix* is about the creation of the new soundtrack. None of these is very long or incredibly fascinating, but they do show the hard work and passion that went to fixing this film for DVD release.

The rest of the supplements (aside from four theatrical trailers and a TV spot, each with optional English subs) are image galleries, which are throwbacks to the age of laserdisc. It's a very meaty set of images, however, worth every moment you spend perusing them. Called the *Production Materials* section, galleries are broken up into 36 chapters, which correspond to the 36 chapters on the film disc. Here you can see storyboards, character model sheets, color models, and cel inserts for each and every scene in the film. There's days' worth of stuff here and it's all pretty incredible to see. You'll also find some unused storyboards and background art, as well as the initial character designs. Rounding out the extras is more artwork – comic and magazine cover art from the various translations of the *Akira* manga around the globe, movie poster and promotional art (including previous VHS, laserdisc, and music packaging), and a text-based glossary of people, places, and things from the film.

It's a bit disappointing that the special edition material is soft on the original manga work Otomo did. The original, must-own Criterion laserdisc had reams of the original manga material collected together. Sadly, all we get here are covers. A commentary of some sort would also have been nice, considering the age and importance of the film. It would have been great to hear a scholarly discussion of the film and its impact. But these are small sore spots on an otherwise great batch of extras.

Akira was one of the first Japanese cartoons to turn American audiences on their heads. We've seen plenty before and plenty since, but for some reason, *Akira* charmed more of us than any other anime of its day. For most fans, *Akira* will stand for all time as the film that defines the anime genre. Those are pretty big shoes to fill . . . but *Akira*'s a pretty big flick.

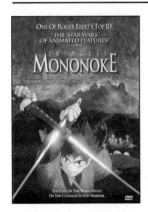

Princess Mononoke

Studio Ghibli/Miramax (Buena Vista)

In a time of gods and demons, Man is finding it harder and harder to exist in harmony with the spirits of Nature. As mankind's hunger for "more and more" grows, Nature's ability to hold onto itself is growing weaker. The gods of nature fight on, but their numbers have dwindled. It's in this time of strife that a young prince named Ashitaka does battle with a demon made of pure, unabashed hatred. Living off the body of a once great Boar God named Nago, the demon, consisting of thousands of black worm-like tendrils, scars the right arm of Ashitaka before he is defeated in a furious battle. Although he "won," Ashitaka is now tainted and cursed. He must live with the evil inside him. And it's his newfound fate to head into the land of the West, where the gods are strongest, to find out what happened to cause the infection of Nago in the first place. There, and only there, may Ashitaka find a cure for his ailment. Or he might find his death.

Taking his faithful steed, Yakul (a badass red elk — one of the best steeds since Trigger), he runs to meet his fate head on. And on his journey, he discovers the truth behind a war raging in the West, on the border of the Deer God's forest. Entrepreneur (and potential warlord) Lady Eboshi is building an iron mill and weapons foundry. To be successful, she has to take from Nature. And the more she takes, the more she needs. Eboshi isn't so much evil as she is headstrong and stubborn. She wants what she wants, and she'll fight to take and keep it. But standing against her are the spirits of the forest, led by three giant Wolf Gods and a young girl named San. They will stop at nothing to kill Eboshi, because she's now destroying their forest home.

When Ashitaka arrives at the iron town, he finds himself in the middle of an epic battle and, not wanting to choose sides (because he can understand both), he seeks to get Man and Nature together somehow. But he has a lot of people to sway. There's Jiko, a mysterious monk working for the Emperor (who is seeking a way to live forever and thinks he's found it in the blood of the Deer God), Moro the lead Wolf God (and San's "mother"), and Okkotonushi (the leader of the boars). Ashitaka bears the mark of hatred to show them that fighting isn't the answer. But in the end, many will die, lives will be changed forever . . . and a warrior and a princess will find love amid the chaos.

DVD fans should be especially proud of this disc. The online DVD community made a big difference when this disc was first scheduled for release. Buena Vista was planning to release this DVD without the original Japanese soundtrack . . . and no Japanese anime should *ever* be released without its original track. When they heard this, DVD and anime fans spoke out, voicing their dissatisfaction. Thankfully, Buena Vista listened.

With *Mononoke*, you'll find one of Buena Vista's best film transfers — ever. This anamorphic widescreen video is incredibly beautiful. There's no print or digital garbage to be found anywhere on this disc. The source is surprisingly flawless and there's not a lick of edge enhancement anywhere. It's utterly amazing. The audio track, in both English and Japanese (as well as French) Dolby Digital 5.1, is equally dynamic. Explosions boom right off the screen, wind bellows through your home theater, and the bass rumbles your bones. It's a pretty expansive sound field and it's quite playful, especially for a dubbed animated film. The English dub pales in comparison to the Japanese track, of course, but if you don't like reading your movies, you'll have a good time with this track. On the subject of the English adaptation, comic writer and novelist Neil Gaiman does the translation duty here, and his adaptation is as lyrical (and in some ways more illustrative) as the original Japanese screenplay. Any way you choose to watch this great film, you'll have a good time with it.

The extras, on the other hand, are disappointing. This really should have been a multi-disc box set (we can only hope that Disney comes back to this film in a couple of years). What you do get is a short "making of" featurette, with interviews and footage from the trailer. There's also the American theatrical trailer, which isn't very stirring. We should also note that the disc uses seamless branching to allow either English or Japanese titles and credits to appear on the film, depending on your audio track selection. A cool touch.

Directed by Hayao Miyazaki, who is an animation legend in Japan (and here in the States), *Princess Mononoke* is a thrill ride. Anything Miyazaki touches becomes beautiful, and this film is no exception. It's action packed, philosophically and emotionally deep, and one of the greatest films of the last decade. You'll love this movie . . . and you'll love the Kodamas too (you'll know what Kodamas *are* once you've seen the film). Don't miss it.

The Films of Hayao Miyazaki on DVD

Princess Mononoke isn't the only Buena Vista release of a Miyazaki film on DVD. If you enjoy this, you'll definitely enjoy *Spirited Away* (winner of 2002's Best Animated Feature Oscar), *Kiki's Delivery Service*, and *Castle in the Sky*. All feature beautiful transfers and great sound (with nice English voice acting if you choose the dubbed audio option). All are well worth a look.

Ghost in the Shell

Manga

The year is 2029, and the world is now one nation, under God, with a unified Internet holding control over everything. The police are broken into Sections, each in charge of policing the one under it. Major Motoko Kusanagi is in charge of "The Shell Squad," a sort of secret service in this new world. Long story short, Kusanagi and crew stumble on a computer virus turned super criminal called the Puppet Master, which believes it's a living being and wants political asylum. Of course, Section 6—a political-based Section—doesn't want this to happen. The story is complicated yet peppered with enough action that you won't have to think throughout the entire film. But then again, as intelligent as the story is, it's not the story you will be watching this anime for—it's the awe-inspiring visuals.

Taking full advantage of all the animation techniques then available, *Ghost in the Shell* is more an art film than it is a cartoon. That probably explains why, during its American theatrical run, it stormed through established art house theaters over the more mainstream theaters. As beautiful as the film is, it's also a bit frustrating because of the way it ends. It just ends — no bells, no whistles, just an end. Fans will be able to explore the film further with a sequel, from *Shell* director Mamoru Oshii, that's due to be released in 2004. Titled *Innocence*, the story is based on creator Masamune Shiro's newest manga series and promises to be as thrilling as the original.

Manga Entertainment has put out a beautiful DVD version of *Shell*. The anamorphic widescreen transfer features sharp colors, strong blacks, and virtually no digital noise. It's a wonder to behold. The audio is also strong, in English 5.1 and Japanese 2.0 Surround. Both are leaps and bounds above previous incarnations of this film on video. The extras are light (including trailers, a "making of" short, and the production report), although they're interesting enough. Hopefully, Manga will revisit this title in the near future as a more loaded special edition. In any case, you just can't miss a great film like this one. The future may not look bright for Kusanagi and company, but with anime like this on DVDs, it's bright for us.

Ninja Scroll
Ninja Scroll: 10th Anniversary Edition

Manga

Ninja Scroll is the exciting story of a masterless ninja named Jubei. While wandering the earth, Jubei bumps into Dakuan, a small, cryptic emissary from the emperor who is investigating a load of gold that's turned up missing. But the gold is not the only mystery in this story. An entire town's citizens have mysteriously turned up dead — and a warning from a dying woman to keep away from the town only causes a panic. Searching for an answer, a team of ninja from the Koga clan try and find out what's going on. But en route to the town, they bump heads with the henchmen of Gemma, the suspected leader of a gang of thieves and assassins known as the Eight Devils of Kimon. One of Gemma's henchman, Tessai, a huge man/monster that can turn himself

into stone (and throws a huge boomerang/ sword that cuts through everything in its path), proceeds to rip everyone to shreds. Everyone that is, except for the beautiful and tortured Kagero. Kagero is the clan's poison taster and, because of her profession, her entire being is left poisonous — and any man who dares make love to her, dies. Tessai and Kagero play a literal game of cat and mouse, until she is "saved" by Jubei, who now finds himself smack dab in the middle of these two mysteries . . . and his own past. Together Kagero, Jubei, and Dakuan must fight to stop Gemma and the other Eight Devils from carrying out their nefarious plan.

The cast of villains here, and the battles with them, is what makes this film so great. Along with Tessai, the stone monster, and Gemma, there is Yurimaru, a hermaphrodite who can electrocute people using a thin metallic string he keeps tied to his hand, and Mujuru, an honorable but deadly blind swordsman. There are two evil women, Zakuro, whose power is to cause things to explode, and Benisato, who plays host to a living tattoo of snakes. There's also Mushizo, a hunchback who keeps a swarm of killer bees in his back, and Shijima, a puppet master of the dead who can replicate himself and hide in the shadows. All in all, it's not a good thing that Jubei has gotten on the bad side of this crew.

Yes, the story, with its swirling battles and dense subplots, is pretty complicated — but it's one of the best pieces of animation we've ever seen. It's beautiful stylized, has a great story, and features some really cool and stylishly designed monsters.

DVD is the only way to watch *Ninja Scroll*, and for now this movie-only edition is the way to go. Before the end of 2003, however, a new 10th Anniversary edition should be released that's packed to the gills with extras. The picture on the current edition is in full frame, and the colors are cool and crisp. The audio is high caliber, and this disc is currently the only way you get the film with the original Japanese audio and English subtitles. It sounds so much better in Japanese than it does in English, if only because the guy playing Jubei in English sounds too jovial. Supplements on this edition include nothing but a trailer, a summary of the film, and a listing of all the characters.

For the record, the new edition will be a two-disc set, featuring the same full-frame transfer of the film on one disc, and a new anamorphic widescreen transfer (1.85:1) on the other. The audio will be included in English DTS, as well as newly mastered English, Japanese, and Spanish Dolby Surround mixes. On the extras side, the story summary will return, as will the trailer and newly produced character biographies. You'll also get new U.S./U.K. *Ninja Scroll* video trailers, as well as a video interview with director Yoshiaki Kawajiri and a look at the English voice-over cast. Should be a killer release, so keep your eyes peeled for it.

If You Like These Films . . .

. . . you should also check out *Blood: The Last Vampire*, the *Macross Plus* films, *Neon Genesis: Evangelion*, the *Armitage* series, both the new and original *Vampire Hunter D* and, as we mentioned earlier, virtually anything with Hayao Miyazaki's name on it.

Documentaries on DVD

To show us ourselves. To show us those we wouldn't normally look at. To show us things that happen in our backyard, which didn't seem to be interesting . . . but are. To show us things we wouldn't see otherwise in a million years. These are just a few of the goals of documentary filmmaking. The examples we've chosen here are not only successful at achieving those goals, they're incredible DVDs to boot.

Editor's Note: *Bits* writer Brad Pilcher contributed to this section.

For All Mankind

The Criterion Collection

For All Mankind is probably the greatest single documentary ever made on the Apollo missions to the moon. Its beauty lies in its simplicity. As a brief title card at the outset explains, *For All Mankind* is composed almost entirely of film footage taken by the Apollo astronauts themselves during the missions. The film is narrated by them as well, with commentary that creates a running, first-person account of the experience. Add to that an ethereal score by composer Brain Eno, and you have a visceral, illusory, and at times even hypnotic film experience, made all the more amazing by the fact that everything you're seeing is real.

As director Al Reinert explains in the audio commentary (which he shares with astronaut Gene Cernan — the last human being to set foot on the moon), virtually all of the film was shot in 16mm. Much of it had never been seen before. The director chose to combine footage from all of the missions to create a single, simulated spaceflight from start to finish. The original camera negatives were obtained from the NASA archives and were carefully blown up to 35mm. The footage was then cleaned up and stabilized, resulting in image quality that's often better than the original.

On DVD, it looks very good indeed. Keep in mind, however, that this isn't video on the same quality scale as you're used to seeing on DVD.

Given the fact that the source was originally 16mm, there's plenty of film grain visible, and print quality varies from clip to clip. But it rarely distracts, and both color and contrast are almost always excellent. Thankfully, there's also very little digital artifacting visible – the imperfections you see almost always lie in the source, and not the DVD's production. The audio is also well done, remixed in full Dolby Digital 5.1 surround sound. More importantly, the remix has been tastefully done and is respectful of the material. There are no cheap sound thrills here – most of the time all you hear is ambient fill. But occasionally, as in the launch and the staging sequences, the surround sound really comes into play, perfectly accenting the visuals without overwhelming them.

The supplemental material on this DVD is very nice (much of it repackaged from the laserdisc version). As mentioned, a commentary track is included, in which the director talks about the process of making the movie, while Cernan describes the experience itself. There are two subtitle tracks – one used to identify the astronauts you're seeing on screen, and another more traditional track (that thankfully identifies which astronaut is narrating at any given time as well). There are several brief audio highlights of historically important moments in the American space program (including, of course, Neil Armstrong's immortal words). There's also launch footage of each major rocket booster used in the program through Apollo. But our favorite extra by far is the gallery of paintings by astronaut-turned-artist Al Bean. Bean was the fourth man to walk on the moon and has spent his days since then working to document the experience on canvas. There are some 24 paintings shown in all. Bean introduces himself in his friendly Texas drawl, and then each image is accompanied by his audio commentary. Bean's a real character – sort of the everyman of the astronaut corps – and his thoughts and musings are welcome and fascinating to listen to. He gives a human quality to the subject, bringing it "down to earth" one might say.

Anyone who has seen *For All Mankind* probably shares our enthusiasm for it. And if you haven't seen it, this disc is simply a must. Even the animated menus on this DVD, which depict perhaps the most enduring single image of the twentieth century – the Earth rising in the blackness of space over the surface of the moon – are classic. Thank you, Al Reinhart, for making this important film. And thanks to Criterion, for doing it justice on DVD.

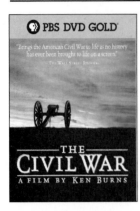

The Civil War: A Film by Ken Burns

PBS (Warner Bros.)

Ask anyone to name a great documentary film-maker, and you're going to get back the name Ken Burns. He's so well known for what he does, *The Simpsons* even mocked him in an episode . . . and they don't mock just anyone. Burns has become arguably the preeminent film documentarian of American history, and he's done it because of style, storytelling, and the fact that he dives fully into every subject he turns his camera eye towards. He could do a six-part documentary about bowling, and we'd sit through every minute of it. We're sure he'd make it interesting somehow.

But of all the films he's done, and all of the subjects he's tackled, none have been as acclaimed or lauded as his 11-hour look at the Civil War.

The Civil War dissects one of the biggest blemishes in American history, closely examining its origins, its participants, and its politics. We learn who died, who fought, who argued, and who lived. Burns shows us how much slavery had to do with the war, how patriotism for the states tore the nation apart, and what battles turned the tide. But more than just examining the details, dates, and places, Burns humanizes the experience through archival photos, letters from soldiers, and diaries from the women left back home. It's utterly fascinating from the first moment, and you'll have a hard time not getting up and switching discs until the whole thing is over.

The Civil War is a five-disc set packaged in a Digipak case and is presented in its original full-frame television aspect ratio. The transfer is gorgeous, with great detail, nicely deep blacks, and beautiful color throughout. The audio is presented in Dolby Digital 5.1 track, but the surround channels aren't terribly active. Given that this is a documentary, you wouldn't expect it to be an intense audio experience, but there are quite a few reconstructed battle audio clips where a more dynamic sound field would have added to the experience. The audio levels are also a bit inconsistent, with certain sections very soft and others blaringly loud. Keep your finger on the volume of your remote, 'cause you'll be using it.

As for the extras, as if simply having an 11-hour documentary wasn't enough, Burns provides audio commentary for all the episodes. He sheds much insight into the making of the film, the history of the production, and its participants. It may be hard to listen to the whole thing, considering the length, but it's a good idea to jump back a chapter here and there and switch over to it.

Disc One houses the video supplements. First up is a look at the process of preparing *The Civil War* for DVD, in a featurette called *The Civil War Reconstruction*. You get to look behind the scenes at the effort put into preserving the archival footage and making the DVD look as good as it does. It's a bit self-congratulatory, but it makes its point well. There are also interviews with the filmmaker and participants of the documentary, as well as a featurette called *Making History* about the technique and style of Burns's work. In addition, in the TV interview, *A Conversation with Ken Burns,* the filmmaker talks about his passion for what he does.

For you Civil War historians, each of the discs has its own battlefield map pertaining to battles fought on the episodes on that particular disc. The war was huge, so to be able to jump over to these maps for a little more context is a very valuable tool. Each of the maps also contains statistics and casualties reports as well. And if you'd like to test your knowledge of what you've just learned from the documentary, head over to the Civil War Challenge trivia game on each disc.

Horrible though it was, the Civil War is certainly an important and interesting piece of our collective American history. This documentary does a great job, not only in showing you what happened and why, but actually making the experience interesting and engrossing at the same time. For this reason and many others, *The Civil War* is close to being *the* perfect documentary. If only all education could be this fascinating and entertaining.

Triumph of the Will: Special Edition

Synapse Films

Triumph of the Will is one of those movies where the production quality is so superb and memorable, and yet they're still dwarfed by the power of the story behind them. In this case, the immense talent of filmmaker Leni Riefenstahl is on display in a film that utilized a huge crew and tremendously elaborate camera setups. Even cameramen on roller skates were used, which only illuminates the inventiveness and experimentation of

Riefenstahl. In the end, she edited 61 hours of footage into 2, and created what is easily the finest documentary/propaganda hybrid film of all time.

The problem for Riefenstahl is that this film was made for and about the Nazis, specifically their 1934 Nazi Party Convention in Nuremberg. That she made several other films for Hitler, who personally selected her after viewing her 1931 release, *The Blue Light*, certainly hasn't helped her cause. So despite her enormous talent for filmmaking, "Hitler's favorite actress" has never been able to shake her Nazi connections and was effectively blacklisted following World War II.

To be sure, the degree and nature of sympathy Riefenstahl had for the Nazi party is a question of some debate. She's made numerous efforts to distance herself from the Nazi ideology, if only out of professional necessity, but has remained unrepentant for the films she made. The whole controversy is as interesting as any of her films, but it's had one indelible effect: Documentary filmmaking has forever been changed by *Triumph of the Will*. Riefenstahl's fusion of propaganda with documentary techniques has skewed the assumption of objectivity afforded the documentarian.

Getting back to the film itself, the documentary opens with Hitler flying into Nuremberg for a week of events. These are made up mostly of marches and speeches by various officials, which makes sense given that this was designed to both introduce the new Nazi leadership to the people of Germany and also to pump those people up with displays of national unity. It probably did both very well for the 10 years it ran in Germany, but to say it gets repetitive would be an understatement. Two hours of saluting crowds, marching uniformed Nazis, and Hitler's ranting can grate on the nerves, and the subject material can be as boring as much as it's repulsive.

However, the structure and style of this film is as mesmerizing as Hitler's speeches were to the Germans of his time. Riefenstahl's cinematography is stellar, to say the least. From religious iconography, such as Hitler's cross-shaped plane shadowing the ground and the halo of light around him, to the early shot of a night rally through a sheer Nazi flag, the movie is still visually amazing all these years later. It set new standards for documentary cinema when it was first released. You simply can't escape the talent involved in the creation of *Triumph of the Will*. The saddest part of all of this is that Riefenstahl's gift has never been fully appreciated due to the sheer depravity of her subject material.

So let's all thank Synapse for bringing this film to DVD. Almost 70 years after its original release, and with the knowledge that history has not been kind to this print, we have a stunningly high-quality piece of video here. Of course, there are still source defects—some grain and even water damage here and there. But you can't expect a black and white film of this age to be pristine. Culled from a master negative held by Robert A. Harris's Film Preserve, *Triumph of the Will* just looks great on DVD and, given the film's artistic and historical value alone, this is a blessing. The image is relatively crisp, with solid blacks (as solid as they get considering the age and period of the print) and surprising depth. We've seen many "classic" films and, on DVD, they've all looked wonderful. But here we feel we must go beyond the customary, "it looks better than it ever has" because it looks damn good.

It also sounds good too, considering that this is a mono track from an era when recording technology introduced plenty of quibbles, pops, and snaps to the audio. The speeches are powerful in their force—surprisingly so. The musical score, an element of her films for which Riefenstahl is famous, comes through nicely as well. Even background noises like the cheering crowds are pretty distinct, all things considered. Overlook the audio defects that remain, as they're simply unavoidable after this much time, and focus on the relative clarity here. It's simply impressive.

Right up front, it's important to note that much more could have been done here in terms of supplements, considering the history of this film. How cool would it have been to get the documentary on Riefenstahl, *The Wonderful, Horrible Life of Leni Riefenstahl*, with this? That said, there won't be any Synapse bashing for not seeking out more supplemental material (especially since Kino has the rights to that documentary). What they *have* included is top-notch. First of all, you get Riefenstahl's short film *Days of Freedom*, which runs about 17 minutes. It can be said that this is nowhere near as good as *Triumph of the Will*. Still, it serves as a further exploration of Riefenstahl's style and is referenced in the commentary track. It turns out that this film, which documents the German armed forces, was only made because they wouldn't stop whining about getting short shrift in *Triumph*.

The other major bonus item is, in fact, the commentary track, which really shines and makes this DVD worthwhile. Featuring a professor of history, Anthony Santoro, it starts out as a sort of play-by-play, with no allusions as to the real opinions of our Ph.D. Then, as we reach a portion where various Nazi officials are shown in snippets, Santoro starts ripping in with how much he thinks these guys are garbage. We learn about the pornography collection of one guy, the person who should've been executed (according to Santoro) but offed himself before they could put him on trial, and so on. The shift sort of throws you at first, but then you just smile at the genius of this commentary. It's full of information, and San-toro's even, nonchalant delivery makes the opinionated zingers all the better. Watching this film without the commentary can be tough, but you can sit happily through the whole two hours of Santoro.

Given the subjective nature of this film, it should be noted that we also have our own opinions (zingerless though they may be). Hitler was a major league bunghole . . . but he sure knew how to pick his filmmakers. Riefenstahl displayed tremendous ability here—ability that forever rede-fined the documentary form and set a new standard for filmed propa-ganda. If you look past the vile nature of this material, you'll find a rock-solid and important DVD. It's absolutely worth a look.

9/11: The Filmmakers' Commemorative DVD Edition

Paramount

Documentaries don't get any simpler, or more awe-inspiring, than this.

Happenstance and fate were working in the favor of two Frenchmen (brothers Jules and Gedeon Naudet), who were making a docu-mentary about a fireman in New York City. They were following a new trainee, from the Acad-emy to being a working member of a firehouse located seven blocks from the World Trade Center. It was the morning of September 11, 2001 and each brother, armed with his own camera, went about business as usual. One went on a routine call about a gas leak in a downtown intersection, while the other stayed back at the station. Out on the street, everything was going in standard fashion, until the sound of a roaring jet engine cut the air. The camera pans up to the image of the towers jutting straight up into the shot. And suddenly we see the only known footage of the plane impacting the first tower. As many have said, it's a surreal moment—something straight out of a Hollywood film. Sadly, we're all familiar with the image at this point.

From there, the Naudet brothers did what anyone with video cameras working with rescue personnel would do—they headed into the belly of the beast. One camera follows a group of firemen into the first tower and captures footage that has been called heart stopping, jaw dropping, and astonishing. It's footage you don't usually get to see from a world-changing event. We watch what happened in the lobby as the towers fell and see how these brave firemen didn't skip a beat, running head first into a situation no one expected would play out the way it did. We see life, we see death . . . we see everything. It's all in first-person view, which makes it all the more disturbing. And throughout it all, we hear the thoughts of the participants in interview footage and voice over. *9/11* is a piece of history that all of us wish never happened. It's something we all wish to forget. Thankfully, this documentary gives us a reason to remember.

9/11 was originally broadcast on CBS, commercial free and hosted by actor Robert De Niro. The film on this disc is a bit longer than the original broadcast version and some changes were made to the program. These changes are few and nothing too significant. Since the video was pulled from consumer-grade video cameras, don't expect reference quality. It's certainly solid, but it's presented in non-anamorphic widescreen. The audio is in Dolby Digital 2.0 and sounds fine. Those looking for extras will find some additional interviews that weren't included in the feature. It's not much, but then again, this documentary really doesn't need features to support it. It stands sufficiently on its own strength.

9/11: The Filmmaker's Commemorative Edition is a great piece of documentary filmmaking. In this age of broadcast TV news and electronic journalism, we as a collective audience are often witness to history in new and powerful ways. But it's not often that we get to see those moments from such an intimate and visceral perspective as this. This film is tough to watch, but it's a powerful and absolutely extraordinary work.

Lost in La Mancha

IFC (Docudrama)

"This project has been so long in the making and so miserable that someone needs to get a film out of it . . . and it doesn't look like it's going to be me."

What happens when a production crew, brought in to chronicle the making of a film for a featurette on an eventual DVD release, actually documents a piece of cinema history, in effect creating the actual movie instead of some throwaway bonus feature? The answer to that question can be found here, in *Lost in La Mancha*.

In August of 2000, director Terry Gilliam (best known for his days in Monty Python, and his creation of *Brazil* and *12 Monkeys*), after 10 years of stops and starts, went to Spain to finally shoot one of his dream projects: *The Man Who Killed Don Quixote*. The film was to be a modern revisioning of the Cervantes' classic story Don Quixote. Johnny Depp would star as an ad exec who is thrown back through time, landing in the seventeenth century to fill in as a confused Sancho Panza — sidekick to the old, crazy, but morally dedicated titular hero. Gilliam had his crew in place, he had his locations set, and he was ready to roll. Unfortunately, fate had other plans. Jet fighters, rain, flood, and a major star having several health problems were just a few of the bigger problems Gilliam's faced. But there were many, many more, and all of them are here in this must-see documentary.

Doomed productions with chronicles almost more fascinating that the actual films are not new. Francis Ford Coppola's *Apocalypse Now* has *Hearts of Darkness: A Filmmaker's Apocalypse*, Werner Herzog's *Fitzcarraldo* has Les Banks's *Burden of Dreams*, and Gilliam's own *12 Monkeys* has *The Hamster Cycle and Other Tales of 12 Monkeys*, which was made by the same fellas behind *Lost in La Mancha*: Keith Fulton and Louis Pepe. The big difference is all of those documentaries illuminate the productions of films that exist, while this documentary is about a film that probably never will — at least with the vision seen here. And although a film about an unfinished film has been done before (1965's *The Epic That Never Was*, focusing on Josef von Sternberg's 1937 version of *I, Claudius*, which would have starred Charles Laughton — available on the DVD of the 1976 version of *I, Claudius*), there hasn't been one that took place while the unfinished film was being filmed. In this case, Fulton and Pepe were there, on the set, catching everything as it went down. As a result, we're given a rare and unusual view on the whole thing.

When the last frame of this intriguing documentary flashes in front of your eyes, you will have a very definite opinion of Gilliam as a filmmaker, how producers work to pass the buck, how clueless investors are, and how corporate bureaucracy can kill artistic vision. Don't be fooled by reviews of *Lost in La Mancha* that paint Gilliam as some quixotic filmmaker chasing down his movies with a certain madness evident in his eyes. Yes, the guy thrives in chaos, but as the whole production begins to fall apart around him, he's the *only* one who doesn't lose his head in the same fashion. Gilliam is an artist with a vision — to blame his directorial style for the downfall of the production is silly.

The quality of this documentary on DVD is pretty damn good when you consider it was shot on hand-held digital equipment. Once you allow for that, you'll find that the picture quality is actually very clear, with bright colors and nice detail. Audio is presented in Dolby Digital 2.0 and sounds fine when you once again consider the source. This was location sound and was never meant to be reviewed as an actual film. As such you get dropouts, and a few harder to understand sequences, which are filled in with subtitles. It's not rich, it's not active, but you get the point and that's what's important.

Although the extras aren't packed, this is a two-disc special edition, so there is plenty to see. All the extras are on the second disc. First up, and substituting for a commentary, we get a series of five interviews with the cast and crew. There's Gilliam discussing the film, Depp talking about his relationship with Gilliam, both members of the directing team of Fulton and Pepe talking separately, and producer Lucy Darwin giving her impressions. All five combined are very informative and fun—a commentary would have been great, but this works in its stead.

Next up are the deleted scenes. There are nine deleted scenes, each with nice little text introductions by the directors explaining why they were cut. Usually it was for length, or the fact that the shots heralded too much gloom or did so too early. These are interesting for the most part, but even *more* interesting is the look at a jettisoned stylistic attempt called "video portraits" that were laid into the film. Apparently, test audiences didn't like it so it was ditched, but it's actually not too bad. Maybe in another documentary it would have worked better.

Included on this disc are two very interesting pieces of video, of about an hour each. The first originally appeared on Telluride Community TV during the Telluride Film Festival in 2002, and features reclusive writer Salman Rushdie interviewing/conversing with Gilliam about the film and a variety of other topics. The other is a career-spanning discussion with Gilliam by *New York Times* film critic Elvis Mitchell, held during the L.A. County Museum of Art retrospective: "Tilting at Windmills: The Fantastical Worlds of Terry Gilliam." An edited version of this interview appeared on IFC, but here we get the whole thing. Both of these interviews are pretty incredible, and absolutely worth your time.

The *Sound Bites* section contains six video featurettes made from unused interview footage (which are a little more informative than entertaining), detailing aspects of the film production and why it fell apart. For those who want to know a little more about the project, these interviews are a nice companion piece to the film and other extras.

The *Storyboards and Production Stills* section includes lots of production drawings and Gilliam's storyboards from the film, without much context—which is a shame and makes it not too worthwhile as an extra. The storyboards, in particular, suffer from poor presentation here. Rather than showing you one storyboard frame per screen, you get four . . . and the sketched-line style of the artwork is hard to appreciate without the closer inspection of the detail. There are only three frames where we get to see the complete storyboard. On the other hand, Benjamin Fernandez's production designs and Gabriella Pescucci's costume designs are given the full screen treatment and they look good. But don't be confused by the title of the section—there are no actual "production stills" to be found, just concept art and storyboards.

Rounding out the disc are trailers. The first disc contains a large collection of cross-promotional trailers for other Docudrama releases, but more important is the theatrical trailer for *this* film on Disc Two.

If you love movies, Gilliam's especially, then *Lost in La Mancha* is a great documentary and a must-own DVD for your collection. Without DVD, this film would simply not exist, making this disc a perfect choice for inclusion in this book. Any production that starts life as an added value featurette, only to go on and represent a fascinating moment in cinema history all by itself is a huge feat. And to be able to show us the inner machinations of filmmaking is a priceless gift to film fans everywhere.

If You Like These Films . . .

. . . we also recommend *The Filth and the Fury, Grey Gardens, Salesman*, the *IMAX Space* series from Warner Bros., the BBC's *Walking with Dinosaurs*, the acclaimed *Joseph Campbell and The Power of Myth*, the *IMAX Super Speedway: Mach II Special Edition* and virtually anything else by Ken Burns on disc, including *Jazz* and *Baseball*.

Great Movies You Might Have Missed on DVD

So many movies get released on DVD each year that it can be hard to keep up with them all. Sometimes a movie was independent and so escaped notice due to lack of promotion. Or it's so classic (that is, old) that its appeal is limited. That's where DVD reviews come in. *The Digital Bits* is the net, and these discs are the best fish that most of you missed. In other words, here's a few DVDs we think everyone should see. They're well worth your time. Trust us on this.

Editor's Note: *Bits* writer Adam Jahnke contributed to this section.

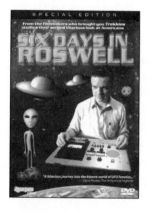

Six Days in Roswell: Special Edition

Synapse Films

Six Days in Roswell is funny. Wet your pants funny, in fact. We first saw this film at the Newport Beach Film Festival a few years back and found ourselves laughing so hard we had tears in our eyes. It's one of those films that just sets off an audience — the whole crowd was in stitches. Afterwards, while speaking with the film's producer, Roger Nygard, we decided we wanted to help get the film released on DVD so everyone could enjoy it. Roger's earlier film, *Trekkies*, had been released by Paramount on disc, but only as a bare-bones, movie-only edition. We wanted to make sure that this film got done right. So we contacted Don May, Jr. of Synapse Films. This special edition is the result.

Six Days in Roswell follows the adventures of one of Minnesota's most unusual native sons — the intrepid Rich Kronfeld. You may remember Rich from *Trekkies* — he was the guy with the motorized Captain Pike wheelchair (Rich really does own that chair, and Nygard assures us that *Trekkies* and *Roswell* are only the first two films in what will probably become a *White Trash Trilogy*). Rich, it seems, has a dream. He wants to be abducted by aliens. But there's a problem, as he says in the film: "Nobody from Minnesota ever gets abducted." So Rich sets out on a spiritual journey of sorts to the virtual Mecca of UFOlogy — the infamous Roswell, New Mexico.

Roswell, as some of you may know, is the sight of an alleged 1947 crash of a UFO — a crash that supposedly yielded alien bodies that the government supposedly covered up. Now it's become ground zero to a strange world of believers and abductees, UFOlogists, and skeptics. Rich arrives in Roswell to take part in the fiftieth anniversary celebration of the crash, and what he finds is a massive carnival atmosphere, rampant "cashing in" by local merchants, and enough bizarre theories about UFOs to fill . . . well, a UFO. In the end, Rich may not have uncovered the truth about alien abduction, but he's certainly learned a thing or two about human nature. And he's given us 81 minutes of great laughs along the way.

Rich Kronfeld is just a true gem — he's one of the funniest guys we've seen on film in a long time. His deadpan reactions to the rogue's gallery of bizarre individuals he meets on his trip are perfect. They provide the jokes and Rich's dry, aloof commentary is the punchline. In this film, Rich enters a "flying saucer" pancake-eating contest, rides his Captain Pike's chair in a small town parade, chats with self-proclaimed UFO "experts" like Stanton Friedman and Whitley Strieber, visits the "crash site" and the Roswell UFO Museum, and even attends "Roswell the Musical" (a community theater retelling of the events of 1947). He must convince his mother that getting abducted is an okay thing to do, and he even gets alien self-defense lessons from a pair of shotgun-toting rednecks (which will come in handy if aliens ever attack Rich with . . . well, shotguns). Along the way, he interviews everyone he meets, hoping for tips that will lead to his own "close encounter." This is very funny stuff.

And this special edition DVD is a great way to experience the film for the first time. The film was shot on Super 16 and was blown up to 33mm for theatrical release, so what you get here is the original full frame. You're going to see some film grain, but the quality of the video transfer is quite impressive. There's very little dust and dirt visible on the print and only minor artifacting. Roger and company spent a great deal of effort on color timing, so the hues here are bright and accurate. The audio on the disc is a standard Dolby Digital 2.0 stereo mix — no surround here. But the track is of perfectly good quality and it's more than adequate for the film. Dialogue is clear and natural and the film's musical track is light and well represented in the mix.

Thankfully, the folks at Synapse let Roger, Tim, and Rich go nuts in terms of the extras. First of all, you get a funny commentary track with the three of them together. Roger sort of sits back and talks about the stories behind each scene and steers the ship, while Tim talks in a straight-laced style about the process of filmmaking and Rich interjects amusing comments here and there. It's very engaging and delivers a few big laughs. There's also a 20-minute documentary on the making of the film, which is a really great piece on the process of low-budget filmmaking. Then there's some 35 minutes of deleted scenes, and this stuff is every bit as good as what was in the final film — there are lots more good laughs here. You also get 30 minutes worth of earlier works from Roger, Tim, and Rich: from their very first Super 8 films as kids to more professional-quality works.

And there's lots more. You get a gallery of about a dozen production photos, production notes, UFO trivia, quotes from other folks who have reviewed the film, and on and on. There's also three Easter Eggs hidden in this disc, which include a trailer for one of Nygard's other films, a video clip on "the atomic age and its influence on household appliances," and an extended version of a piece seen in the film.

Here at *The Bits*, we're proud of having a hand in making this DVD happen, so we're probably biased. But this is a great little film, and Rich is gonna make you laugh. If you liked *Trekkies*, or even think you might be interested, then you owe it to yourself to pick up this disc. It's a fun little spin, packed full of extras that are going to entertain you as much as the film itself. This is one of those flicks that we think is gonna pick up a big cult audience on DVD. So get your friends together, order a pizza, and give *Six Days in Roswell* a try. We're pretty crazy about it and we think you will be too.

The Films of Wes Anderson on DVD

Ah, Wes Anderson. You've gotta love the guy. He's made three feature films and all of them are great. We hear from aspiring filmmakers all the time at *The Bits*, and the one thing we hear the most is discussion about how they all want to make movies like Anderson. His characters, his style, his dialogue . . . they're all so ripe.

Anderson, so far in his career, has made a partnership with writer/actor Owen Wilson. The two started their journey together with *Bottle Rocket*, a movie not a whole lot of people saw, mostly because the cover art, which intentionally apes the *Reservoir Dogs* genre, features its stars pointing guns. That's not what the movie is about. Not by a long shot. Instead, it's a beautiful character study, a road picture, and a heist flick, all rolled up into a single awesome film that more people should see. It's out

as a movie-only DVD, with no extras to speak of, but the rumor is a new special edition is on the way. When it arrives, you can bet we'll be right there in line to pick it up.

Next, Anderson and Wilson moved on to *Rushmore*, a surreal, coming-of-age comedy about young Max Fisher, an extracurricularly addicted high school student. He'd rather "do" than study, which causes some troubles now and then. But the crux of this tale is Max's infatuation with a young teacher at his school (Olivia Williams) and the love triangle that pops up between the two of them and Max's new best friend, a self-made millionaire (Bill Murray) who hates himself and his family (not necessarily in that order). This love triangle drives the film, the comedy, the drama, and everything that lives in between.

Rushmore is not necessarily for everyone, but it's a great film. If you decide to check it out, you'll find it available as a very nicely done special edition from Criterion. The transfer is gorgeous, the commentary is funny and informative, and, best of all, it includes the funny *MTV Movie Award* interstitials that were themed around this film. There's also enough great production features and other extras to make this a DVD worth hunting down.

Wes Anderson's most recent movie, though, is his best . . . *The Royal Tenenbaums*.

The Royal Tenenbaums

Miramax (The Criterion Collection)

The Tenenbaums are a family of eccentric geniuses. Eldest son Chas (Ben Stiller) is a budding tycoon, still suffering from the death of his wife. Adopted daughter Margot (Gwyneth Paltrow) was a renowned playwright but hasn't written a word in years. Younger son Richie (Luke Wilson) was a tennis pro, until his career was curtailed by a very public meltdown. All of them have lived their extraordinary lives in the shadow of patriarch Royal Tenenbaum (Gene Hackman). Flat broke and with nowhere to go, Royal tries to win back the family he's been separated from for years by claiming he has just six weeks to live. The timing for such a scheme couldn't be worse. Royal's wife Etheline (Anjelica Huston) has just received a marriage proposal from her accountant Henry Sherman (Danny Glover), and Margot's marriage to cultural anthropologist Raleigh St. Clair (Bill Murray) has hit the skids, thanks in large part to her affair with Richie's friend, neighbor, and Tenenbaum-wannabe, Eli Cash (Owen Wilson). And those are just the main characters.

There are plenty of movies with terrific individual elements. Things like a great performance, amazing cinematography, or a moving musical score. But the best movies are those in which every single piece of the puzzle gels to create a seamless organic whole. *The Royal Tenenbaums* is one of those movies. Start with that amazing ensemble cast led by Gene Hackman, who can comfortably add Royal Tenenbaum to his list of indelible screen characters (like *The French Connection*'s Popeye Doyle and Harry Caul from *The Conversation*). Hackman is never less than interesting onscreen, and this movie gives him a chance to flex his comedy muscles (often underused by directors, but no surprise to anybody who remembers *Superman* or his unforgettable cameo in *Young Frankenstein*). It's a brilliant performance, unfairly snubbed by the Academy Awards.

And speaking of Oscar oversights, how is it possible that production designer David Wasco and costume designer Karen Patch were ignored for this movie? The Tenenbaum house on Archer Avenue is as much a character in the story as anyone, bursting with so many details that a dozen viewings of the film couldn't catch them all. And the look of the characters is utterly bizarre, but seems perfectly natural in this context. All you have to do is look at these people to get a sense of their lives. Of course, all of these aspects are in service to a rich, layered screenplay by Anderson and Owen Wilson (he's the blond Wilson). Anderson has said that he wanted the movie to feel like it was based on a book that never existed and he fully succeeds. The chapter headings, the frequent use of flashbacks (some of which last just a couple of seconds), and the unobtrusive narration by Alec Baldwin, all help make this feel novelistic in the best sense of the term.

This literary feel is only heightened by the terrific two-disc set from the Criterion Collection. The movie comes packaged in a dual-disc Amaray keep case, contained in a handsome (and, we might add, surprisingly sturdy) cardboard slipcase designed to look like an old, dog-eared book. While we're on the subject of packaging, we should mention the two inserts provided. The first will be familiar to anyone who's ever purchased a Criterion release before. It features a lengthy essay on the film (this one by *Film Comment* editor Kent Jones) as well as film and disc credits. The second insert is an amazingly detailed floor plan of the Tenenbaum house, illustrated by Anderson's brother Eric. The idea behind this was to guide the production designer but, as Wes Anderson explains in his own liner notes, Eric's drawings were so meticulous that they weren't finished until after most of the work on the film was already done. In any case, this insert is a very nice addition to the set.

Oh yeah, there's a couple of DVDs included in here too. The movie is on Disc One, and it looks simply amazing. Presented in the ultra-widescreen format of 2.40:1, and anamorphic enhanced for your viewing pleasure, this is a gorgeous transfer with virtually no flaws. Colors are warm and vibrant, but never over saturated. Even shots like Ben Stiller in his bright red tracksuit, standing in a deep red hallway, are handled extremely well. The image is never overly sharp, nor does it dissolve into a soft sea of grain. It's a perfect balance, conveying the texture of the film. The audio is presented in both Dolby Digital and DTS 5.1, as well as a 2.0 Dolby Surround track. Since this is a dialogue-driven film, most of the action is front-centered and the surrounds are not too active. However, they certainly come into play with the film's music, and this is one of the best soundtracks of non-original music we've heard in quite some time. Very little difference between the DTS and Dolby Digital are detectable on this one. Both are subdued, but natural, and fit perfectly with the spirit of the film.

The only extra on the first disc is a very interesting commentary by Wes Anderson. Anderson is amusingly self-deprecating at times and discusses a wide range of topics, including the music, some of the details in the design work, and his eclectic array of influences . . . some obvious (Orson Welles' *The Magnificent Ambersons* is a clear touchstone for this movie) and some considerably less so (an episode of *The Rockford Files*). The other supplements are all on Disc Two and an impressive batch they are. *With the Filmmaker* is an intimate peek behind the scenes at Anderson's creative process, co-directed by the great Albert Maysles. No offense to DVD producers, but you can't beat having a "making of" piece done by one of the guys who made the brilliant documentaries *Gimme Shelter* and *Salesman*. The only complaint with this piece is that, at approximately 30 minutes, it's far too brief. A nice counterpart to this feature is a series of video interviews with the cast. These are all good, and cutting the interviews together with behind-the-scenes footage beats the talking-head syndrome of similar interviews. Rounding out these is *The Peter Bradley Show*. Sharp-eyed viewers will realize that *The Peter Bradley Show* is the program Eli Cash walks off of in the movie. Bradley is a dim-witted, Charlie Rose-type, well played by Larry Pine. And on Disc Two, Bradley interviews five cast members, all longtime friends of Wes Anderson. You're not really going to learn anything substantial about the movie here, but it is awfully funny.

Perhaps the highlight of Disc Two is the extensive set of still galleries, featuring artwork, book covers, and storyboards. It's odd to call a still gallery a highlight, but there is simply so much art in this movie, much of it barely glimpsed, that it's a pleasure to be able to linger over it in detail. The large gallery of still photos by set photographer James Hamilton, is a step above the usual collection of posed publicity shots. These are all behind-the-scenes pictures and they're very well done. Wrapping the whole thing up are a pair of theatrical trailers and a pair of cut scenes. Oh yes . . . there's also several Easter Eggs on the second disc, at least four of 'em. They're quick, cute, and pretty easy to find, so we won't spoil them for you here.

All together, this is a fantastic DVD release of a brilliant movie, that's definitely worth at least one (and probably many) viewings in your home theater.

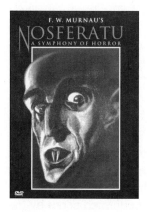

Nosferatu: A Symphony of Horror

Blackhawk Films (Image Entertainment)

Ladies and gentlemen, we give you *Nosferatu: A Symphony of Horror* — historically the most significant horror film ever made. No other horror film has been more analyzed, copied, or praised. F.W. Murnau's 1922 silent masterpiece is a study of terror, with its cryptic imagery and frightening intelligence. Enigmatic German actor Max Schreck (whose name literally means "fear" in German) not only portrays Count Orlok (a.k.a. Count Dracula) in this film, but has forever burnt into our collective subconscious a vision of utter terror, with his long, pale face, pointed ears, and claw-like fingers.

We won't bother explaining the plot of 1922's *Nosferatu*, because almost everyone out there is familiar with the traditional story of *Dracula*. This film is a bit different technically, because Murnau could not secure the rights from Bram Stoker's estate to create a movie based on the novel. What Murnau changed are the names of the characters and the location of the story (from London to Wisbog, Germany). After the film was finished, Stoker's widow sued the producers of the film, and the settlement stipulated that all copies of the film were to be destroyed. Thankfully, that didn't happen.

On the surface, *Nosferatu* seems to be a very straightforward vampire film, with all the requisite characters and events. However, Murnau constructed a poetic tale overflowing with intelligence and symbolism. Everything from camera angles to set architecture to dialog has some hidden meaning that adds to the mystery and psychology of the movie. Murnau's consummate dedication to artistry and experimentation are present in every frame of this film, and it will be studied and talked about as long as the world possesses the cinema as an art form.

Image Entertainment, in association with Blackhawk Films, has released what we consider to be the definitive version of this film on DVD. It's been restored and remastered from 35mm archival material, with all the color tinting fully restored, along with the correct playback speed. The color tinting is important to the film, because different colors are meant to signify things like time of day and candlelight ambiance. The playback speed having been corrected gives the action a more human pace; not fast and erratic like that of a *Keystone Cops* short. The image quality is outstanding given the rather abysmal condition of the original source print. Sure, there are a lot of scratches and blemishes, but this is a 79-year-old film that was probably never stored correctly to begin with. Given the circumstances, the picture looks quite good. The intertitle cards are brand new, but their newness is not distracting.

There are two audio choices. The first is the same recording of the original organ score that accompanied a previously released DVD, compiled and performed by Timothy Howard. It's presented here in Dolby Digital 2.0 and it sounds grand, boasting great fidelity and range, with some very deep registers that will put your subwoofer to the test. The second audio option is a newly composed and performed score by the Silent Film Orchestra, presented in Dolby Digital 5.0. This score is well recorded, with a crisp and clean character. Unfortunately, the Silent Film Orchestra score is just plain silly. At times, it sounds like music from *The Wonder Years* and, other times, it sounds like really bad new age music. The composition uses too many clichés and seems out of place with the tone of the film. Stick with the classic organ score—it's eerier and truer to the original creative vision.

New materials have been added to make this a nice special edition. *Nosferatu* isn't loaded with extras, but what Image and Blackhawk have included is worth a look. The disc includes an audio essay by Lokke Heiss that makes its return from the original release. It runs over the film like a commentary but is not always screen-specific. Heiss is obviously reading the text, and he's not the most exciting person to listen to, but the information he provides really helps to enhance your enjoyment of the film. Heiss discusses the history of the film and explains a lot of the subtleties of Murnau's creativity. Definitely give it a listen. Next up is a brand-new extra, *The Nosferatu Tour*, which is a 10-minute "then-and-now" photographic tour of the locations used by Murnau to shoot the film. We are shown the locations first as stills from the film, and then in modern-day photographs to compare them over the years. All the while, Heiss discusses bits of relevant trivia. It works well as a continuation of the audio commentary. The next new feature is a brief deconstruction of the carriage ride to the castle from the film, which highlights one of Murnau's unusual methods of filmmaking. Rounding out the supplements is a large gallery of conceptual art and poster advertising created by the film's producer and art director Albin Grau, along with a large collection of early nineteenth century paintings that inspired Murnau's vision. And peppered throughout the art are production notes. This last extra was also a holdover from the original edition.

Any self-respecting film buff owes it to themselves to make *Nosferatu* a part of his or her DVD library. It's an intelligent, haunting piece of art that's worthy of study and discussion. The audio and video quality of the new special edition is about as good as it will ever get. And if you think quality over quantity when it comes to the extras, you should be very happy. This new edition comes very highly recommended from your friends here at *The Bits*.

Rififi

The Criterion Collection

You wanna see one of the greatest films you never saw? Well . . . this is it. *Rififi* is a tried and true American heist film, but told in the French new wave style (in fact, it was the earliest breakout hit of the French new wave). American expatriate Jules Dassin, (who gave us classic fare like the prison yarn *Brute Force* and two of the blackest noir films ever, *The Naked City* and *Night and the City*) wrote and directed *Rififi*, loosely based on the pulp novel by Auguste Le Breton. *Rififi* simply claws into your gut and climbs its way into your brain — where it will live forever.

Set in and around Paris, the film follows Tony (Jean Servais), pasty and tubercular, who's coming home after serving five years in prison for jewel robbery. Out for good behavior, Tony hooks up with his old pals Jo (Carl Möhner) and Mario (Robert Manuel). As we find out, Tony actually went to jail to take the fall for his young protégé, Jo, because the kid was a newlywed with a young son. Hoping that Jo has finally set his life straight, Tony quickly learns that Jo and Mario are still dancing on the wrong side of the law. And they have a nice little homecoming planned for Tony — a diamonds score from a nearly impenetrable jewelry store downtown. Finally deciding to go straight, Tony turns the deal down and heads to the L'age d'Or club, where he's heard that his former flame and loyal girlfriend, Mado (Marie Sabouret), is now working. But Tony sees that she hasn't been so loyal lately. Unwilling to wait for Tony's release from jail, Mado has taken up with a vicious gangster named Grutter (Marcel Lupovici). Brokenhearted, Tony handles this the only way a grizzled ex-con knows how . . . he takes her jewelry and fur coat, roughs her up, and kicks her out of his apartment. And, of course, he calls Jo and agrees to do the heist. After all, what's he got to be straight for now? Tony agrees to do the job on two conditions (that will keep the group alive and hopefully prevent him from ending back up in jail). First, no guns. The game is always played by different rules when a rod is involved. And second, they have to rob the jewelers' vault instead of the planned window display. That's when a charismatic Italian ladies man and safecracker named Cesar joins the team (played by Jules Dassin a.k.a. Perlo Vita). And so begins the tale of a flawless heist that goes wrong about every way imaginable.

Actually, the heist itself is flawless—a beautiful piece of cinema. It's become one of the most talked about sequences in all of film, and deservedly so. The four men meticulously go over every aspect of the job. They case the shop, detailing every bit of information possible, from how long the shops surrounding the jewelry stores are open to delivery schedules for those shops. and even how long it takes to walk around the block. They also construct an exact duplicate of the store's "state of the art" alarm system to figure out the best way to disarm it. But all of this, grand though it may be, is nothing when compared to the actual heist — a 30-minute, dialogue-free jaw-dropper that is so detailed, the film was actually banned in several countries because of how realistic it was. If you don't fall in love with *Rififi* at this point, check your heartbeat — you're probably dead.

Once again, Criterion did us all proud. The film is gorgeous on DVD. Luscious in all its original black and white glory, *Rififi* practically sparkles on disc. We kid you not. A DVD fan couldn't gush enough, especially after having to suffer for years with horrible video transfers of this film plagued by crappy subtitles. Here the video is presented in the original full frame, with deep dark blacks, nice solid grays, and edible whites. There's nary a digital artifact to be found. Just check out the detail in the opening credits alone. This is a "wow" transfer. The audio is also pretty flawless. We get the choice between an English dub, which sounds fine, and the original French, which is equally well presented. Both are in the original mono. But just trust us here . . . you need to see this film in French.

The extras included on this disc are pretty simple, but when you're done with them, you'll find them invaluable. The first is a video interview with writer/director Jules Dassin, who tells stories about his life in Hollywood, his life on the run as an American expatriate (trying to make films in Europe with little success), and his meeting with Le Breton (who, after reading the script, wanted to know, "Where's my book?"). It's beautiful. Mr. Dassin is a great storyteller. Anyone who owns the Roan Group laserdisc of *Brute Force* knows this from that disc's three-hour commentary track. Hell . . . anyone who's seen *any* of his films knows he can tell a story. But here, instead of just going on about the film, it's like catching up with an old family member and listening to what's been going on since last you saw him. The interview is very personal and is well worth checking the disc out for all by itself. There are also some production notes, a stills archive of production photos, and the American release trailer.

When we gush over movies like this, all we really want to do is have everyone see it. If we all went to see more films like *Rififi*, instead of giving $130 million to brain-dead, summer shoot-'em-ups, then the cinematic world would be a better place. *Rififi* is a great film, made even better with the love and care given to it for this DVD release. Criterion hasn't failed us yet and, because they took such good care of this film, the world of cinema is all the better for it.

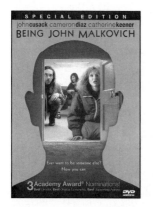

Being John Malkovich: Special Edition

(USA Films) Universal

Being John Malkovich was the best, most interesting, and original film of 1999. The film's story is simple, in principal. John Cusack plays Craig Schwartz, a down-on-his-luck puppeteer. Craig struggles unsuccessfully to make a living at his chosen craft, until his wife Lotte (Cameron Diaz) suggests he get a real job. So Craig answers an employment ad that leads him to the Murton Flemmer building, and the seventh-and-a-half-floor offices of LesterCorp. A brief and bizarre interview later, Craig's a newly hired file clerk. Before long, Craig finds a surprise in his office . . . a hidden portal leading into the head of actor John Malkovich. No kidding. It works like this — you simply crawl down a dark tunnel, and suddenly find yourself inside the head of Malkovich, experiencing his life for 15 minutes, after which time you get dumped out on the side of the New Jersey Turnpike. Simple right?

Craig shares his discovery with a woman he's met on the seventh-and-a-half floor, Maxine (played by Catherine Keener, in a performance that won her a Best Supporting Actress nomination). Craig is bored with his life, and he's hot for Maxine, so he's happy to go along when she suggests that they make a buck or two off the discovery. Together they form JM, Inc. and charge $200 a pop for people to experience the thrill that is Malkovich. But there are problems. First of all, when Lotte finds out and tries the portal, she can't get enough of the experience. She also falls in love with Maxine, who loves her back . . . but only when she's inside Malkovich. What's a jealous Craig to do? And what does John Malkovich have to say about all this? Plenty . . . and that's just scratching the surface of this bizarre and funny film.

It's really hard to believe that *Being John Malkovich* was former music video director Spike Jonze's first film (you may actually know Spike better for his acting turn as that hick American soldier in *Three Kings*). This is a really amazing piece of work. The story, by screenwriter Charlie Kaufman, is well written and completely original. The film features great performances by Cusack, Keener, and Cameron Diaz. You have to give Diaz credit — her appearance in this film is surprisingly good, and she was brave to take the role (take one look at her in this film and you'll understand). And naturally, John Malkovich steals the show. Malkovich's puppet show/dance, in which he flails around wildly, is worth the price of this disc alone.

So how is the DVD? Pretty cool. The anamorphic widescreen video is solid, with accurate (if very muted) colors, deep blacks, and excellent shadow detail. This isn't reference quality, but it looks very good overall. The audio is about equal to the video — the Dolby Digital 5.1 track doesn't put your system to the test, given that this is a dialogue-driven film, but it's plenty good. It creates a good ambience and the dialogue is clean and clear.

Now . . . let's talk extras. What you get on this disc seems like a lot, but it's got to be the most bizarre set of bonus materials to be included on

DVD to date. To start with, you get the film's theatrical trailer and four of the most unusual TV spots ever seen for any film. You also get a pair of really great short videos, which were produced for use on camera during the film. The first is the orientation on the seventh-and-a-half floor that Craig is made to watch after he gets hired at LesterCorp ("at least there will be one place on God's green earth where you and yer cursed kind can live in peace . . ."). The second short is the fake documentary on Malkovich, which is seen late in the film, featuring his second career as a puppeteer (and a funny cameo by Brad Pitt). In addition, there are a pair of "behind-the-scenes" videos. One focuses on the art of puppeteering and the other is an interview with someone who was hired to drive her car back and forth in the background of the New Jersey Turnpike scenes. Are you scratching your head yet? How about this—you also get a page with nothing on it (literally) and an interview with director Spike Jonze where he . . . well, he gets sick and pukes on the side of the road. Rounding things out are a gallery of production photos and an unadvertised feature that can't quite be called an Easter Egg—as you skip from menu to menu, you'll get to listen to five or six complete audio tracks from the film's soundtrack in the background, including the film version of Bjork's "Amphibian." Just skip around to find them all, and then don't touch anything if you want to listen—the entire track will play.

But there is one complaint to be had with this DVD, but it's a *big* one. Where's the commentary track? A film this unique and interesting absolutely screams for a director's track, but you won't find one here. It also would have been nice to hear Malkovich talking about this film. Like, what he thought when he first got this script—a script with his name on the cover? The lack of commentary is really disappointing. But again, that's our *only* real complaint with this disc.

Being John Malkovich is as offbeat a film as you will ever see, and see it you must. If you missed this one in theaters, you really owe it to yourself to catch it on disc. And while the DVD isn't the best you'll ever see, it's almost perfectly matched to the spirit of this film. It's absolutely a must-spin (and for serious film fans, a must-own as well).

If You Like These Films . . .

. . . we also recommend *12 Monkeys, Titus, Three Kings, Wings of Desire, Pi*, the original Norwegian version of *Insomnia, The Decalogue, Pitch Black*, and *Waking the Dead*.

TV on DVD

During the early days of DVD, the format was never known to be a show-case for TV shows. The ones that did come to disc came as "best of" sets, and no one seemed to be interested in that. Fans wanted every episode, and who could blame them? Then Fox made a bold move. They started putting out the cult favorite *The X-Files* in complete season sets . . . and DVD fans gobbled them up. With that success, and the advent of other cult shows and film-like cable series, today we're seeing a true explosion of TV product on DVD. Along with the likes of *The Sopranos* and *Friends*, you can now start collecting complete seasons of *All in the Family*, *Space: 1999*, *Mary Tyler Moore* . . . the list is long and constantly expanding. That just tells you, wholeheartedly, that TV is big on DVD.

Editor's Note: Bits writer Greg Suarez contributed to this section.

The X-Files

Twentieth Century Fox

As fans of *The X-Files* since its first season, it's a lot of fun to look back at the earliest episodes of the series and rediscover the things that hooked us in the first place. Striking and noir-ish cinematography, against-the-grain charac-ters, unforeseen plot twists, aliens, government conspiracies — they're all here from the show's pilot through the nine seasons that make up its broadcast run. *The X-Files* is a show that dared to be different, and different it was. We often wonder if the real magic of the show is that series creator Chris Carter didn't have a clue as to where all this was going when he created it — all the story arcs and character revelations that have come over the years as the show has developed. Because if he did . . . wow. We suppose that when you create a hit show, you're entitled to allow everyone to think you're a genius, right? Now Fox has brought the *entire* series to DVD as a number of multi-disc boxed sets. *The X-Files* on DVD is, simply put, a fan of the show's dream come true.

So how do these look and sound, you may be wondering? Pretty damn good. Not reference quality by any means, but definitely very, very good. Let's start with the video. If you're used to the way the show looked during its original broadcast, you'll be blown away at the quality of the image on these discs. Unless you've visited a post-production suite at Ten Thirteen Productions when one of the episodes was being edited, you've probably never seen *The X-Files* looking this good before. That's not to say that the video doesn't have its problems. The video often has an overly soft, digital-looking quality to it — but then the show has always looked a little "processed." You'll see plenty of film grain, but then that's just the style in which the show is shot, low lighting and all. You'll see some edge enhancement, but that's par for the course for any video not originated in high-definition these days. But what you'll also see is very nice contrast, solid and accurate color saturation (if often muted by choice of style), and very good blacks. That last bit is important. *The X-Files* is a show where so much takes place in the shadows, making good black-level detail critical. And these DVDs definitely deliver it. In addition, all of the episodes from *Season Five* on are presented in anamorphic widescreen, as they were originally produced — a very nice touch.

The audio is presented in Dolby 2.0 Surround, and it's a perfect match for the video. *The X-Files* has never sounded this good in your living room. It's surprisingly encompassing, despite using only light surround effects for ambience. It creates a solid acoustic environment that draws you into the action on screen nicely. The dialogue is crisp and clear, and there's decent bass in the mix. Every now and again, it surprises you. Overall, the audio doesn't jump through any hoops to impress, but it works just fine. And it comes in French flavor too.

Now we'll talk extras, 'cause there is plenty included in these sets. There are featurettes, interview clips, and "behind-the-scenes" segments, some of which were run before each episode in syndication on the FX cable network. But we're not done yet — not by a long shot. You also get the TV promo spots that appeared on Fox, "international" clips, tons of deleted scenes, and even DVD-ROM material.

The X-Files on DVD is a welcome treat, and it was one of the first TV series to be given deluxe treatment on the format. Sure, this set isn't exactly cheap (the SRP is $149.98 per set). But if you're a fan and you can get it for a good price, it's a must own. The truth may be out there . . . but for those of us who love *The X-Files*, the fun is definitely in these discs.

24

Twentieth Century Fox

Here's a great concept, and we can't believe someone hasn't tried to pull it off before now: A government agent has 24 hours to prevent something bad from happening . . . something that can utterly change the world as he now knows it. At every step, something or someone stands in his way. Oh . . . and to make it all the more interesting, the story is told in "real time" over the course of 24 one-hour television episodes. Brilliant. But can it work logistically? Yes and no.

24, as a show (both seasons), is full of plot holes you can drive a truck through. It's got major logic jumps, hackneyed story lines, and even a few real-world math problems, such as . . . how the hell does a guy get through L.A. traffic in 10 seconds without a helicopter? But you know what? The whole thing works because it's so friggin' fun. Star Kiefer Sutherland not only plays lead agent Jack Bauer, but simply *is* Bauer. We loved his character from the minute we first saw him on the screen. We didn't think he could do it — be the tough guy we actually root for. But after just the first few episodes of the show, he's in your hearts and you want him to win.

Now . . . half the fun of the show lies in going into it without knowing anything. So if you don't know *24*, and you haven't seen it yet, good. Watch the DVDs. If you did watch the show, then regurgitating the story for you here won't do a lick of good. You either loved it or hated it. But here's a summary of *just* the first episode from *Season One*, so you can get an idea of what *24* is all about.

Season One begins at midnight on the day of the California presidential primary. Jack Bauer, head of the government's *Counter-Terrorism Unit* (CTU) based in L.A., is called in to investigate the possibility that Senator David Palmer, the first African-American with a shot at being elected president, is the target of an assassination plot. Meanwhile, Jack's teenaged daughter, Kimberly, sneaks out of the house with her friend, Janet, for a date with two young guys. When Teri (Jack's wife) calls and lets him know that Kim's gone, he tells her not to worry. We then establish Jack's team at CTU: Nina (his right-hand "man"), Jamey (the computer geek), and Tony (a.k.a. "Soul Patch" Tony, the shifty-eyed bad boy who may or may not be up to something). We also meet George Mason, a co-worker who has it in for Jack because of a past situation involving dirty agents that Jack turned in. Jack doesn't trust Mason, because he's been tipped that the assassination plot might involve someone working on the inside at CTU. Meanwhile, at his campaign headquarters, Senator Palmer gets a mysterious phone call from a reporter who claims to know something dangerous about his son's past. Back at Jack's house, Teri's hooked up with Janet's father to go looking for the missing girls . . . a plan that soon leads them to believe that the girls are in trouble. And high above all of this, in a plane bound for California, Mandy (a.k.a. "Naked Mandy") joins the mile high club with a photographer named Martin in hopes of getting his credentials to access a press conference that Palmer is expected to attend later in the day. All of this ends with an explosion high over the Mojave Desert. And where all these events might be going is anyone's guess.

Hooked yet? So are a *lot* of people. No doubt about it . . . *24* is cool. When watching this series during its broadcast run, each episode keeps you hanging on pins and needles for the next. Thankfully, now you can watch *24* at home without the wait.

On DVD, the video quality is quite good. First of all, the episodes are all presented in anamorphic widescreen—this show was *definitely* meant to be seen in widescreen. Viewing it that way makes the experience seem like a continuous, 24-hour long film that you just can't stop watching. The colors are rich and accurate, with excellent contrast. The look of the series is dark, ominous, and moody, reminding us in many ways of *The X-Files*, and the DVDs capture this look perfectly. There is a certain graininess to the episodes by design, and this is noticeable on the video. You will also see a certain amount of MPEG-2 compression artifacting, particularly in overly bright or washed-out areas of the picture. But none of this is distracting in any way. Fans who watched the original broadcasts will experience video on these DVDs that is far and away superior to what they remember.

The audio on these discs is also good, but not outstanding. English-only audio is presented in Dolby Digital 2.0 Surround, and the track generally services the video and the story well. Dialogue is always intelligible, and the series' tension-ratcheting score is well presented without overwhelming the action. The rear channels provide some light ambience and fill, but don't expect much in the way of true surround play beyond this.

The extras from *Season One* were a bit disappointing, in that all you got was an alternate ending to the final episode of the season (with optional audio commentary by the show's executive producer) and a video introduction to the set by Sutherland. Sutherland also briefly hints at the events of the second season (all without spoiling anything for either season). As far as the alternate ending, it goes without saying that you shouldn't watch this until after you've seen all the episodes. We think you'll agree that the ending that *was* used is much more dramatic and effective.

Naturally, we would have liked more in the extras department, but the lack of them was necessitated by the quick release of the season to DVD, just a scant few months after its original Fox network run. *Season Two*, thankfully, is much more packed, featuring no less than 44 deleted scenes, along with alternate endings and audio commentary tracks for several episodes.

All in all, *24* isn't perfect. It's not the best dramatic television you'll ever see. But it *is* original, wild, fun, and very, very cool. On DVD, the series is interesting if for no other reason than *24* marked the first time a TV show had been released on DVD right after the completion of its first season (and before the premiere of its second), allowing fans to catch up and/or discover the show for the first time. *24* went on to become one of the most successful shows of the 2002-2003 season, in parts thanks to DVD. In any case, we highly recommend this series, and DVD is absolutely the perfect way to enjoy it.

The Sopranos

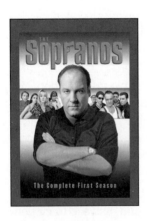

HBO (Warner Bros.)

Since David Lynch's *Twin Peaks* departed ABC in 1991, television drama has been adrift in a sea of lawyers, doctors, and police detectives . . . and one cop show is virtually indistinguishable from the next. What happened to originality? What happened to the quality and craftsmanship of stories and character development? And most of all, what the hell happened to good, old-fashioned acting? Thankfully, HBO's original series *The Sopranos* is a cool, refreshing drink of water. Some argue that what makes this show unique is its frank depiction of reality. This isn't kids' stuff—there's graphic violence, strong language, drug usage, and nudity. To be fair, a cable network like HBO can

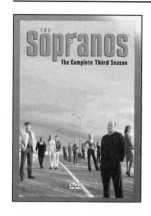

freely air such content without much worry of the dreaded censor, while the four major broadcast networks would never recover from the fallout of angered viewers. However, let's be experimental. Delete the crimson spray of gunshot wounds to the head, the trio of strippers in Tony Soprano's nightclub, and the barrage of four-letter words uttered throughout an average episode. What you're left with is a show that's fit for NBC or CBS, yet still remains completely original and expertly written. It's a program that's arguably better than any other drama on the air, and that would surely draw primetime viewers away from courtrooms and ERs. But given that even a censored version of *The Sopranos* would still be a cut above most broadcast television cookie-cutter fare, that extra wallop of violent, graphic reality ties the series together and makes it that much more powerful.

Quality programming deserves a quality audio/video presentation on DVD, and while *The Sopranos* might fall shy of the "reference quality" label, HBO has gone above and beyond the call to make these discs special. HBO presents the shows in anamorphic widescreen (framed at 1.78:1), and overall, the picture is impressive. Evidence of compression artifacting does pop up here and there, and the video can also be a bit noisy in places. That said, most of the video on these discs appears very smooth and detailed, portraying a convincingly cinematic look. *The Sopranos* is shot on film, and the creators make it a point to shoot each episode like a mini-movie. Color and contrast are also generally accurate. The cinematography on every episode is first-rate, and the DVD format is definitely the best way to experience it.

For DVD, audio for *The Sopranos* has been remixed in a quality 5.1 Dolby Digital. While it always sounds smooth and clear, it's not an especially active mix. The surrounds are used almost exclusively for music fill, with only occasional ambience heard during the onscreen action. Much of the soundtrack is very screen-oriented, but the front soundstage is nice and spacious. A Dolby 2.0 Surround mix is also included (the 5.1 version edges it slightly in dynamics, as expected).

While these boxed sets are not as feature-laden as we would have liked, HBO did throw in a few extras for each season. There are behind-the-scenes featurettes, which tend to be less than five minutes each and were used on HBO to advertise the show. They offer almost no insight into the series and are barely worth mentioning. Audio commentary tracks with the show's creators run through select episodes. These are always fairly informative and fun. The real treat of the supplements is contained on the first season: a 77-minute interview with David Chase by filmmaker Peter Bogdanovich. This interview covers almost everything a fan of *The Sopranos* would want to know, from the genesis of the show to its development, writing, and music . . . you name it. While we normally like to see more of a variety of features, this interview provides enough depth and information to cover all the seasons. The only missing element on these sets is more involvement by the cast. We're pretty sure we're not the only *Sopranos* fans out there that would like to hear an interview or commentary with James Gandolfini or Edie Falco. Gandolfini has so much presence and power on the show, it would be great to hear him talk over a few of the episodes.

While the DVD format is a collector's medium, it's not often that a studio will release a title (or titles) that can truly be considered special. It's even more rare when that kind of treatment is given to a TV series. Rest assured, you can definitely add *The Sopranos* to the elite list of the best the DVD format has to offer. *The Sopranos* is not only a benchmark of television drama, it's also one of the most provocative shows to debut in years. If you've missed it on cable, this is the perfect opportunity to see what everybody's been talking about for the last few years. Highly recommended.

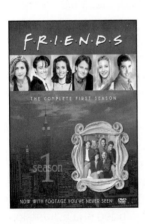

Friends

NBC (Warner Bros.)

You love 'em on TV, but now they'll be there for you on DVD—anytime you want 'em. Warner Bros. has been putting out these complete season sets of *Friends* for a while now, and they're actually pretty great. Originally, Warner was releasing episodes of the show in a single-disc "best of" format, but it wasn't working for the fans. So Warner gave in to demand and began releasing these complete season sets. As expected, they're flying off the shelves. If you love all the trials, tribulations, breakups, and make-ups of your friends (Rachel, Ross, Monica, Chandler, Phoebe, and Joey), then you'll love these DVDs.

Friends is even more a show about nothing than *Seinfeld*. Six Gen-Xers live and love in New York, cracking wise and helping each other

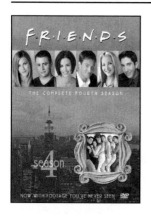

through life all the while. They started as late-twentysomethings and have kept going from there. The characters are modeled after all the types of friends we have in our own lives. There's Rachel, the spoiled rich girl, learning that there is life without pampering; Ross, the goof with a heart of gold; Monica, the picky one (with a 250-pound ghost in her closet); Chandler, the spastic and sarcastic friend we all have; Phoebe, the ditzy one; and Joey, the brainless but loveable hunk. The comedy is sharp, the writing crisp, and new and the characters are all well acted by a cast of superstars. What's not to love?

Presented in full frame as exhibited on TV, these episodes look great. In fact, they look better than what we saw on the original broadcasts. Colors are bright, artifacts are nil, and detail is tight. The sound is a very nice and full Dolby Digital 5.0 (lacking a booming LFE, but who needs it here?). In terms of audio and video, there are no complaints.

The extras are fun, but as we're finding with these TV sets, somewhat minimal. There are commentaries on select episodes (although so far, no appearances from the big-name cast — only producers and writers). There are also "tours" of the various environments the show takes place in; *Season One* has Central Perk, *Season Two* has Monica and Rachel's loft, and *Season Three* features Chandler and Joey's digs. There are also trivia games, character and actor bios, and other bits, such as an uncut video of the song *Smelly Cat*.

While not loaded with bonus material, *Friends* is definitely a show worth owning on DVD.

The Simpsons

Twentieth Century Fox

The Simpsons — one of the greatest social commentaries on life in America — is now on DVD, and we couldn't be happier. Sure, it'll take forever for us to get the entire series on disc, but what we've gotten so far has been worth the wait.

The series is being released in complete season sets, but the quality of what we've

gotten so far is a mixed bag. Video-wise, the colors are beautiful. Yellows are warm and solid, blues are as blue as the sky and all the other primary colors used in the show are well represented. Unfortunately, the blacks are lacking on the earlier episodes and there's some heavy edge enhancement used here and there throughout all the releases thus far. There are also artifacts from digital compression throughout the video on these episodes. Just look at the credits for proof of that. Overall, the video quality is fair, but we expect it will get better as Fox releases newer seasons.

The audio, on the other hand, would make Homer give a big "Woo-hoo!" There's a very nice Dolby Digital 5.1 track on each of these episodes, and they all sound great. There's a little bit of play in the rear channels, clean and centered dialogue, and an overall crisp and clear soundstage. You'll also find the original Dolby Digital 2.0 audio here, as well as a 2.0 French track for our French-Canadian friends.

So what about the extras? Mmmmm. Extras. There's not a whole lot of meat on these discs, but there are a couple of things. First up, select episodes feature the original scripts. This is pretty neat, as there are doodles in the margins that lead you to believe that they were probably culled from creator/producer Matt Groening's personal library. So far, nearly every episode on these DVDs has scene-specific commentary by members of the production team. You'll find folks like Matt Groening, producer James L. Brooks, director David Silverman, and writers Al Jean, Mike Reiss, and Jon Vitti to name a few. There's a lot of information on these commentaries, but most of them are really just these guys coming back together to watch the shows after many years. We can only hope they continue the trend for upcoming releases.

Other extras include outtakes with optional commentaries, animatics (preliminary animated drawings) with commentary, short documentary material, behind-the-scenes footage, foreign language clips, promo clips, music videos, galleries of production art, and more. And let's not forget the Easter Eggs scattered amongst the discs. It's fair to say that the discs aren't packed, but what you get is cool and nice to have.

The Simpsons is one of the only shows on TV that many of us here at *The Bits* bother to watch religiously. So having it on DVD is a treat. Hopefully, Fox will get more in depth with the extras in future sets. But they'd better hurry, because at the rate Fox is going (about one season per year), our kids will be completing our *Simpsons* DVD collection as we sit in the old folks' home writing angry letters to TV producers.

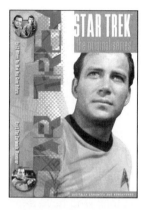

Star Trek on DVD

Paramount

We definitely fall in the category of *Trek* fans at *The Digital Bits*. We spent most of our youth following the syndicated TV adventures of Kirk, Spock, and the crew of the *Starship Enterprise*. Long before *Star Wars* took us to a place long ago and far away, *Star Trek* fired our collective imaginations like nothing else could.

So we're pleased as punch that Paramount's delivered the goods when it comes to *Star Trek* on DVD. Not only are they working to release special editions of each and every *Star Trek* feature film, including a *Star Trek: The Motion Picture – The Director's Edition* two-disc edition that *Trek* fans can really sink their teeth into, they're releasing the TV series episodes faster than Tribbles can multiply. There were 79 episodes of the original series and we know every one of them. So it's with fond memories that we're now able to go back and revisit this small-screen adventure, digitally remastered on DVD where it belongs.

Trek never looked (or sounded) so good as it does on DVD. The quality of these *Original Series* discs is first rate. Paramount undertook a complete digital remastering of the entire series. Fully digital, high-definition transfers were done of all the episodes in the series, using the original film interpositives from the studio's vaults. Then the digital master tapes were "cleaned" using special digital filters to reduce the amount of dust and dirt visible. The result, on virtually every episode, is a crisp and vibrant picture, with terrific contrast and lush, accurate colors. The video isn't exactly reference quality, but for a series from the late 1960s, it looks *way* better than we can ever remember having seen it before. If you're a fan of the series, and you've only seen these episodes previously on some syndication network, you'll be blown away.

But if you dig the look of these discs, wait until you hear them. Specifically for DVD, Paramount remixed the audio of all of the episodes in Dolby Digital 5.1, and we can say with utter assurance that it makes a *big* difference. All of the mechanical tape hiss in the original masters has been eliminated, along with bad music and sound effects edits and the like. And the original mono tracks have been digitally extrapolated to create a three-dimensional sound environment for home theater. The result is wonderful ambience and some nifty panning and surround sound effects. When the opening credits appear, you'll hear the *Enterprise* woosh over your shoulder. The enhanced surround sound adds just the right amount of gee-whiz to the DVD experience and helps to make the episodes fresh again for those of us (and I know you're with me on this) who have seen them dozens of times over the years. Very cool.

The only extras provided on *The Original Series* DVDs are preview trailers for the two episodes on the disc, plus two more trailers for the episodes on the next volume. You can access the trailers for each episode in that episode's submenu page, or all four trailers on the disc can be accessed by highlighting the "Enterprise arrowhead" symbol on the main menu (it's that shiny thing Kirk wears on his uniform).

There are 40 volumes of *The Original Series* on DVD, each with two episodes per disc. And when you're done with that, you can check out all seven seasons of *Star Trek: The Next Generation* that have already been released on DVD. Paramount wisely learned from other studios and released *The Next Generation* in complete season box sets. And each set has numerous behind-the-scenes featurettes on various aspects of the series and its production. You might get a look at the special effects, a bio of a particular character, interviews with the writers, actors, and producers – you name it. Still not enough *Star Trek* on DVD for you? Okay . . . by the end of 2003, Paramount will have released all seven seasons of *Star Trek: Deep Space Nine* in complete season box sets as well. These discs have even more features than were on *The Next Generation* discs. And just to keep you busy for another year, Paramount plans to release *Star Trek: Voyager* on DVD in 2004. That's enough to keep even a Klingon happy.

It's a blast watching *Star Trek* on DVD, and our favorite is still *The Original Series*. You'll see William Shatner's ham-handed Kirk swagger, and

Nimoy's unflappable Spock arch his eyebrows. Scotty will bitch and moan about his engines, McCoy will remind you that he's a doctor (as opposed to something else), and there are plenty of short skirts and funny-looking aliens (who look surprisingly like humans with painted skin and funny ears). Lots of red-shirts will die on away missions, several computers will be logic-looped by Kirk, and hordes of extras flood the bridge just in time to be tossed around during red alerts. So what are you waiting for? Set your phasers on stun, and get yer *Trek* DVDs pronto, space cadets!

If You Like These Shows . . .

. . . we also recommend checking out *Buffy the Vampire Slayer*, *Smallville*, *M*A*S*H*, *Farscape*, *Stargate SG-1*, *The Twilight Zone*, *Family Guy*, *The Outer Limits*, and *Twin Peaks*.

Bad Movies . . . Great DVDs

Sometimes, the worst films can make the best DVDs. Great DVDs can put the film under a microscope and show us all what was really going on behind the scenes. Where did the filmmaker go wrong? What was the original idea, and how did it get corrupted? Was it the fault of some producer or studio pressure? Was it just a bad idea to begin with, or did it get worse as it went on? Good DVD special editions can answer questions like these, sometimes without even meaning to. When you're listening to audio commentary or watching interviews, a good filmmaker will usually own up to the problems, while a bad filmmaker will just be totally oblivious. Nothing will reveal the problems better than a director's own words. And that's just great entertainment, any way you slice it.

So, submitted for your perusal, here's a list of flicks we find wanting . . . but which (surprise!) make great DVD special editions.

Editor's Note: *Bits* writer Brad Pilcher contributed to this section.

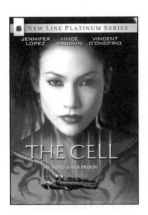

The Cell: Platinum Series

New Line

When you first saw the trailer for *The Cell*, you had to think to yourself, "Oh yeah, baby! This film is gonna rock." New Line has been known for bringing us some of the greatest young directors of this generation: David Fincher, P.T. Anderson, F. Gary Gray — these are guys that impress the hell out of us every time we see their films. And so when New Line brought us the feature film debut of Tarsem Singh, a commercial and video director that first made himself known with the video for R.E.M.'s "Losing My Religion," we were all expecting something great — especially when held up next to the trailer. We should have known better.

In *The Cell*, Jennifer Lopez plays a psychologist who rides the cutting edge of a new technology that allows her to enter the minds of people and help them solve their problems through experimental therapy. When a serial killer/necrophile, played by the always kooky Vincent D'Onofrio, falls into a coma due to a rare neurological virus complicated by his acute schizophrenia, the police learn that he has one last victim still out there and they must race the clock to find her. The FBI, led by Vince Vaughn, turns to Lopez and her team, hoping she'll enter the killer's mind and pull out the information they need. The problem is a man's mind is his castle and there's no telling what kind of world this killer has created for himself in his head. And there's an added risk—if Lopez begins to believe in his world, she might get caught inside and never find a way out. Freaky.

With images straight out of some of the greatest gothic paintings, *The Cell* looks wicked cool. In terms of visuals, this is one hot picture—eye candy on the highest level. It's just too bad the characters are so poorly drawn. The killer is a cartoon, no matter how hard D'Onofrio tries to inject him with depth. Lopez is a blank page. Just because she looks concerned, sulks around her apartment in men's shirts and a pair of hot underwear, and smokes blunts, it's not going to make the audience care for her as a character. Vince Vaughn is yet again wasted, but here more than he usually is. Part of the problem lies in a poorly executed script with paper-thin characters and bad dialogue between them. Part of the problem lies in the direction the actors were given or, better put, the lack thereof. *The Cell* is a very poor film. It's a great music video, but a poor film. And that's a shame, because it could have been really cool.

But somehow, this title has become one of the must-own DVDs for anyone with an interest in filmmaking. One reason for this is because it's such a visually rich film, having it on a medium like DVD somehow makes it easier to swallow . . . slightly. The other and more important reason is this DVD gets to the bottom of everything. The DVD actually helps to explain why *The Cell* fails as a film. Director Tarsem, as you'll find on the director's commentary and in the deleted scenes, is a fountain of film references. He's got it all down. If fact, this guy's knowledge is impressive. He knows what he's talking about and his comments are rapid fire. The problem is Tarsem doesn't seem to understand context very well, such as *why* a scene is cool. Tarsem references films like *Coma*, *Lost Highway*, and David Mamet's *Homicide* . . . but he doesn't seem to know *why* the images he steals work in the context of the original story he pulled them from. Yes, the bodies hanging from the ceiling in *The Cell* look like the ones in *Coma*, but it's certainly not freaky. The freaky factor comes in knowing that the people in *Coma* are being harvested. Just because he had the bio-suits designed to look like flayed bodies doesn't mean that the scene is going to set you on edge. Horror works when it's accessible and mental. Gratuitous visual stimulation is nothing but masturbation—it's an empty experience.

The brilliance in this disc is that everything we need to know about this film is preserved on the commentary track for us to listen to. Tarsem unknowingly condemns himself over and over again, as he spouts off his thoughts like some fan boy. He even admits that the test audiences "forced" his planned pacing of the film to change due to "uneasiness." It's all such a smack in the gut. The hope is that film students will pick this disc up and witness for themselves exactly why a film like this doesn't work. What's apparent here is filmmakers like Tarsem, who have incredible visual genius, also have a third-grader's capacity for psychology and an inability to get actors to act (as evidenced on the deleted scenes commentary, when Tarsem audibly shudders at the idea of improvisation on set). Even the featurette is brilliant, because while everyone is waxing Tarsem's ass in the interviews, we see footage of Tarsem directing the camera and frames and putting actors on their marks or showing them how to move so that it'll be cooler looking. If only we had a nickel every time we watched a filmmaker play puppeteer with talented actors. You know . . . for a commercial and video director, this guy really, honestly, and truly *is* a genius. But Tarsem needs to learn the nonvisual side of his craft and go work with actors in a theater, where he can't be so reliant on camera setups. If this guy could really direct, he'd frickin' rock. He has a lot to offer the cinematic world . . . but he's got some work to do first.

The supplements in total are very well done. Because there was just no way of making this disc cool relying on the film, the producers of this DVD went for the only route they could — the visuals. Besides the commentary by Tarsem and the film's isolated score (which actually makes watching the film a hundred times better), we have a commentary with the key production members behind the film (director of photography Paul Laufer, production designer Tom Foden, makeup artist Michelle Burke, costume designer April Napier, special effects supervisor Kevin Haug, and composer Howard Shore). They all have some enlightening things to say and, as this film shows, they have a lot of talent. You get deleted scenes and a featurette dedicated to Tarsem, entitled *Style as Substance*, as well as a breakdown of six scenes from the film with accompanying production team interviews, storyboards, and behind-the-scenes footage. What's cool about this is that you can watch the interviews, storyboards, and behind-the-scenes stuff all at the same time picture-in-picture style, or you can switch between them via alternate angles. There's also the film's trailer, the international teaser trailer, a map of the brain (with explanations for everything in there), a psychology test based on your level of empathy, and a handful of DVD-ROM features (like script-to-screen and a demo of a PC game that has nothing to do with anything).

For the record, the film is presented in a beautiful anamorphic widescreen transfer at 2.40:1. The colors are stunning with deep, detailed blacks. Grain and artifacting are not to be found. The sound is also pretty sweet, really backing up the visuals wonderfully. This Dolby Digital 5.1 track is actually one of the better we've listened to, creating a nicely wide and atmospheric soundstage. Just check out Lopez's first meeting with the killer in Chapter 12 – creepy. You'll also find Dolby Digital 2.0 on board as well.

The Cell sucks, no doubt about it. It's silly, needless, and stupid. But, man, does it look cool. And that's the problem with so many films today. Too bad Hollywood doesn't hold this film up and examine why it failed so miserably. We hope people give the DVD a chance and see it for themselves, because it's brilliant. It's a great lesson for film students, and so it's worth picking up for that reason alone . . . if for no other.

Mallrats: Collector's Edition

Universal

Mallrats is the simple and tender story of two ne'er-do-well slackers who try and get their girlfriends back after losing them both on the same day. Brodie and Quint (*Jaws* anyone?) are friends who have problem relationships. Brodie would rather stroke and caress his Sega game system than his beautiful girlfriend Rene (played by Shannen Doherty). She gets fed up, leaves his lazy ass, and heads to the mall to shop. Quint plans on running away with his girl Brandi (played by Claire Forlani) to Universal Studios Florida to get married in the *Jaws* attraction. His plans are smashed though when he accidentally causes a woman to die and has to watch his girl take her place in her father's game show. Her father would rather see Quint dead than as a son-in-law, and before the day is out, he may just have his way.

Mallrats is fast paced and filled with wonderful characters and viciously funny lines. So there are no problems there. Absolutely no problems there. So why do so many fans consider this the lesser of Smith's films and nonfans consider it crap? There are two words that explain why some could easily not like the flick: Jeremy London. His acting is wretched, his delivery is awful, and the overall feeling the film gives is that he's the main character. And he's horrid as the central character. Now some would say that this is the fault of the director. It's not in this case. Look at everyone else. Look at Jason Lee, who never acted before, and see how good he is in this film. Jeremy London is just plain bad, and the film is less than good because of it.

So, if you accept this as a "bad film," then you have to accept this as a really great DVD. It features a whole stack of outtakes, one of the funniest commentary tracks *ever* recorded (it's even better than the *Chasing Amy* track on the Criterion DVD), and a healthy smattering of other treats. One of those tasties is related to the aforementioned (and already stellar) commentary track. At various points in the track, you can access "live" footage of the guys talking. Smith and company (actors Ben Affleck, Jason Lee, and Jason Mewes; the producer Scott Mosier; and View Askew historian Vincent Pereira) are so funny that at points in this track you'll be laughing out loud while listening to it. The deleted scenes consist mainly of a cut storyline showing a "failed assassination attempt," but loads of other things were cut as well—something like an hour in all. You have to see it to completely understand it.

There are so many fun bits on this disc! Aside from the commentary and deleted scenes, there is a video featuring Jay and Silent Bob for the song "Build Me Up Buttercup" by The Goops. You also get several production photos, cast and crew bios, the trailer, and a production featurette that's pretty damn good and even reveals more information about the production on top of what we learned on the commentary track.

The video quality itself is also pretty stellar. The video is anamorphic enhanced, and it's a great transfer. The colors are nice, and there's virtually no digital crap anywhere in the picture. The only issues with the video are related to the print itself—some occasional visible film grain and slight softness now and again—but these are *very* minor. The Dolby Digital 5.1 sound is sweet as well—nice, natural, and well rounded. You will hear much more background dialogue in this mix.

As Smith gets more popular, more and more people have started to embrace *Mallrats*, and well they should. It's not a great film, but it's very funny, and a necessary addition for tourists in the View Askew Universe. This is, without a doubt, an absolutely must-own DVD, for fans of both

Smith and special edition DVDs in general. And it includes one of the best Easter Eggs of all time. We'll let you find that on your own.

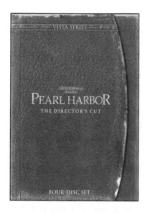

Pearl Harbor: The Director's Cut— Vista Series

Touchstone (Buena Vista)

Oh . . . to live in a world envisioned by director Michael Bay. Every image looks like it fell off a Norman Rockwell canvas. All the men are brash, red-blooded, handsome, and strong (except that one awkward kid with the red hair and the big Adam's apple), and all the women are perky—sweet and rosy-cheeked—the type you'd take home to Mom. The sun is always smiling on this America. Somewhere down the street, there's a baseball game going. Old Glory's waving in the breeze and the heady smell of apple pie hangs in the air. That is . . . until the asteroids start falling, or, in this case, 500-pound bombs from those Japanese suckers' Zeros.

Pearl Harbor is a film that very much aspires to follow the pattern of success of James Cameron's *Titanic*. The structure of the film itself is even the same, featuring a two-dimensional love triangle that dominates the first half of the story, with an action-packed second half that's super-charged by the best special effects Hollywood can offer. But while this pattern mostly worked for *Titanic*, it falls flat here. This feels like two completely different films, tacked together rather uncomfortably. The attack on *Pearl Harbor* itself is visually amazing . . . more realistic than the real thing. But it starts getting a little uncomfortable when any filmmaker makes history look like a video game.

Most of the actors in this film seem like they're walking through their roles, just reacting to what's happening around them. That's not to say that they're necessarily bad—just that this script is extremely contrived. Ben Affleck and Josh Hartnett play Rafe and Danny, a pair of boyhood friends who aspire to be fighter pilots and eventually get their wish as World War II looms across the Atlantic. Rafe falls in love with a pretty nurse named Evelyn (played by Kate Beckinsale) just before he volunteers for an assignment with the RAF defending England from the Nazis. Meanwhile, Danny and Evelyn find themselves shipped off to balmy Hawaii, about as far from the war as they can get. When they get word that Rafe's been killed in action, the two soon find themselves inexorably drawn together, first to ease their sorrows and then romantically. But surprise — Rafe's alive and well, and returns to complicate things on December 6, 1941 . . . just in time for all three to find themselves caught up in the surprise ILM — I mean, Japanese — attack.

Pearl Harbor is pretty much a wash as a piece of entertainment. The good stuff here is so weighed down by the bad that it only slightly breaks even — like matter and antimatter annihilating each other . . . just without the bright flash and all the energy. Still, there is a certain amount of guilty pleasure to be found here. Filling out the film's variously thin subplots are Jon Voight (in a great supporting turn as President Roosevelt), Alec Baldwin (in a cheesy but fun performance as Colonel James Doolittle), and Cuba Gooding, Jr. (good but barely used as a cook on board the doomed *U.S.S. Arizona*). These are all real historical characters (unlike the leads), and they're far more interesting than the story of Rafe, Danny, and Evelyn. Unfortunately, they're largely used as filler in an already bloated film. All in all though, if you lower your expectations, there are far worse ways to spend three hours of your life than watching this sort of gung-ho, premasturbatory stroking of our collective American egos. *Armageddon* comes to mind.

Given that *Pearl Harbor* runs some 183 minutes, the film has been broken up into 2 parts (over 2 discs) for this DVD release to maximize image quality. On the whole, the anamorphic widescreen video is excellent, reference quality even. The source print used for the transfer occasionally exhibits moderate grain, but that is no doubt intended — an artistic choice. The contrast is also excellent, with very deep blacks and yet good detail throughout the image.

On the audio side, the discs feature both English Dolby Digital and DTS 5.1 soundtracks, along with an additional French Dolby Digital 5.1 track. All are excellent, featuring active surround sound fields, a wide front stage, and deep low-frequency support. Dialogue is almost always clear, which is an achievement given the energetic nature of these mixes. The DTS soundtrack is just a hair better than the others, sounding slightly more smooth and natural, but it's a very minor improvement that will probably only be appreciated by those with high-end home theater systems. This is good audio all around, whichever option you choose.

There's one other audio feature on these discs that represents a first on a major studio DVD — a 2.0 Dolby Headphone track. Dolby Headphone is basically a process that simulates the effect of listening to full, multi-channel surround sound, using *any* pair of average stereo headphones. It's designed for those who watch their discs on the go, using portable DVD players and laptops with DVD-ROM drives. We first saw this process demonstrated at CES a few years ago, and we were totally blown away by it. When we saw the demonstration, it required that a special chip be installed in the device, which then processed any existing Dolby Digital 5.1 track. We couldn't wait to see portable players hit the market with this feature. When few did (we assume because manufacturers didn't wish to bear the added cost of the chip), Dolby changed tactics. What they've done for this DVD is to take the film's 5.1 soundtrack and preprocess it into a 2.0 Dolby Headphone track, which is then encoded onto the disc. That means that no additional hardware processing is necessary—just plug in your headphones and go. The effect is amazing. It's not true 5.1 of course, but the process adds tremendous depth and realism to the listening experience. Dolby Headphone is one of the coolest DVD features we've seen yet.

Kicking off this incredible DVD special edition are three feature-length audio commentary tracks. The first is by director Michael Bay and film historian Jeanine Basinger. Bay takes his film very seriously in this track and Basinger chimes in from time to time with information about how the film and events contained therein relate to the true events they are based on. The second commentary is a bit more fun. There's producer Jerry Bruckheimer, together with actors Ben Affleck, Josh Hartnett, and Alec Baldwin. Affleck and Hartnett (recorded together) play the track for comedy for the entire three-hour run time of the film. There are, of course, the few moments where they are serious and reflective with a few references to the September 11th attacks and the parallels between it and the attack on Pearl Harbor. And Bruckheimer and Baldwin bring the track back to the more serious side of filmmaking. The final audio commentary features cinematographer John Schwartzman, production designer Nigel Phelps, costume designer Michael Kaplan, art director Martin Laing, and composer Hans Zimmer. It's the more technical of the three tracks, but is worth the listen.

Aside from the three commentaries, there're literally hours of supplemental material here on these discs. The only real extra on Disc One is an Easter Egg comparing the widescreen and pan-and-scan versions of the film. Press right on "main menu" on the disc's audio setup menu. Disc Two features the *Journey to the Screen: The Making of Pearl Harbor* documentary, brought back from the theatrical DVD release. It's a 50-minute

EPK piece with some slightly valuable information. You'll also find Faith Hill's "There You'll Be" music video, and a preview trailer for the *National Geographic Beyond the Movie: Pearl Harbor* DVD, which is available separately for those who find they'll need more info about the film and its true history. You'll also find another Easter Egg on Disc Two as well—a gag reel.

Disc Three splits the extras into two sections: *The Film* and *The History*. *The Film* features some nine featurettes, breaking the various key aspects of the film down. You get to see everything that went on with the production. You'll also see how the final footage compared with what was being shot and get more commentary from Michael Bay. Additional featurettes like *The Soldier's Boot Camp* and *Officer's Boot Camp* bring you in on the elaborate, four-day training the actors received in military procedures and daily life. There's also comparisons of the film to stock footage of the bombing, showing just how far the film went to be authentic. In *The Film* section, you'll also find the film's teaser and theatrical trailers.

The History contains an excellent pair of documentaries about the real events depicted in the film, produced for the History Channel. *One Hour Over Tokyo* includes interviews with the surviving members of the Doolittle Raiders. *Unsung Heroes of Pearl Harbor* highlights some of the individual stories from survivors of the actual attack. Each is made up of stock film, modern interviews, and reams of archival photos. It really helps to put you in the shoes of the brave men and women who were there. The last feature on Disc Three is *Oral History: The Recollections of a Pearl Harbor Nurse*, which is simply a modern reenactment of the testimony of a war nurse who was stationed at Pearl at the time of the attack, featuring more historical photos and stock film.

After what already amounts to several hours of supplements, we head on over to Disc Four, where we find a multi-angle version of the entire attack on Pearl Harbor sequence. Going from animatic to CGI wireframe to the final footage, we explore what it took to replicate a 60-year-old event. There's also the original animatic commissioned by Michael Bay and used to sell the film to both the U.S. Navy and to Buena Vista. You'll be utterly amazed at how close ILM came to the final footage in the film. Michael Bay and visual effects supervisor Eric Brevig also take us through the entire process of putting the effects sequences together in the featurette, *Deconstructing Destruction*, as well as DVD-ROM material and a gallery of photos and artwork broken into various subjects for easier digestion. Finally, you get an "interactive timeline" documentary called *When Cultures Collide: From Perry to Pearl*, which details the history of American/Japanese relations as well as many political events that led to changes in policy on both sides.

Pearl Harbor: The Director's Cut — Vista Series stands as one the finest special editions ever produced. The movie blows, but this is definitely how a big epic film should look on DVD.

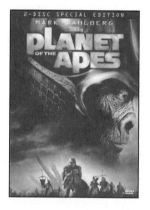

Planet of the Apes (2001): Special Edition

Twentieth Century Fox

When we heard that Twentieth Century Fox was doing a remake of *Planet of the Apes* . . . well, what else could you do but groan loudly and shake your head? Seriously, how dare they remake one of the most loved classics of sci-fi? Then news came that director Tim Burton was taking the reigns, and well, just think of it — a Tim Burton *Apes* movie! That would *have* to be cool, right? Well . . . it almost was. Almost.

The film's revised story centers on Captain Leo Davidson (Mark Wahlberg), a United States Air Force pilot stationed at a primate research facility/space station near Saturn (we think). It seems that the primates, including the stouthearted Pericles, have been genetically engineered to make them better space pilots. Suddenly, there's an electromagnetic storm that's right outside the space station windows. So Leo tucks Pericles into a rocket pod and sends him out to see what the storm's all about. Surprise . . . he disappears. Surprise again . . . Leo disobeys orders and goes after his little banana-lovin' buddy. And wouldn't you know it, he crash lands on a planet where apes rule and humans all look like Estella Warren and Kris Kristofferson. Well okay, not all of 'em, but you get the idea. Leo is quickly captured by the nasty Thade (Tim Roth) and taken to Ape City, where he's befriended by the kind-hearted Ari (Helena Bonham Carter), who believes that humans are people too. She helps Leo escape, and before long, humans and apes square off in a battle to rule the planet.

It's easy to enjoy almost everything that takes place in this film, from the time Leo crashes on the planet of the apes until just before the end.

With the exception of some totally unnecessary camp references to the original film (including a silly cameo by Charlton Heston), pretty much everything is "believable." The cast is excellent, the performances are good, the atmosphere is appropriately Burton-esque, and the production design is damn cool. But the film's setup is so weak as to be silly. And the ending feels so tacked on that not only do you not get the payoff you're looking for, you're left scratching your head trying to figure out what possible twists of logic could allow for it. Suffice it to say, it feels like a cheap copy of the original film's ending. And while Burton obviously understands it (as he explains in his audio commentary) he isn't about to tell us what it means (as he also explains in his commentary).

So that leaves the setup. Why does the USAF have a space station orbiting Saturn for which the only purpose seems to be training chimps to fly rocket pods? Is there some need for chimps that can fly rocket pods? We seem to do pretty well these days sending robot probes to check out distant objects in space — have we lost that capability in the future? Why would Davidson risk his own neck going after Pericles? Is there some special bond between the two of them? These are all questions that go largely unanswered, which we suspect the filmmakers would say fall in the category of things that don't need explanation — you're just supposed to accept them as part of the suspension of disbelief demanded by all movies. But that's just lazy storytelling. How hard would it be to explain these things? Not hard. Imagine that the military decided that it wanted to experiment with genetically engineered primates . . . but genetic tampering's been outlawed on Earth. So the USAF sets up a secret genetic research facility on the edges of our solar system. When the electromagnetic storm happens, Pericles, being the curious and highly intelligent genetically engineered chimp that he is, decides to go check it out on his own. So he hops into a rocket pod and blasts off into the storm. Davidson must go after him, because his superiors are worried that if Pericles somehow makes his way back to Earth, their secret research will be compromised. And everything else about this story could play out like it already does. See? It's as easy as that. The biggest problem with Hollywood sci-fi films these days is that you're expected to just accept plot elements on face value, even if they make no logical sense within the context of the story. And that's just bad storytelling.

So if the film itself is a miss, what about the DVD? Simply put, this two-disc set is one of the most interesting we've seen. Let's start with Disc One. The film is presented here in an excellent anamorphic widescreen transfer, which preserves its 2.35:1 aspect ratio. *Planet of the Apes* is as dark and atmospheric a film as one would expect of Burton. And the DVD presents that perfectly. The contrast is excellent and the images are clean without being edgy. There's plenty of detail in the darker areas of the picture. Color is rich and spot-on accurate. And, surprisingly, there's very little compression artifacting visible. This is excellent DVD video that, if not quite reference quality, is pretty close.

And that's impressive given that this disc also features dual 5.1 audio tracks in Dolby Digital and DTS flavors. It seems like we spend a lot of time saying this, but as expected, the DTS track provides a greater measure of clarity and naturalism, resulting in a somewhat fuller and more unified soundfield. But the Dolby Digital track is no slouch either. Both tracks feature active surround channels during action scenes (the crash, for example), which remain active, if more subtle, throughout the rest of the film to maintain atmosphere (during, say, scenes set in the jungle). The dialogue is clean and audible, there's plenty of bass in the mix, and Danny Elfman's score is ever present without being overwhelming.

Disc One features a pair of audio commentary tracks, one with director Tim Burton and the other with Elfman. Burton's commentary ranks among the better he's done, but there are still plenty of gaps where he simply stops talking. There are also a number of things we'd love for him to talk more about that go unexplained (the ending again stands out). Elfman's track features the film's terrific score isolated in Dolby Digital 2.0 Surround, with occasional sound effects and Elfman speaking in between the music cues. Also included on Disc One are cast and crew bios, THX-Optimode test signals, features that will work only in Nuon-enhanced DVD players (like "viddies" and various zoom points during the film), and DVD-ROM extras (including the script-to-screen viewing mode and web links).

But the really cool thing about Disc One is the ability to watch the film in "enhanced mode." If you select this option, at various points during the film, a small picture-in-picture window will appear in which various cast and crewmembers will talk about the making of the scene you're watching at that moment. Or you might see behind-the-scenes footage of that scene being shot. The video plays right over and along with the film itself (these segments are inserted via seamless branching). Additionally, at various times during the film, an icon will appear on the screen (à la the "Follow the White Rabbit" feature on *The Matrix*). Pressing Enter on your remote at that time will take you out of the film for a few minutes to view a separate piece of behind-the-scenes video. So, for example, when you see the space station, an ILM staffer will show you the full-size miniature in more detail and talk about how it was filmed. Then you're sent right back into the movie without having missed anything. Both of these features are cool ways to convey the usual "here's how we did that" kind of information in the context of viewing the film, and they're used very effectively here.

Disc Two provides even more behind-the-scenes material. And if this material lacks any kind of perspective (the DVD extras were completed within a few days of the film's theatrical release, allowing little time for retrospective thinking), it's all so interestingly presented that you'll hardly notice — a very neat trick indeed. You get an HBO documentary on the making of the film, which is pretty standard. But you also get five cool behind-the-scenes featurettes, some quite substantial, on various aspects of the production. There's a look at the recording of the score, a video on the "ape movement" classes the cast attended, the creation of the makeup effects and costumes, what it was like to shoot on location, and so on. You also get a number of video "screen test" segments, where similar video material appears in quad-screen, and you can select which audio track you want to listen too. And we're just getting warmed up. You get five extended scenes, the film's trailer and teaser, eight TV spots, and an extensive gallery of production photos, storyboards, and conceptual artwork indexed by subject. There's an additional gallery that provides a look at the film's poster and print campaign, along with images from the press kit. You even get a music video remix of the film's theme. Best of all, however, eight different multi-angle featurettes give you a look at the production of various scenes in the film. Each featurette has up to four different video angles and up to three different audio tracks that you can select in whatever combination you wish. A little icon at the top tells you which angle you're on at that moment and lets you select another. And a "command bar" ever present at the bottom of the screen allows you to dive out of the featurettes momentarily to view production artwork created for that scene, to view the final scene in the film, or to read a text excerpt of that scene from the screenplay.

So there's lots of substance in this two-disc set. But it's the way the material is organized and presented that is so completely impressive here. This is one of the most interesting and efficient uses of DVD interactivity we've seen. *Planet of the Apes* is arguably the most technically advanced DVD ever produced, in terms of both authoring complexity and interactivity. It's just too bad the flick wasn't better.

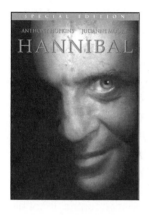

Hannibal: Special Edition

MGM

Often, when a movie is based on a well-known novel, film critics will spend a great deal of space in their reviews discussing whether or not the picture is a faithful adaptation. This can be kind of interesting from a purely academic perspective, but in the long run, it's completely irrelevant. What works in a novel doesn't necessarily work in a movie. Sure, you can stay true to a good book and make a good movie (as in *The Silence of the Lambs*) and you can deviate from a good book and make a bad movie (as in David Lynch's *Dune*). But you can also deviate from a good book and still make a good movie, like Ridley Scott did with *Blade Runner*. You can take a bad book and turn it into a good movie, like Steven Spielberg did with *Jaws*. Or you can keep everything simple and just turn a bad book into a bad movie, like Ridley Scott did with *Hannibal*.

Hannibal is the 10-years-later follow-up to Jonathan Demme's Oscar-winning *The Silence of the Lambs*. Julianne Moore takes over for Jodie Foster as Clarice Starling, now a full-fledged FBI agent who, as the movie opens, finds herself taking the flak for an arrest that went horribly wrong. She escapes punishment thanks to the machinations of Mason Verger, who wants her reassigned to the Hannibal Lecter case. Verger is the only one of Lecter's original victims to survive, though his face is now nothing more than a hideous mask of scar tissue. Verger's plan is to use Starling as bait to draw Lecter out of hiding. Once Lecter is located, Verger will capture him and feed him to a specially trained herd of man-eating pigs.

Okay, so clearly we're not in the same genre as *Silence of the Lambs* here. You can call *Hannibal* a black comedy, a dark romance, or a horror movie, but you can hardly call it a psychological thriller. There isn't a single moment in *Hannibal* that generates an iota of the tension found throughout *Silence of the Lambs* (or Michael Mann's *Manhunter*, for that matter). Lecter is no longer a believable threat. Anthony Hopkins seems content to play him as sort of a gourmet Freddy Krueger, tossing off silken witticisms and demonstrating his fine taste in music, art, and perfumes. The movie has lost all handles on the character by the time an attack dog cowers in fear at Lecter's mere presence. The movie's pace is off too. A huge chunk in the middle of the film is taken up by Italian cop Giancarlo Giannini's growing suspicion that he tracked down Hannibal Lecter. Of course, we already know what he's going to discover, so there's no suspense in the sequence at all. And by the time we get Lecter and Verger face to scarred face, it's as if the filmmakers have lost faith or interest in the man-eating pigs idea. We're rushed through these scenes as quickly as possible, denying the audience a really interesting confrontation between the completely twisted Verger and his old tormentor.

For all its flaws, *Hannibal* is still worth at least half a look. For starters, Gary Oldman seems to be having a grand old time under all that makeup as the demented Mason Verger. Oldman is always worth watching and his performance here certainly ranks among his most eccentric, if not his most richly layered. Also, Scott and crew should be commended for not shying away from the more violent aspects of the story. *Hannibal* is a lot more graphic than its predecessor. Of course, this doesn't make it a better movie, but if you're going to lessen the intensity, you'd better provide something to compensate for it. The gore is surprising, over the top, and realistic enough to make you squirm, but treated lightly enough to make you laugh. Just add a couple of injured eyeballs and replace Hans Zimmer's score with a pounding rock soundtrack by Goblin, and this could have been Ridley Scott's homage to Lucio Fulci.

The main reason to watch this movie, particularly on DVD, is the gorgeous cinematography by John Mathieson. This is a beautiful-looking movie and MGM's anamorphic transfer more than does it justice. There are virtually no flaws in this picture, apart from a very slight shimmer to a couple of shots. Otherwise, it looks fantastic. The colors are rich, deep, and rock solid, with almost no noticeable bleeding. Scott and Mathieson make good use of light and shadow, and the transfer highlights this perfectly, keeping the blacks nicely detailed. The stroboscopic effect Scott's become so enamored with in action scenes lately (see the opening shoot-out here, the huge battle in *Gladiator*, and pretty much all of *Black Hawk Down*) can get downright annoying, but even that looks solid on this disc. As for the sound quality, MGM provides dual DTS and Dolby Digital 5.1 audio options of equal brilliance. Both of these tracks are outstanding, not just in scenes with active and obvious surround sound, but throughout. There doesn't seem to be any difference between DTS and Dolby Digital, which would either point to this being a slightly weak DTS track, or a particularly full-bodied Dolby Digital mix—either way, they both work well enough.

MGM pulled out all the stops, supplements-wise, with extras for this two-disc release. It's worth noting that the extras were produced by Charles de Lauzirika, who has helmed the production of many of Ridley Scott's films on DVD. The first disc contains the movie and a full-length audio commentary by Scott. His comments are fairly interesting, though he does seem to run out of steam after the first hour and doesn't get back on track for about 20 minutes or so. The centerpiece of the second disc is *Breaking the Silence*, a 76-minute "making of" documentary that is viewable either as a full-length feature or as 5 individual featurettes. The documentary is surprisingly in-depth, though perhaps not as candid as one would like. However, at least the attempt is made to discuss *Hannibal*'s tumultuous development, including the non-participation of Jodie Foster and Jonathan Demme, Hopkins' startling announcement that he was retiring from acting, and the decidedly mixed reaction everyone had to Thomas Harris' novel. Unfortunately, there are still plenty of unanswered questions at the end of these features. Like what exactly did credited co-screenwriter David Mamet do when it seemed that Scott, Hopkins, and everybody else considered Steven Zaillian to be the sole writer of the script? And what about the deviations from Harris' novel, particularly the controversial ending? Either the commentary or the documentary would have been the place to address these subjects.

Disc Two also contains a trio of multiple-angle features. One breaks down the opening shoot-out scene, allowing you to toggle between the four simultaneously running cameras. This feature is like a miniature film school, as each angle also gives you technical information (like what lens and camera mount was being used)—very interesting stuff for film buffs. The second multi-angle feature compares the storyboards drawn by Ridley Scott himself to the final footage, while the third analyzes Nick Livesey's title sequence with an optional commentary by Livesey. Other bonuses on the disc include a whopping 14 deleted and alternate scenes (with optional commentary by Scott), the teaser trailer (which only uses footage from *Silence of the Lambs*), the full theatrical trailer, 19 TV spots, a gallery of still photos, cast and crew biographies, and production notes taken from the original press kit. By far our favorite feature was the gallery of poster concepts, displaying a huge variety of unused ad campaigns for the movie, some of which were quite beautiful. There is also a nifty little

Easter Egg hidden in the documentary submenu — a video montage of countless "flash frames" trimmed from the dailies, set to original music.

Hannibal is an odd film — a gore-drenched exploitation movie with a veneer of elegance and sophistication. It's doubtful anyone was really satisfied by it, but it made a fortune nonetheless — enough to justify Universal's *Red Dragon*, with Hopkins playing a younger, still imprisoned Lecter. Whether or not you like *Hannibal*, it's hard not to be impressed by this DVD. This is a bona fide special edition, with extras examining virtually every aspect of the film, all the while presenting the feature in absolutely pristine condition. It's kind of a shame that the worst thing about this DVD is the movie itself.

A.I.: Artificial Intelligence—Special Edition

DreamWorks

There are dozens, if not hundreds, of films that never get made every year. *A.I.* could easily have fallen into that category. Based on the short story "Super-Toys Last All Summer Long" by Brian Aldiss, the film was the brainchild of Stanley Kubrick. He bought the rights to the short story in the early 1980s with the intention of eventually bringing it to the silver screen. But the legendary director was never able to fulfill that intention. Despite years of work, including some consultations with Steven Spielberg and considerable storyboard work by comic book illustrator Chris Baker, Kubrick passed away before he could bring his vision to life.

That hurdle alone would've likely killed most other films, but Spielberg picked up on the concept of a robotic boy forming an emotional connection with a human mother and decided to complete Kubrick's work. In two months, he banged out a screenplay and began working on the film's development, incorporating Baker's storyboards into the production design. Haley Joel Osment, who became everybody's favorite child actor after performances in *Pay It Forward* and *The Sixth Sense*, signed on to play the robot child and Spielberg suddenly had all the pieces needed to make Kubrick's film.

Still, more than a few wondered immediately if Spielberg, the sappy filmmaker who had a hand in fairy tales like *E.T.* and *The Goonies*, could pull off the irony of Kubrick. Dystopias where humanity builds robotic fantasies to substitute for true emotion, then abandon them when they become too inconvenient, just didn't seem quite Spielberg's style. Nonetheless, that is exactly what the famed director embarked upon.

In *A.I.*, David is a robotic child created with the capacity to show emotion. In the home of a couple who had lost their son to a coma, David seems to fit in nicely at first. Things soon become disturbing though, and David, who is programmed only to show unconditional love, is incapable of understanding why he's becoming more of a burden than a blessing. When the couple's child revives and returns home, David no longer serves a purpose in this happy home. His mother abandons him in the forest and, because he has no choice but to love her, he embarks on a journey to find his way back to her.

At this point, the film transitions into a quest fraught with dangers and adventures. Remembering a fairy tale his mother once read, David decides to try and find the "blue fairy" that can turn him into a real boy, so that he can go home to his family. Along the way, he meets another robot named Gigolo Joe, designed to be a lover for lonely women. Fueled by Jude Law's inspired performance, Gigolo Joe leads David to Rouge City and eventually to what is left of New York City, buried underwater from the melted polar ice caps.

It would be an understatement to say the film has a great deal going for it. The initial vision of a filmmaker like Kubrick, the direction of Spielberg, and the acting of talented performers like Osment and Law all join with the groundbreaking special effects that finally allowed the vision to be made real. Yet despite all of these elements, the film never comes together, perhaps because of the old adage about two many chefs in the kitchen. This was Kubrick's film, not Spielberg's. But Spielberg resurrected it, and, though he accomplished more in the realm of irony than perhaps he was ever thought capable of, those undercurrents of hope that so defined his Peter Pan phase still linger.

Maybe it was simply that the film was too grand in scope. It starts as a straight-ahead domestic drama, switches to a sort of road picture, and then loses its way. All of that is compounded by the fact that this is a science fiction film exploring philosophical questions about the nature of emotion and what makes us real. By the time we get to the final minutes, where it all comes swerving to a disconcerting end, you've been bombarded with film noir, comedy, and farce, all woven into a visual tapestry of sometimes awesome beauty. But it's not overstepping to say that this is just too much for a single film to bear.

Whatever we think of *A.I.* as a film, the DVD that supports it is stunning. The film is presented in anamorphic widescreen and ranks up there with the best transfers out there, defining the term reference quality. Color levels and detail are spot on, with heavy black tones without any grain or artifacting to be found. The audio is given to us in both Dolby Digital 5.1 and DTS 5.1. Both are incredible mixes, with good play in the surrounds, a well-centered dialogue track, and a very nicely distributed LFE.

Most of the extras on this disc are presented in documentary form (Spielberg doesn't do commentary tracks for some frustrating reason). They start on Disc One with *Creating A.I.*, where the information is a bit general and serves more as an appetizer for what's on Disc Two. The second disc breaks down by subject, starting with the documentary featurettes *Acting A.I.* and *Designing A.I.* When you're done there, you can move on to the *A.I. Archives*, which are galleries of storyboards, sketches, production and behind-the-scenes photos, and two trailers. Next up is *Visual Effects/ILM*, consisting of a group of featurettes that examine the work of ILM and Stan Winston's effects group. If you're interested in the photography of the film, look no further than *Lighting A.I.*, where director of photography/cinematographer Janusz Kaminski talks about his lighting choices and what he was trying to achieve. The last of the documentaries is *The Sound and Music of A.I.* where sound designer Gary Rydstrom and composer John Williams discuss what they did to create a very distinct aural world for the film. Lastly, Spielberg comes back to discuss the reality of the film in *Closing*. The disc also includes text-based cast and crew bios and production notes, but you'll hardly find any real need of going there by the time you're done with all the featurettes.

The idea of *A.I.*, embodied in those treatments and storyboards, was a lovely and powerful concept. As realized by Spielberg, however, that concept lost something in translation. Thankfully, the DVD turns this less than stellar piece of cinema into a brilliant exploration of how great filmmakers like Spielberg and Kubrick approach such grand ideas, and how, sometimes, they're hamstrung by their own grandeur. For that reason alone, *A.I.* becomes a must-have special edition.

If You Like These Films . . .

. . . first of all don't feel bad. Everyone has right to their own guilty pleasures. In that spirit, we also "recommend" *Armageddon*, *Wild Wild West*, *Men in Black II*, *The Emperor's New Groove*, *Minority Report*, *Robin Hood: Prince of Thieves*, and *Legionnaire*. Just roll with the punches.

Part V

The Future
of DVD

Of course, no guidebook to DVD, especially ours, would be complete without addressing the future of the format. After all, if you're going to invest your hard-earned money in a good DVD-based home theater system and a nice library of movie discs, you want to at least know that your investment is going to have some sort of lasting value . . . instead of, all too quickly, becoming obsolete.

As you're no doubt aware, technology is always improving. So the best thing for you (as a DVD consumer) to do is arm yourself with knowledge concerning the latest developments in DVD technology. There are also important questions to ask, like: What new features can you expect in the months and years ahead? How can you make smarter purchase decisions right now to give yourself an edge in taking advantage of those features? Which of these players makes my butt look bigger, the black one or the silver one?

With these questions and concerns in mind, let's take a look at the cutting edge of DVD, and give you our advice on what each of these developments means to you as consumers.

Do keep in mind, however, that our look at the future of DVD, and home entertainment in general, is based on the best available information at the time of this writing. Since the world of technology is fluid and ever-changing, we recommend that you check *The Digital Bits* web site regularly for the latest, updated information.

We know our stuff, no doubt, but even *our* crystal ball isn't *that* good . . .

Recordable DVD

In the same way that you've long been able to record your favorite movies and programs from broadcast TV and cable onto blank videotapes, special DVD players have now become available that allow you to record such programs digitally, onto blank DVD discs.

Recordable DVD players have only recently become widely available. As a result, they're still quite a bit more expensive than your standard DVD players. In addition, the blank DVD discs you need to record your favorite shows in these players are only now beginning to approach the low cost of current CD-R and CR-RW media.

All of these issues aside, Recordable DVD players generally work very well. They usually allow you to record in varying degrees of video quality. As a rule of thumb, the better the quality of the recording, the less video (in terms of length) you can store on the disc. Recordable DVD players employ a variety of copy-protection technologies that make it difficult to record copyrighted material and, like current VCRs, they only record analog video signals—not digital ones. So forget about making perfect copies of your favorite pre-recorded DVDs. It's not going to happen.

Unfortunately, there are a number of different (and competing) disc formats for recordable DVD currently on the market. They each allow for different recording options and have their own strengths and weaknesses, but not all of them are compatible with regular DVD players. In other words, the discs you record will not always work in other DVD players—a major concern if you're planning to send discs to family and friends, or store your favorite shows on disc for years to come.

These different formats include DVD-R, DVD-RW, DVD+R, DVD+RW and DVD-RAM. DVD-R is a one-time recording disc but is also the cheapest (at about $5 per blank disc) and will work in most DVD players and DVD-ROM drives. DVD+R is a single-use format compatible with about 85 percent of existing players, but it's not technically part of the official DVD spec. DVD-RW is a rewritable format, like videotape, that can be used up to 1,000 times. It's compatible with about 70 percent of existing DVD players. DVD+RW is also a rewritable format, but again isn't technically part of the DVD spec. And DVD-RAM is a rewritable disc format that can be used up to a 100,000 times and stores much more data than standard DVD recordable discs, but it's incompatible with the vast majority of existing players.

Confused yet? Don't worry; you're not alone. We fully expect that, at some point, one or two of these recordable disc formats will become the most commonly used. When that day will arrive, however, is uncertain. In addition, there's a larger issue to be concerned about . . . none of these current formats allows you to record digital, high-definition video.

With digital and high-definition television broadcasting set to replace current analog broadcasting over the next few years, our recommendation is that most consumers should wait to buy recordable DVD technology. After all, your current VCR works just fine right now for recording standard-definition programs, so why invest in technology that's going to be obsolete in a few years?

At *The Digital Bits*, we do believe that some format of disc recorder *will* one day replace the VCR. That said, until the technology is compatible with digital and high-definition broadcasting, and the physical disc format has been standardized, your money is better spent elsewhere.

High-Definition DVD

Imagine video so clear and detailed looking that watching TV is almost like looking out a window. That's the promise of high-definition, and it's quickly becoming a reality, as HDTV prices continue to drop and high-definition programming becomes more readily available. But while you can already enjoy lots of great films and TV shows in high-definition quality, both through over-the-air broadcast and special high-definition packages from satellite services like DirecTV, the Holy Grail for most home theater fans is a digital disc format that contains *pre-recorded* films and other programming in glorious high-definition. It makes sense. The logical progression of current DVD technology *is*, after all, an eventual HD-DVD (or *high-definition* DVD) format.

Just think of it: Five or six years from now, you buy a new HD-DVD player that lets you watch all your favorite movies in true high-definition video quality, with the same kind of Dolby Digital and DTS surround sound you already enjoy on current DVD. And best of all, it still plays all of your existing CDs and DVDs, so your collection of movies and music retains its value for years to come. Eventually, you'll even be able to record your favorite shows in high-definition quality onto blank discs. That's the place most home theater enthusiasts, and many in the film and consumer electronics industries as well, want to reach in the years ahead.

In fact, work is already well under way toward achieving that goal. The various members of the DVD Forum have, for several years now, been working on various elements of the basic technology needed. But there are a lot of roadblocks that have to be overcome first.

One of the biggest concerns that the Hollywood studios have is ensuring that the content stored on the discs is secure. Remember, a high-definition digital version of a film is virtually a master-quality copy, and digital information, if left unprotected, can be duplicated perfectly every time. That's a very attractive target for video pirates. So any disc format that's intended for the distribution of films and other copyrighted content in high-definition quality must have very robust copy-protection schemes built into the system.

In addition, the many studios and consumer electronics companies involved have very different ideas about what the next-generation high-definition disc format should look like. Economics play a large role here. Right now, a number of companies receive royalty payments every time a DVD is manufactured and sold. That's because they hold patents on the existing DVD technology. Companies that receive smaller royalty payments from existing DVD would very much like to have a larger piece of the pie with the next-generation format. Several different groups are crafting their own high-definition disc formats in the hope that their particular format will eventually dominate the market the same way current DVD has.

Right now, three major format concepts are vying for the high-definition disc crown: Advanced Optical Disc, Blu-ray Disc, and Extended Red Laser. Actually, there are several others as well, including proposals from Microsoft (based on Windows Media 9) and Pixonics (pHD), but these three are the heavy hitters at the moment.

Advanced Optical Disc (AOD) is being developed primarily by Toshiba and NEC. It uses blue lasers and physical discs that can store many times the data of current DVDs. One nice feature of AOD is that these discs could be manufactured on existing DVD production lines, making it an attractive option in many ways. Of the three major proposals, AOD *seemed* to have the edge, in that it was favored by the DVD Forum for many months. However, in June of 2003, the Forum's Steering Committee officially voted against adopting AOD as the basis for an eventual HD-DVD format. It's telling that of the 17 manufacturers that hold Committee votes, the majority of the 11 companies that voted against the AOD proposal, or abstained from voting, are members of the Blu-ray Disc group.

Blu-ray Disc is also a blue laser-based format, again using a physical disc that can hold much more information than current DVD discs. Blu-ray Discs are enclosed in special cartridges, so existing manufacturing lines would have to be retooled in order to produce them. It's important to note, however, that Blu-ray is also designed to be a high-definition recording format in addition to just playing prerecorded software.

A high-definition, Blu-ray format disc. Note that the disc is contained in a special cartridge.

Panasonic USA/Matsushita Electric Corporation of America

Blu-ray Disc is being developed by Sony, Matsushita (Panasonic), and Philips (along with several other manufactures), and is already well on the way to market delivery. This group hasn't even bothered to present Blu-ray Disc to the DVD Forum as the format of choice for HD-DVD. Instead, they're simply doing an end run around the Forum, advancing their proposal as *the next big thing*. This means Blu-ray Disc is being positioned as a direct competitor for eventual HD-DVD. And since the group has managed to kill momentum for AOD within the Forum, the development of HD-DVD is effectively stalled.

Warner Bros. has also suggested a high-definition DVD format that would be based almost entirely on existing technology using the same red lasers found in current DVD players. The difference is that their "Extended Red Laser" concept would simply employ greater digital video compression to cram high-definition video onto regular DVD discs. This idea has the advantage of being immediately compatible with existing hardware and manufacturing equipment, but it's drawn the ire of home theater enthusiasts because its increased compression would almost certainly result in compromised video quality.

So which format will eventually win the high ground? Unfortunately, it's getting increasingly hard to say. Blu-ray is obviously moving forward. Meanwhile, Toshiba and NEC have vowed to continue their effort to develop AOD, even if they have to go around the DVD Forum to do so, as the Blu-ray Disc group has done. Although it's still a little early to say for sure, and significant industry forces are working to prevent it, it's starting to appear that a format war is inevitable.

If we had to weigh in with our preference at *The Digital Bits*, we'd probably go with the Blu-ray Disc concept. Its blue laser technology would ensure the highest-quality video delivery, and the capability to record in high-definition right from the start is very attractive.

Warner's idea is just ridiculous, at least by our reckoning. Why launch a high-definition format at all if you're intent on compressing the video to death? The idea here should be *superior* video quality, not compromised video quality.

One critically important thing we should note here is that *whichever* high-definition disc format eventually wins the high ground, most everyone in the industry agrees that the new players *must* be backwards-compatible with existing DVD software. Even the manufacturers who support these different formats acknowledge that the capability of future hardware to play current DVD movies, in addition to the new high-definition discs, is virtually a given. That's very good news for you as a consumer. It means that you don't have to worry about your existing DVD collection becoming instantly obsolete when the new HD format arrives.

So how soon will all this happen? Prior to the June 2003 Steering Committee vote, we'd been told that the DVD Forum expected to have version 1.0 of the HD-DVD specification finalized in 2004, with the first actual HD-DVD players appearing as early as 2005. Of course, that seems unlikely now. The first consumer Blu-ray Disc hardware has *already* been launched in Japan, and Sony reps have said that a U.S. debut could happen in 2005 (probably with a few Columbia TriStar titles available for the launch).

The bottom line is that high-definition discs *are* coming . . . eventually. But the bigger issue in our minds is this: Is the market really ready for them yet? Sure, early adopters—the same folks who were among the first to buy current DVD players—are already chomping at the bit for high-definition video on disc. But most consumers are just now really diving into DVD, and few of them are likely to want to replace their brand-new DVD players in the next few years.

Our feeling at *The Digital Bits* is that we'd rather see the industry get together and haggle out a *single* high-definition format, as was done with current DVD, rather than launch competing formats and (in the process) confusing consumers. When a single format is ready (whether it's HD-DVD based on AOD, Blu-ray Disc, or whatever), with robust copy protection and all the bugs worked out . . . *then* put it on store shelves.

Sadly, that ideal scenario seems unlikely to happen at this point. Our best guess is that none of the high-definition disc formats is really going to be viable until much later in this decade—in the 2007 to 2010 time-frame. Remember, you can't watch high-definition video on disc if you don't have a compatible Digital TV. More and more people are buying them, but it's going to take a while for the majority of consumers to do so.

In the meantime, our recommendation to all of you is not to worry. Existing DVD is going to be the best game in town for at least the foreseeable future, and your discs should all still work on the new high-definition players . . . whenever they do finally arrive. And with the potential for a format war looming, it's best to take a "wait and see" approach rather than rushing in too soon and potentially wasting your money.

D-VHS: High-Definition Films on Tape

It's worth mentioning that there *is* a way to watch prerecorded, high-definition Hollywood films in your home theater right now. A new format has become available called D-VHS, which was developed and introduced primarily by JVC. D-VHS stands for Digital VHS, and it's basically a digital version of the regular old VHS tape format we've all been using for years.

D-VHS can record high-definition signals off the air and play back high-definition signals from D-VHS tapes. The format can also play all the existing VHS and S-VHS tapes you already have (although not in high-definition quality). D-VHS players look and function almost identically to a regular VCR. Unfortunately, D-VHS players are several times more expensive than regular VCRs—in the $500 to $1,000 range, which is more expensive even than most fully loaded DVD players.

A handful of the Hollywood studios have agreed to support the D-VHS format with prerecorded feature films using a special encryption format of D-VHS known as D-Theater. These include Twentieth Century Fox, Universal, DreamWorks, and Artisan Entertainment.

Sarah Hunt

Typical D-VHS movie titles.

Currently, about fifty movie titles are available on D-VHS, priced at about $30 to 35 each. These include such hit films as *X-Men*, *Terminator 2*, *Independence Day*, *Planet of the Apes*, *U-571*, *Galaxy Quest*, and others. We've tested D-VHS personally at *The Digital Bits*—there's no doubt that the movies look and sound great. After all, the tapes contain high-definition video, created directly from the studio archive masters, and encoded at an MPEG-2 video data rate of 28.2 Mbps (megabits per second). By comparison, broadcast high-definition video is only 19 Mbps, while standard-definition DVD video has a maximum data rate of 9.8 Mbps (with 4 to 5 Mbps being the average). In addition, D-VHS tapes feature Dolby Digital 5.1 sound encoded at a whopping 576 kbps (kilobits per second), as opposed to the 384 kbps (or sometimes 448 kbps) usually found on DVD. The quality advantage offered by D-VHS should be obvious.

But, since D-VHS is a tape-based format, it has all the disadvantages you'd expect from tape. It's prone to damage, it's large and bulky to store, you can't have all the supplemental bells and whistles you'd find on DVD . . . and you have to rewind the damned things. In addition, very few (if any) other studios are likely to join Twentieth Century Fox, Universal, DreamWorks, and Artisan in releasing movies in this format. The reason is that the rest of Hollywood seems content to wait until the next-generation disc format is ready for prime time before releasing their films in high-definition. And once HD-DVD is ready to go, you can bet that even the D-VHS supporting studios will support it sooner or later.

To us at *The Bits*, the future of high-definition is clearly on disc . . . not tape. One of the reasons that so many people love current DVD is that the disc-based format is just so much more durable and convenient than tape. So unless you absolutely just can't wait to watch high-definition movies at home, and you've got money to waste on a format that's almost certainly going to be obsolete in a few years, D-VHS is better left to high-end aficionados.

Broadband DVD Players

In terms of standard DVD, one of the features that we expect to see added to players over the next few years is broadband connectivity. Simply put, the DVD player would have a built-in, high-speed modem that allows it to download data from the Internet. Using this capability, your DVD player could, for example, download new animated menu screens each time you watch the disc—menus that would change over time. New special features could be accessed directly from a studio's server, such as featurettes, deleted scenes, and other programs—features that are made available *after* you've purchased the disc and that aren't actually included on the disc itself. Broadband connections could even allow you to use your DVD as a special key that would access online web sites or live events, such as real-time video commentaries with a film's director that you watch right on your regular TV in your home theater.

Unlike most DVD-ROM features that you currently access with your computer, such broadband features would have to offer real, tangible value to consumers in order to be worth experiencing. At *The Digital Bits*, we have yet to see a DVD-ROM feature that was worth buying a disc for, much less a whole new DVD player. That said, the potential of broadband DVD players *is* there. Whether or not the option ever becomes common in DVD players, much less reaches that potential, will depend in large part on what the Hollywood studios and other content providers do with it.

High-Resolution Audio

With much of the cutting-edge technology we've described in this section of our book, we've encouraged you to take a cautious, "save your money for now" approach. But there is one new technology that we're very excited about at *The Digital Bits*, and if you love music as much as we do, it's absolutely worth looking at right now. That technology is high-resolution audio.

There are two new high-resolution audio formats on the market: DVD-Audio and *Super Audio CD* (SACD). The experience of hearing music in high-resolution for the first time is really amazing. Live recordings actually *sound* live. High-resolution brings back the warmth and *flavor* of the music you've loved for years—it's like listening to your favorite music for the first time again. All of the subtle nuances that were present in the original master recordings are suddenly there for you to hear. This kind of high quality was always present in the original recording, but vinyl record albums, cassette tapes, and even CDs didn't have the ability to let you hear it all.

Each of these new high-resolution audio formats has its strengths and weaknesses, but both are steadily winning the support of the music industry. Let's take a closer look . . .

DVD-Audio

The DVD-Audio format was developed by the same group of companies that created DVD-Video. It's designed to do the same thing for music that DVD-Video has done for movies. DVD-Audio discs feature high-resolution audio encoded using a scheme developed by Meridian Audio, called MLP (or *Meridian Lossless Packing*). This high-resolution music can only be accessed with DVD-Audio-compatible players. In addition, DVD-Audio discs often contain the same music in Dolby Digital or DTS format as well. That means that DVD-Audio discs are at least partially compatible with *all* existing DVD players. We should note that, while these additional tracks aren't true high-definition, they generally sound as good as the movie soundtracks on regular DVD discs. DVD-Audio discs are not compatible with existing CD players, although a special hybrid format of DVD-Audio disc is currently in development that *would* allow the discs to play a CD-quality soundtrack on most current CD players.

DVD-Audio is designed primarily for the surround sound presentation of music, so most discs contain 5.1 mixes (in both high-resolution and Dolby Digital or DTS). In addition, DVD-Audio discs usually contain a number of interesting, video-based extra features, like DVD-Video discs do. These can include documentaries, artist interviews, music videos, song lyrics, artist discographies, and more. These extras are viewed on your TV screen and are accessed with your player's remote via the same kind of interactive menus you see on DVD-Video discs. That means that DVD-Audio discs require authoring in order to function, just like DVD-Video discs do.

DVD-Audio discs typically come packaged in standard CD jewel case or Super Jewel Box packaging. The Warner Music Group is the main supporter of this format, although many other companies are supporting DVD-Audio as well (or have announced that they'll support it in the future).

Super Audio CD

Super Audio CD (or SACD), on the other hand, was developed by Sony and Phillips—the same companies behind the introduction of the standard compact disc (CD) format in the 1980s. SACDs feature high-resolution audio encoded using a scheme called DSD (or *Direct Stream Digital*), which was first developed for the master-quality, archival storage of audio recordings in digital form. DSD is designed to preserve the sound quality of the original master recordings in exacting detail. Because of this, the SACD format was originally marketed toward audiophiles, although it has since garnered a much wider audience.

SACD discs are intended to present the music in quality that is as close as possible to the original recording, simply in high-resolution. As such, most SACD discs feature stereo or even mono audio. New recordings, however, are occasionally presented on SACD in 5.1 multi-channel, surround format. In addition, hybrid SACD discs are already available that play high-resolution audio on SACD-compatible players, and CD-quality audio on existing CD and DVD players. SACDs do not include video-

based extras, but some discs do include bonus audio tracks not found on the original CD releases (video options may become available on future SACD releases).

SACDs come in standard CD jewel cases, Digipak, and Super Jewel Box packages—virtually every kind of packaging that's used for current CDs. The Sony Music Group is the biggest supporter of the SACD format, although (as with DVD-Audio) many other companies are supporting SACD as well (or have announced that they'll support it in the future).

Literally hundreds of DVD-Audio and SACD releases are available right now for you to listen to, with many more on the way. These include both new and catalog albums from many of your favorite artists, as well as music in a variety of genres, including rock, jazz, classical, and others. Fancy listening to the likes of Miles Davis, John Coltrane, The Rolling Stones, Pink Floyd, the Police, R.E.M., and Peter Gabriel in high-resolution? No problem—they're all available right now. Discs can easily be purchased from online retailers, and they're starting to become more widely available at stores like Tower Records and Best Buy as well. And most of them are priced affordably at under $20 each (some are even priced the same as regular CD releases).

DVD-Audio and SACD discs don't *necessarily* require a separate, stand-alone player. If you happen to be in the market for a new DVD player, there are many new models that also support either DVD-Audio or SACD. Better still, there are a few models of DVD player that support *both*

Typical high-resolution audio titles. SACDs are on the left; DVD-Audio discs are on the right. Sarah Hunt.

high-resolution audio formats while still being relatively inexpensive (well under $500, and some as low as $250). If you're interested in high-resolution audio, this is definitely our recommendation.

So those are the basic details on the new high-resolution audio formats. But you've probably already got a few important questions you'd like answered. Is one format better than the other? Is there a format war between them? Will one of these formats eventually dominate the market while the other disappears into oblivion?

The good news is that both DVD-Audio and SACD sound great. If we (at *The Bits*) were forced to make a choice between DVD-Audio and SACD . . . right now, we'd pick SACD. The whole point of high-resolution audio is, after all, the high-resolution audio. All the other bells and whistles that DVD-Audio discs offer (the video interviews, onscreen lyrics, and so on) can already be found on music DVD-Video discs. Since SACDs were originally designed for audiophiles, a lot of careful thinking went into engineering the format to sound as good as possible. We've listened to literally hundreds of hours' worth of DVD-Audio and SACD music. Simply put, of the two formats, SACD sounds better. That's not to say that DVD-Audio sounds bad—not by any stretch of the imagination is that true. But, as we've mentioned, DVD-Audio tends to emphasize the *surround* sound presentation of music. And while we certainly enjoy the possibilities that surround sound can offer, 5.1 presentation changes the character of the music and usually involves actual alterations of the original audio mix. That's like colorizing a black and white film, in our mind. We'd rather just hear our favorite music in the original stereo (or even mono) format in which it was recorded . . . but just in the highest resolution possible, so

Pioneer's DV-563A plays both SACD and DVD-Audio music discs in addition to DVD-Video movies and regular music CDs (suggested retail price: $249).

you're hearing the music in the same quality that the artists did in the recording studio. And that is *exactly* what SACD is all about.

That said, the great thing about DVD-Audio and SACD is that you don't *need* to choose just one or the other. Rather than competing in a format war, our feeling is that these two audio formats can coexist nicely. In the same way that DVDs are often made available with Dolby Digital or DTS sound, some discs will be available in DVD-Audio, some in SACD and some (eventually) in both. Virtually all of the major record labels have announced support for one or the other, and at least one major label, Universal Music, is starting to support both. And by taking our advice and choosing a combo player that works with both formats, on the off chance that one of the formats does end up dominating the market, not only are you *not* stuck with a useless player, but all your discs from the other format will still work just fine on your machine. It's a no-risk proposition.

The bottom line is that, just like high-definition is the future of home video, high-resolution audio is the future of music. If you love listening to music as much as we do, high-resolution audio is absolutely worth checking out right now.

DVDs and Videogaming

There's one last new wrinkle of DVD technology that we should mention before we close, and that's the connection between DVDs and videogaming. Two of the three videogame console systems on the market right now, Microsoft's Xbox and Sony's PlayStation 2, feature the capability to play DVD-Video discs in addition to game software discs. With the Xbox, you need to purchase a separate remote control to take advantage of the system's DVD compatibility. The PlayStation 2 (PS2) allows you to control DVD playback with the system's own controllers, but you can purchase a separate, DVD-specific remote (with more functionality) to control the system as well. And in 2004, Sony is expected to introduce an upgraded version of the PS2, dubbed the PSX, which is specifically designed to act as the center of your home entertainment system. It will come with a built-in TV tuner.

Nintendo's GameCube system currently does not allow DVD playback as a feature. However, it's worth noting that both the GameCube and the Xbox *do* give you the capability to play many videogames in progressive scan video, so if you have a new widescreen Digital TV, now you've got something else to kill time with.

We can't vouch for the home-theater quality of these game systems, but the fact that so many of them are in people's homes means that you may *already* have a device capable of playing DVD movies and you don't even know it. You can definitely expect the trend of DVD compatibility in videogaming to continue in future, next-generation systems. It's even likely that some of them will eventually be compatible with future high-definition disc formats in the years ahead.

Final Thoughts

So there you go, folks. Thanks to the power of the written word, you should all now be knowledgeable in the ways of DVD. Do you know more than us? Well . . . no. But that's part of being a DVD insider, right? You have to hold back just a little bit to keep people coming back. And the goal this time was to keep things simple enough so that everyone can understand. That way, when you give this book as a gift to Grandpa, he can impress all his VFW buddies by dropping terms like layer switch, telecine, and aspect ratio. Maybe we should have included a section on how to set your VCR clock. Oh well.

Anyway, insider or not, you should all now be able to converse with your friends about DVD like experts. You can wax poetic about the benefits of anamorphic widescreen . . . and correct that guy at the video store who still believes that full frame is the best way to go. You can show off with pride that spiffy new home theater system you've put together for a song and dance (using the advice we've given you, of course). And we've provided you with a killer list of must-own DVDs to take with you the next time you visit your favorite video store, so you can fill out your movie library in style.

The single most important thing we hope you've learned from reading this book is that DVD is here to stay. The format is alive and well, with more than 26,000 titles already available, and more released every week. Among them are many of the great films and TV programs, both new and classic, that we all love so much. Some of these titles have never been released on video before, and they're more affordable than ever. Best of all, thanks to the anamorphic widescreen feature (and the backwards-compatibility that future HD-DVD is likely to provide), you'll be enjoying your current DVD discs for many years to come.

Of course, long after this is book published, there will be new developments. DVD will grow and change in ways we can only dream of at the moment. But we've given you a good foundation of understanding, so that you can grow and change with it. And don't forget—you can always get the latest news and information from us online, at *The Digital Bits* web site. If you ever have questions, concerns, or suggestions, we'll be there to listen and help guide your way.

Hugs and kisses . . .

Bill Hunt & Todd Doogan
Editors of *The Digital Bits*
www.thedigitalbits.com

Glossary

The following is a glossary of some of the most common terms you'll be confronted with as you explore the world of DVD and home theater. We've broken them down and explained them in simple, easy-to-understand terms.

1.33:1—This is the aspect ratio of a standard television screen, also referred to as full frame. Most television shows, cartoons, family programming, and pre-1950s films were shot in 1.33:1 format. The number indicates that the image is 1.33 times as wide as it is tall. This aspect ratio will completely fill a standard TV screen, but the image will require gray side bars on a widescreen, digital/high-definition television screen.

1.78:1—This is the widescreen aspect ratio of most digital/high-definition television screens (DTV/HDTV). The number indicates that the image is 1.78 times as wide as it is tall. Displaying this aspect ratio on a standard TV will require black bars on the top and bottom (letterboxing), but the image will completely fill a widescreen, digital/high-definition television screen.

1.85:1—This is one of the two most common widescreen aspect ratios in use today for theatrical films. The number indicates that the image is 1.85 times as wide as it is tall. Displaying this aspect ratio on a standard TV will require black bars on the top and bottom (letterboxing), but the image will completely fill a widescreen, digital/high-definition television screen.

2.35:1—This is one of the two most common widescreen aspect ratios in use today for theatrical films. The number indicates that the image is 2.35 times as wide as it is tall. This is one of the widest aspect ratios you will normally see. Displaying this aspect ratio on a standard TV will require large black bars on the top and bottom (letterboxing), and smaller black bars on a widescreen, digital/high-definition television screen.

3-D—Stands for three-dimensional. A film gimmick of the 1950s (revived in the 1980s), 3-D is making something of a comeback. There are a few home 3-D systems available and a handful of DVD titles that use the format. Special glasses are required to convert the 3-D video signal into a three-dimensional image.

16x9 or 16:9—*See* anamorphic widescreen.

Analog—A non-digital signal format for storing information (video, audio, and other data). Analog generally provides good quality, but not the highest possible quality. In addition, duplicating an analog signal results in a copy that is of poorer quality than the original.

Anamorphic Widescreen—This is a special feature of DVD that allows a studio to present widescreen films in the best possible quality on both standard and widescreen televisions. The image on the disc itself fills the entire frame but is horizontally "compressed" to fit the square picture area. If you have a standard TV, your DVD player electronically "compresses" the image vertically to produce the proper widescreen aspect ratio. Black bars are then added to the top and bottom of the screen to fill in the unused space. If you have a widescreen TV, the TV "uncompresses" the image horizontally, until it fills the screen (small black bars may still be required for the widest aspect ratio films). The result is a high-quality image with full vertical resolution. If the film on DVD is in letterbox widescreen but isn't anamorphic, a widescreen TV would have to electronically "expand" the image on the disc, resulting in a significant loss of resolution and thus picture quality. Most DVD fans believe that all widescreen films should be anamorphic to provide all viewers with the best possible video quality. To determine if a DVD features anamorphic widescreen video, look for the following words on the disc's packaging: "anamorphic widescreen," "16x9," "16:9," or "enhanced for widescreen televisions."

Artifacts—These are unwanted picture elements that are sometimes visible in the video image of a DVD. These can include dust, dirt, scratches on the original film print, and other such things. In a digital video signal, these can also include erroneous digital errors found within the image because of poor mastering or authoring. Most often these errors are found in darker areas of the digital image; they show up as slow-moving pixels or blocks. Artifacts are more common in older releases but still pop up in newer releases.

Aspect Ratio—Film and TV images are measured by taking into account the width-to-height ratio, or aspect ratio, of the image. For example, many films today are presented in the 1.85:1 aspect ratio, which means that the image is 1.85 times as wide as it is tall. Other common aspect ratios are 1.33:1, 1.78:1, and 2.35:1. Aspect ratio is sometimes also referred to as "screen format."

Audio Commentary—This is a special audio track that, if available on a DVD, you can choose to listen to while watching the film, instead of the film's regular soundtrack. Usually, this is a recording of a filmmaker, actor, or other expert who will discuss the film you're watching and give insights into the production. You can access this feature through the disc's menu screens or sometimes by using the Audio button on your DVD player's remote.

Authoring—Because DVD is formatted just like computer software, it must be programmed. This process is called authoring. Technicians, using special computer programs, carefully assemble all of the necessary instructions to make the disc work (pressing this button plays the movie, pressing that button takes you to the special features page, and so on.) and test it before sending it out to consumers.

Banana Connector—A speaker wire connector consisting of a single, fat shaft bulging on the sides, which inserts into a receiver or speaker's binding post.

Beta (Betamax)—A consumer video format developed by Sony during the 1980s, using a half-inch magnetic tape to record and play back video. Because of Sony's hesitance to license the Beta format to pornographic distributors, VHS eventually became the format of choice for home viewers. Variations of Beta are still used by professionals today as both an archival storage format and in the form of broadcast/production-quality Betacam.

Bit—The absolute smallest piece of digital data. There are 8 bits in a byte and 1,024 bytes in a kilobyte.

Bit Rate—The number of bits transferred in one second by a digital device. Also known as data rate.

Black Level—The level of brightness in the darkest part of a visual image.

Bleeding—An image error where the color from one object in an image seems to cross over or "bleed" into the object next to it.

Blooming—Distortion caused by excessive brightness resulting in exaggerated images having a soft focus.

Blue Laser—A laser with a shorter wavelength than the red lasers used for compact discs and the ruby lasers used for regular DVDs. Using a blue laser to read a physical disc allows the disc to store greater amounts of data.

Blu-ray—Blu-ray or Blu-ray Disc is the name of an optical disc video recording format developed by Sony (and eight other manufacturers) to enable the recording, rewriting, and playback of high-definition video. Blu-ray makes it possible to record over 2 hours of digital high-definition video or more than 13 hours of standard-definition video on a 27GB disc. There are also plans for higher-capacity discs that are expected to hold up to 50GB of data. Blu-ray is a competing format for future HD-DVD.

Bookshelf Speaker—A small speaker, 18 to 24 inches in height, best suited to sit on a small stand, side table, or bookshelf in a living room or office-based home theater.

Brightness—A control function found on a video display that allows you to adjust the black level of the image.

Burning—A slang term meaning to write digital data to a blank DVD or CD disc.

Calibration—In the simplest of terms, this is nothing more than setting up a device or system in order to ensure the best operation. In most cases, video and audio equipment, when first purchased from a store, will have to be recalibrated or adjusted for the best performance in your home.

Captions—The text representation of the audio part of a video presentation. Captions differ from Closed Captioning in that they are often included as a selectable option in a DVD's menu screens. Closed Captioning, on the other hand, consists of a hidden signal embedded within the video program that is read and displayed by your TV.

CATV or Cable Television—This refers to video and audio signals, transmitted through coaxial cables rather than over the air, that enable televisions to display a vast selection of channels without the need for an antenna.

CAV—Stands for Constant Angular Velocity. CAV is a laserdisc format that allows less video to be stored on the disc (about 30 minutes per side), but contains additional features, such as still frame and slow motion, not offered by other types of laserdisc.

CD—Stands for compact disc. The CD is a small reflective disc, like a DVD, that stores digital data in the form of microscopic pits that can be read by a laser. CDs store less data than DVD discs and are commonly used for prerecorded music and software.

CD-R—A recordable optical disc format that can be used to save digital information. A CD-R disc can only be used for recording a single time.

CD-RW—A recordable optical disc format that can be used to save digital information. A CD-RW disc can be recorded, erased, and re-recorded many times, making it reusable.

Center Channel—The primary channel used for reproducing dialogue in movie surround sound. In home theater systems, the center channel speaker is usually placed above or below the video monitor, so that the dialogue it delivers correlates logically with onscreen images.

CGI—Stands for Computer-Generated Images or Imagery.

Channel—A single component of an audio track. In home theater surround sound systems, each speaker is typically driven by its own channel of audio information.

Closed Captioning—This is a text-based feature (similar to subtitles) that describes the audio information contained in a film for individuals who are hearing impaired. It often describes other sound cues than just dialogue, making it distinct from (and usually more accurate than) regular subtitles. Closed Captioning consists of a hidden signal embedded within the video program that is read and displayed by your TV.

CLV—Stands for Constant Linear Velocity. CLV is a laserdisc format where a maximum amount of video storage is allowed (approximately 60 minutes per side of the disc).

Coaxial Cable—The standard form of cable for television reception in the home.

Color Bars—A graphical test pattern used to calibrate the color reproduction on video displays.

Component—This is one of three common types of video signal connection between your DVD player and your TV. The video signal is separated into its three basic color components (red, blue, and green), and each is routed to your TV in a separate cable, where the signal is recombined for display. This allows for the cleanest, highest-quality video possible, compared to composite and S-Video connections. Use component connections whenever possible to get the best video quality from your DVDs.

Composite—This is one of three common types of video signal connection between your DVD player and your TV. It's the most often used connection, but it provides for somewhat reduced video quality compared to S-Video and component connections.

Compression—In order to fit an entire film's worth of video and audio information onto a DVD disc, that digital information must be compressed, or reduced in size, for easier storage. The compression works by removing redundant information from the signal while inserting special instructions for your DVD player or sound system to later use to restore the missing information. If it's done properly, you'll never notice that any compression occurred. MPEG-2 is the most common form of video compression used for DVD. Dolby Digital and DTS are common digital audio compression formats.

Contrast—A video display control used to adjust the white level. It also defines the range of brightness between the lightest and darkest portions of an image.

Copy Protection—An electronic scheme developed to thwart the copying of information contained on DVDs and other media.

CRT—Stands for Cathode Ray Tube. These are commonly found in TV displays. A CRT is a large vacuum tube, with a slightly curved glass surface on one end and an electron emitter on the other. The emitter fires a stream of electrons from the backside of the screen to the front, resulting in a video image being displayed.

CSS—Stands for Content Scrambling System. CSS is the standard digital copy-protection format for DVD.

Data Rate—The number of bits transferred in one second by a digital device. Also known as bit rate.

DBS—Stands for Digital Broadcast Satellite. This is a television transmission format in which the digital television signals are broadcast from a satellite orbiting the earth rather than a regular land-based transmission tower. This allows the same signals to be delivered to a much larger geographic area. The signals are received by small dishes and then decoded for viewing. See also: DirecTV and Dish Network.

Digipak—This is a type of cardboard and plastic packaging that is used for many multidisc DVD special edition releases.

Digital—An electronic signal format for storing information (video, audio, and other data) as binary code that can be read by computer devices. Digital generally provides the highest possible quality. Unless copy-protection has been employed, digital information can be duplicated to create perfect copies.

Digital Audio—A method of encoding analog audio signals into digital format (binary code), which allows for easy transmission and duplication without the distortion and degradation typical of standard analog signals.

Digital Cable—A format of digital content delivery in which the video signals and other data are transmitted from the cable operator to your TV via coaxial cables rather than via satellite or over-the-air transmission.

Direct Digital Transfer—A procedure that allows for digitally produced movies (such as computer-animated titles) to be encoded onto DVD directly, with no video or audio degradation and no generation loss. What you are seeing is almost exactly what the filmmakers saw as they worked on the final cut of the film.

DirecTV—A content provider of digital broadcast satellite signals. *See also* Dish Network and DBS.

Disc Menu—A disc menu is a graphic page that can contain both text and images. It allows you to navigate DVD features using your player's remote. Often, these will employ animation, sound effects, and music, and they're almost always themed to the program on the disc. A disc menu is often the first thing you see when you put the disc in your player and start it up, although sometimes you'll get copyright warning screens and studio logos first.

Dish Network—A content provider of digital broadcast satellite signals. *See also* DirecTV and DBS.

DLP—Stands for Digital Light Processing. DLP is a sophisticated digital video projection technology developed by Texas Instruments. DLP-based systems are replacing traditional film projectors in some theaters.

Dolby Digital—This is the primary sound-encoding format for DVD (also known as AC-3). Developed by Dolby Laboratories, this format is compatible with all DVD players. It's highly compressed, which allows more information to be stored on a DVD disc. Dolby Digital comes in various multi-channel arrangements that allow for different configurations of surround sound. The most common is Dolby Digital 5.1 (which indicates that sound information is being delivered to five speakers: front left, center, and right; rear left and right; and a subwoofer).

Dolby Surround—Also developed by Dolby Laboratories, this sound-processing format creates "simulated" center and surround channel speaker signals from a standard Dolby Digital 2.0 soundtrack to make the audio sound more dynamic and directional. This is something your sound equipment does electronically to the signal, so the sound isn't going to be as dynamic as it would be if it were actually mixed by the studio that way. A variation of this, known as Dolby Surround EX, creates a "simulated" center-rear speaker signal out of a Dolby Digital 5.1 soundtrack.

DSD—Stands for Direct Stream Digital. DSD is a digital coding scheme developed for the master-quality, archival storage of audio recordings in digital form. It's also the primary scheme used by the high-resolution SACD format.

DTS—Stands for Digital Theater Systems. Originally developed for Steven Spielberg's *Jurassic Park*, DTS is a sound-encoding format that allows for multiple channels of surround sound (the most common is DTS 5.1, which sends sound information to five speakers—front left, center, and right; rear left and right—and a subwoofer). A DTS signal is less compressed than Dolby Digital, which in theory allows for a purer, higher-quality sound. Many people describe it as more "natural" sounding. In order to take advantage of DTS, your player and sound system (particularly the receiver) must have DTS capability, and the disc itself must be encoded with a DTS signal. Consult your equipment manuals to determine if your player and sound system are DTS compatible.

DTS-ES—This is a variation of DTS that adds an additional channel of surround sound in the center-rear portion of the sound field. It's noted as 6.1 (meaning that sound information is sent to six speakers—front left, center, and right; rear left, center, and right—and a subwoofer).

DTV—Stands for Digital Television. This is the television transmission format that the United States is currently in the process of upgrading to. It's marked by cleaner, clearer, and more accurate picture and sound quality. Digital television signals can be displayed in two different formats: standard definition and high definition. Standard definition looks basically like your current TV does today. High-definition signals are the ones people usually think about when they hear the term DTV. They're those amazing widescreen images you see at your local electronics store that people describe as "so good it's like looking out a window." They generally have almost double the resolution of a standard definition picture. DTVs are usually in the 1.78:1 or 16x9 widescreen aspect ratio. DTV is composed of 18 formats including 6 high-definition (HDTV) formats and 12 standard-definition (SDTV) formats. *See also* HDTV and SDTV.

Dual-Layered—In order to allow you to view a complete film and special edition materials without having to flip the disc, many DVDs are dual-layered, which means that the information is encoded on two layers of a single side. Usually the film is on one layer and the bonus material is on the other. If the film itself is split over the two layers, it's probably an RSDL (*Reverse-Spiral Dual-Layer*) disc.

Dual-Sided—Sometimes both sides of a DVD disc will have video and audio material on it. This means that you watch everything on the first side, and then you have to take the disc out of your player, flip it over, and start it up again. Dual-sided discs are commonly referred to as "flippers" for this reason.

DVD—The DVD is a 5-inch, CD-sized optical disc, capable of holding digital video and audio information for movies, music, computer games, and more. DVDs hold digital video using MPEG-2 compression and digital audio using primarily Dolby Digital or DTS. DVD stands for Digital Video Disc, although DVD discs can obviously store more than just video signals.

DVD-Audio—This is a special format of DVD that is designed primarily to play high-resolution audio signals (usually music programming), often in surround sound. DVD-Audio discs often include some video-based materials as well (such as music videos, interviews, song lyrics, and so on). Note that you must have a DVD-Audio-compatible DVD player to be able to enjoy the high-resolution audio. Consult your DVD player's manual to find out if your player is compatible with this format.

DVD-R—A recordable optical disc format, based on DVD, that can be used to save digital information. A DVD-R disc can only be used for recording a single time.

DVD-ROM—This is a special format of DVD disc designed for use with computers. Many DVD-Video discs also feature special bonus material in DVD-ROM format that can only be accessed by a computer (usually a PC running Microsoft Windows 95 or higher) with a compatible DVD-ROM drive.

DVD-RW—A recordable optical disc format, based on DVD, that can be used to save digital information. A DVD-RW disc can be recorded, erased, and re-recorded many times, making it reusable.

DVD+RW—A recordable optical disc format, based on DVD, that can be used to save digital information. A DVD+RW disc can be recorded, erased, and re-recorded many times, making it reusable. Unlike DVD-RW, however, DVD+RW is incompatible with many other DVD devices.

DVD-Video—This is the most common format of DVD and is designed primarily to present video material (usually films) at the highest possible quality given today's standard televisions. It also has the capability to present high-quality video on digital/high-definition televisions, but not at true high-definition resolution. DVD-Video discs usually feature some type of Dolby Digital and/or DTS sound encoding to present theater-like surround sound, and they may include many extra, interactive features as well. This format was designed to replace (and is replacing) VHS videotapes and laserdiscs for movies.

D-VHS—Stands for Digital Video Home System. D-VHS is a digital variation of the half-inch VHS magnetic tape cartridge format developed for the home playback and recording of analog video signals. D-VHS can deliver true high-definition video signals for viewing on an HDTV.

DVI—Stands for Digital Video Interface. It's a standardized form of an all-digital cable connection between electronic devices, which allows for clearer onscreen images. DVI is becoming more common with new digital and plasma TVs.

Easter Egg—Many DVDs these days include special hidden features that aren't listed on the back of the disc's packaging. These can include blooper reels, trailers, deleted scenes, and all sorts of fun, film-related items. They're called Easter Eggs because they're usually hidden somewhere on the DVD's menu screens. The idea is for you to find them as you explore the disc. You might have to highlight a symbol tucked on a corner of one page or enter a code on your DVD player's remote when you're at a particular spot. Easter Eggs add to the fun of the DVD experience and are becoming more common all the time.

Edge Enhancement—An electronic process that artificially increases the "sharpness" of the edges of objects displayed in a video signal. Edge enhancement was often needed in analog video transfers, but it is generally unnecessary in digital video transfers.

Encryption—The process by which digital information is "scrambled" so that it is unreadable except by authorized users with the proper hardware or software.

Fiber Optic Cable—A cable that uses light beams to transmit information rather than electrical signals traveling over metal wires. *See also* Toslink.

FireWire—A type of high-performance cable connection designed to transfer digital information.

Front Projection—A video display in which images are projected onto a large reflective screen by a separate projector mounted in front of the screen (just like in a movie theater).

Full Frame—This is the aspect ratio of a standard television screen, also referred to as 1.33:1. Most television shows, cartoons, family programming, and pre-1950s films were shot in full-frame format. When referring to the numerical aspect ratio, the number indicates that the image is 1.33 times as wide as it is tall. This aspect ratio will completely fill a standard TV screen, but the image will require gray side bars on a digital/high-definition television screen.

HD Radio—High-Definition Radio. A new digital audio broadcasting format that is designed to provide FM-quality sound for AM stations and CD quality sound for FM stations. In addition to improved sound quality, broadcasters can also deliver other data, such as song information, news, and weather reports right to your receiver.

HDTV—This acronym stands for High-Definition Television. This is a format of digital television (DTV) that generally displays almost double the resolution of a standard definition picture. High-definition signals are the ones people usually think about when they hear the term DTV. They're those amazing widescreen images you see at your local electronics store that people describe as "so good it's like looking out a window." HDTVs are usually in the 1.78:1 or 16x9 widescreen aspect ratio and are capable of displaying interlaced video with 1,080 scan lines (lines of vertical resolution) and progressive video at 720 scan lines.

Home Theater—A term used to describe a complete audio/video system used for viewing movies at home. A home theater generally consists of a video display, at least one video source (like a DVD Player), and surround sound equipment (a receiver and multiple speakers) that work together to reproduce the movie theater experience in the home. Home theaters can range from simple and very inexpensive systems designed to be included in your living room to those that duplicate a theater environment in exacting detail.

Home Theater in a Box—A slang term used to describe a low-end, inexpensive home theater system that comes complete in one package.

IMAX—The largest commercial film format in use today, IMAX uses specially designed cameras and projectors to shoot and project images of tremendous clarity and resolution. The IMAX film frame is 10 times larger than the 35mm format used in regular movie theatres and three times larger than standard 70mm film.

Interlace—Interlace is a method of video projection that displays only half of the entire image at one time. Half of the image appears on the screen, while the other half is being scanned. It happens so fast that you can't see the process—your brain interprets the result as a complete image. This is the projection method used by nearly all standard televisions.

Isolated Score—This is a special audio track that, if available on a DVD, allows you to watch the film while listening to just the musical score without added dialogue and sound effects. Sometimes a composer will also record his thoughts in between the sound cues, like an audio commentary. This is something many fans of film music will appreciate.

Jewel Case—This is the type of clear plastic packaging most commonly used for music CDs.

Keep Case—This is the most common form of DVD packaging. It protects the disc in a plastic, book-like case. Variations of this case can hold multiple discs.

Laserdisc—A 12-inch optical disc format for viewing movies at home. Laserdisc was the precursor to DVD. It offered superior video and sound quality compared to VHS, but not as good as DVD. Laserdisc was the first home video format to popularize the presentation of films in their original aspect ratios. In addition, laserdiscs often included some of the same kinds of bonus features (audio commentaries and deleted scenes) found on DVDs today.

LCD—Stands for liquid crystal display. This is a type of video display that uses liquid crystals, which change the amount of light they let pass through them when an electric current is applied. LCDs are commonly used in laptop computers and portable videogame systems.

Letterbox—This is a format for presenting widescreen films in their original aspect ratio on standard televisions. When a film is in letterbox format, black bars are visible on the top and bottom of the image. The important thing to note here is that you're not missing part of the picture. Everything that was seen in the film's original presentation in theaters is visible.

Macrovision—A common format of copy protection. Macrovision prevents material from being duplicated through encryption and digital watermarking.

Main Speaker—A large speaker, usually floor-standing, that's generally designed to present the most important channels of sound in a stereo system or home theater. They're often placed to the left and right of the video display.

MLP—Stands for Meridian Lossless Packing. MLP is a digital audio coding scheme developed by Meridian Audio. It's the primary scheme used by the high-resolution DVD-Audio format.

MP3—Stands for MPEG Layer 3. MP3 is an audio compression format for encoding digital music.

MPEG-2—This is the digital video compression format used to encode and store movies on DVD. It works by removing redundant picture information from the signal, thus making the video data small enough to be stored on the disc. It also inserts special instructions in the signal that your DVD player later uses to restore the missing information for display on your TV. If it's done properly, you'll never notice that any compression occurred. Just FYI, MPEG stands for Motion Picture Experts Group.

Multi-angle—This is a special feature of the DVD format that allows you to compare multiple video programs instantly with the push of a button on your DVD player's remote. Although all DVD players have this capability, multi-angle has to be added as a feature on the disc for you to take advantage of it. You'll often see multi-angle used for sports and music programming. It's less commonly applied to films, but you'll occasionally see it used in the special features section of a disc.

NTSC—This acronym stands for National Television Systems Committee. This is the primary television transmission format used in the United States, Canada, and Japan, until the conversion to digital/high-definition television (DTV/HDTV) is complete. It's also the name of the group of engineers that created the format back in 1941.

Nuon—This is a special enhancement to DVD available on some players, which adds additional interactive capabilities to the disc. Only DVD players with Nuon capability can take advantage of Nuon-enhanced discs. Consult your DVD player's manual to find out if your player has this feature.

OAR—This acronym stands for Original Aspect Ratio. Many people believe that a film should always be shown on home video and DVD in its original aspect ratio to preserve the artistic intent of the filmmakers. This sometimes means that a widescreen film has to be shown in letterbox format on a standard TV to preserve the intended image.

Optical Digital—This is a common type of audio signal connection between your DVD player and your sound system receiver. It passes the signal as pulses of light via an optical cable, which in theory allows for greater accuracy and better sound quality.

PAL—This acronym stands for Phase Alternate Line. This is the primary television transmission format used in most of Europe and other countries around the world. PAL was introduced with SECAM in 1967 and both produce images with up to a maximum of 576 scan lines (lines of vertical resolution) compared to the NTSC used in the U.S., which produces 525 scan lines.

Pan and Scan—This is a common method for transferring widescreen material for viewing on a standard TV. When the film is being transferred to video, the video camera actually zooms in on one part of the film image, and electronically pans back and forth to keep the most important part of the action in the center of your TV screen. The important thing to note here is that while this allows the picture to fill your screen, you're actually losing as much as 50 percent of the image in the original film, severely compromising the filmmakers' original artistic intent.

PCM—This is a sound-encoding format that sometimes appears on DVD discs, particularly music titles. It's uncompressed, which in theory allows for a purer, higher-quality sound. PCM stands for Pulse Code Modulation. It's also the commonly used format on current audio CDs.

Progressive Scan—Progressive scan is a method of video projection that displays the entire image in one pass, making for a cleaner and steadier-looking picture. This is how your computer monitor displays its image. Many DVD players feature progressive scan video outputs that can be connected with compatible televisions and display devices to take advantage of the improved quality. Consult your DVD player's manual to find out if your player has this feature.

Rear Projection—A video display that projects images onto a screen from behind it, using a series of mirrors that reflect light. Most rear projection television displays are 40 to 80 inches in diagonal measurement.

Recordable DVD—DVD players are now available that can record TV programs digitally onto special, blank DVD discs. In terms of operation, these are used much like your current VCR, and they'll eventually cost about the same. Players with this capability use some combination of three currently competing recording formats: DVD-R, DVD-RW, or DVD+RW. Eventually, one of these formats will probably dominate the recordable DVD player market, so it's best to wait until then to buy one.

Redbook CD—The document, first published in 1982, that provides the specifications for the standard compact disc digital audio format first developed by Philips and Sony.

Red Laser—A laser of a longer wavelength of light, used in current CD and DVD players.

Region Code—Region coding is a system that is built into nearly all DVD players and that restricts the showing of certain discs in various parts of the world. The world is divided up into eight separate regions (the U.S. and Canada are in Region 1). The idea behind this system is to prevent a film that's been released on DVD in one region from being imported into another region where it hasn't been released yet. This allows the studios to control the home video distribution of their films throughout the world.

Resolution—Resolution refers to the number of bits of information that are used to recreate a signal. It also describes the number of pixels or individual picture elements that make up an image on a video display. Higher resolution generally provides higher-quality images.

RGB—An acronym that refers to the three primary colors of light: red, green, and blue. A typical video signal is composed of electronic representations of these three colors.

ReplayTV—ReplayTV is a brand of digital video recorder technology that allows you to copy and play back television programs using a hard drive. It works much like a VCR, but digitally and with instantaneous access. See also: TiVo.

Ripping—A slang term defining the act of digitally extracting audio tracks from a CD and storing them on your computer, usually as MP3 files.

RSDL—This acronym stands for Reverse-Spiral Dual-Layer. DVDs that are RSDL dual-layered will usually exhibit a slight audio and video pause at some point during the film presentation. This is because the film information on the disc is stored on two separate layers. Your DVD player will read the first layer of the disc and then will switch layers and work its way back through the second layer to play the remainder of the film. Using the RSDL format allows studios to include longer films on one side of a DVD disc, keeping you from having to flip the disc over halfway through the film. The slight pause is caused by the DVD player's laser refocusing on the second layer and changing directions. It's completely normal and doesn't mean that your DVD is defective.

SACD—This acronym stands for Super Audio CD. It's a high-resolution audio disc format created by Sony and Phillips for music aficionados and is capable of presenting both stereo and multi-channel music. It looks like a standard CD or DVD disc, but it uses a different digital encoding method. Only players with SACD capabilities can play SACD discs. Consult your DVD player's manual to find out if your player is compatible with this format.

SACD Hybrid—This is a special type of SACD disc that includes a "redbook" layer that is backwards-compatible with almost all current CD and DVD players. It will also have a high-resolution layer that works only in SACD-compatible players.

Satellite Radio—A digital radio format in which the digital audio signals are broadcast from a satellite orbiting the earth rather than a regular land-based transmission tower. This allows the same signals to be delivered to a much larger geographic area.

Scene Selection—This DVD function allows you to directly access particular scenes in a film from the menu screens, as if they were chapters in a book.

SECAM—This acronym stands for Sèquential couleur avec mèmorie (or Sequential Color with Memory). This is the primary television transmission format used in France, Russia, and a few other countries. SECAM features 576 scan lines (lines of vertical resolution) compared to the NTSC used in the U.S., which produces 525 scan lines.

Sharpness—The detail in clarity, particularly of edges, seen in an image.

Sirius—A satellite-based digital radio service. *See also* XM.

Snapper Case—This is a form of DVD packaging that is half-cardboard and half-plastic. It actually "snaps" open and closed using a long, black plastic flap along the right-hand side of the package, thus the name.

Sound Meter—An instrument used to measure sound levels. Sound meters are very useful in properly calibrating your home theater audio system.

Spade Connector—A flat, U-shaped speaker wire connector that inserts into a receiver or speaker's binding post.

Special Features—This generally means any kind of bonus, "behind-the-scenes" material, other than the film itself, found on a DVD disc. It can include the film's theatrical trailer, a documentary on the making of the film, an audio commentary track, a gallery of photographs or production artwork, or any of a number of similar items.

Storyboards—During the process of filmmaking, sometimes an artist is commissioned to draw on paper the shot-by-shot look of the action contained in the screenplay. These drawings are called storyboards and are often used as a guide to the filmmakers on the set. These are sometimes included on a DVD as a special feature, and you can view them by navigating through them with your DVD player's remote.

Subtitles—A subtitle track is basically a text translation of the onscreen dialogue that appears on the bottom of your TV screen. This allows, for example, English language films to be enjoyed by people who don't speak English. On the other hand, it also allows foreign language films to be enjoyed by those who only speak English. It takes some practice to both read the subtitles and watch the film at the same time, but the process makes the world of international films accessible to the average viewer.

Subwoofer—A special speaker that reproduces the lower portion of the audible frequency spectrum (usually from 80 Hz down to 20 Hz—the lower limit of human hearing). There are two primary types of subwoofer—powered and non-powered. Powered subwoofers have their own built-in amplifier, while the non-powered models need to be connected to the amplifier used for the main stereo speakers. Powered subwoofers are by far the most popular.

Super Jewel Box—This is a type of clear plastic packaging that is sometimes used for DVD-Video discs and is often used for SACD and DVD-Audio discs.

Surround Speaker—A special type of speaker used to reproduce surround channel information, creating ambience and sonic realism in your home theater. The surround channel speakers are usually hung on a wall or attached to the ceiling to the left and right sides of the listener.

S-VHS—Stands for Super Video Home System. This is a variation of the standard analog VHS videotape format that allows for marginally higher quality.

S-Video—This is one of three common types of video signal connection between your DVD player and your TV. It provides for better quality than a composite connection, but is not quite as good as a component connection.

Telecine—The process of capturing the filmed image and converting it into a digital-format video.

THX—This is a rigorous system of standards and guidelines that are sometimes followed when making a DVD to ensure the highest possible video and audio quality. The system was originally created in part by director George Lucas as a way to improve the presentation quality of films in theaters and on home video. A DVD that features the THX logo was produced in compliance with these quality guidelines. It is sometimes mistakenly thought of as an audio format, such as Dolby Digital or DTS. The basic THX process can also be applied to audio and video equipment.

Tint—A color attribute that describes a color in relation to a primary color.

TiVo—TiVo is a brand of digital video recorder that allows you to copy and play back television programs using a hard drive. It works much like a VCR, but digitally and with instantaneous access. *See also* ReplayTV.

Toslink—A fiber-optic cable connection that uses beams of light to transfer digital information between electronic equipment components. The use of light beams avoids potential interference from electrical wiring and other electrical devices. In terms of home theater, this is usually an audio signal delivered from your DVD player to a receiver.

Universal Remote—A remote control device that has the capability to send commands and control features of multiple components in a home theater system, rather than just one.

VCD—Stands for Video CD. This is a disc-based, digital movie format of lower quality than DVD. It has the same form as a standard CD or a CD-ROM, except that instead of music or software, it holds movies using compressed MPEG-1 video. It has a video resolution roughly comparable to VHS.

VCR—Stands for Video Cassette Recorder. This is a video/audio playback device that accepts standard analog videocassette tapes (VHS, S-VHS, Hi-8), reads the data, and passes the signals to your TV. Most VCRs allow for a user to record in addition to allowing playback of the tape.

VHS—Stands for Video Home System. VHS is a half-inch, magnetic tape cartridge format developed for the home playback and recording of analog video signals. For many years, VHS was the most popular format for the home video distribution of Hollywood movies.

Watermarking—Also known as digital watermarking, this is a copy-protection technique where a pattern of imperceptible data is inserted into a digital image, audio, or video file to identify copyright information.

White Level—The level of brightness of the lightest portions of an image.

XM—A satellite-based digital radio service. *See also* Sirius.

Index